Cryogenic Processing

Cryogenic Processing

Advances in the Food Industry

Edited by

Sumit Sudhir Pathak
Karunya Institute of Technology and Sciences
Coimbatore, Tamil Nadu
Bharat

Rama Chandra Pradhan
National Institute of Technology
Rourkela, Odisha
Bharat

Sudarshan Ramanathan
Karunya Institute of Technology and Sciences
Coimbatore, Tamil Nadu
Bharat

WILEY

This edition first published 2026

Registered Offices
John Wiley & Sons, Inc., 111 River Street, Hoboken, NJ 07030, USA
John Wiley & Sons Ltd., New Era House, 8 Oldlands Way, Bognor Regis, West Sussex, PO22 9NQ, UK

For details of our global editorial offices, customer services, and more information about Wiley products visit us at www.wiley.com.

The manufacturer's authorized representative according to the EU General Product Safety Regulation is Wiley-VCH GmbH, Boschstr. 12, 69,469 Weinheim, Germany, e-mail: Product_Safety@wiley.com.

Library of Congress Cataloging-in-Publication Data

Names: Pathak, Sumit Sudhir, 1992– editor | Pradhan, Rama Chandra, 1977– editor | Ramanathan, Sudarshan, 2001– editor
Title: Cryogenic processing : advances in the food industry / edited by Sumit Sudhir Pathak, Rama Chandra Pradhan, Sudarshan Ramanathan.
Description: Hoboken, NJ : Wiley, 2026. | Includes bibliographical references and index.
Identifiers: LCCN 2025054140 (print) | LCCN 2025054141 (ebook) | ISBN 9781394280087 cloth | ISBN 9781394280100 adobe pdf | ISBN 9781394280094 epub
Subjects: LCSH: Frozen foods–Technological innovations | Low temperature engineering | Frozen foods–Analysis
Classification: LCC TP372.3 .C7964 2026 (print) | LCC TP372.3 (ebook)
LC record available at https://lccn.loc.gov/2025054140
LC ebook record available at https://lccn.loc.gov/2025054141

Cover Design: Wiley
Cover Images: © CB studio/Shutterstock, © BravissimoS/Shutterstock

Set in 9.5/12.5pt STIXTwoText by Straive, Pondicherry, India

Dedicated at the feet of

'Arulmigu Perur Patteeswarar'

'Arulmigu Velliangiri Andavar'

'Arulmigu Maruthamalai Subramaniaswamy'

Contents

List of Contributors *xix*
Preface *xxiii*

1 **Cryogenics: An Overview and Introduction** *1*
Sudarshan Ramanathan, Teena Bijo, and Sumit Sudhir Pathak
1.1 Introduction *1*
1.1.1 Simple Concept of Cryogenics *2*
1.1.2 Significance of Cryogenics in General Science and Technology *2*
1.2 Historical Developments of Cryogenics *2*
1.3 Fundamental Concepts of Cryogenics *3*
1.3.1 Behaviour of Materials at Room Temperature *3*
1.3.2 Phenomena and Temperature Scales Related to Low Temperatures *5*
1.3.3 Behaviour of Solids, Liquids, and Gases at Extremely Low Temperatures *7*
1.3.4 Low-Temperature Thermodynamics *8*
1.3.5 Various Cryogenic Fluid Phase Transitions *8*
1.4 Cryogenic Equipment and Systems *10*
1.4.1 Cryogenic Fluids and Their Properties *11*
1.4.2 Challenges in Cryogenic Fluid and Material Selection *12*
1.4.3 Safety Measures – Working with Cryogenic Systems *12*
1.5 Impacts and Application of Cryogenics *13*
1.5.1 Cryogenics in Space Science *13*
1.5.2 Cryogenics in Physics and Material Science *14*
1.5.3 Cryogenics in Health and Medicine *14*
1.5.4 Cryogenics in Food Processing *15*
1.6 Influence of Cryogenics in Food *15*
1.7 Safety in Cryogenics *16*
1.7.1 Hazards of Cryogenic Fluids *16*
1.7.2 Safe Practices for Transporting *17*
1.7.3 Cryogenic System Safety *17*
1.7.4 Safety Aspects: Designing Cryogenic Systems to Prevent Failure and Accidents *17*
1.7.5 Pressure Curtailment and Self-Explosion Management *18*

1.8 Limitations and Challenges in Cryogenics *18*
1.9 Summary and Structure of the Book *19*
1.10 Conclusion *20*
References *21*

2 Molecular Foundations of Cryogenic Processes: A Chemical Perspective *25*
Sakkaravarthy Abishek and Parameswaran Gurumoorthi
2.1 Introduction *25*
2.1.1 Overview of Cryogenic Processing in the Food Industry *25*
2.2 Thermodynamic Principles of Cryogenic Systems *27*
2.2.1 Heat Transfer and Temperature Gradients *27*
2.2.2 Phase Diagrams and Freezing Point Depression *28*
2.2.3 Energy Changes During Cryogenic Transitions *28*
2.3 Kinetic and Molecular Motion in Cryogenic Conditions *30*
2.3.1 Molecular Vibrations and Rotations at Low Temperatures *30*
2.3.2 Diffusion and Reaction Rate Alterations *31*
2.3.2.1 Effects on Diffusion *31*
2.3.2.2 Effects on Chemical Reaction Rates *32*
2.3.3 Glass Transition and Molecular Immobilization *32*
2.3.3.1 Glass Transition in Cryogenic Food Systems *32*
2.3.3.2 Kinetic Consequences of Glassy States *32*
2.4 Intermolecular Forces and Structural Stabilization *33*
2.4.1 Hydrogen Bonding and Cryoprotection *33*
2.4.2 Hydrophobic and van der Waals Interactions in Frozen Matrices *34*
2.5 Cryogenic Effects on Biomolecules *34*
2.5.1 Protein Denaturation and Refolding *35*
2.5.2 Lipid Crystallization and Oxidative Stability *36*
2.5.3 Carbohydrate Glass Formation and Mobility *36*
2.6 Water Behaviour and Ice Crystallization in Cryogenic Processing in the Food Industry *37*
2.6.1 Nucleation and Growth of Ice Crystals *37*
2.6.2 Amorphous vs. Crystalline Ice *37*
2.6.3 Effects on Cell and Tissue Integrity *38*
2.7 Role of Cryogens: Liquid Nitrogen and Carbon Dioxide *38*
2.7.1 Physical and Chemical Properties *38*
2.7.2 Reaction with Food Constituents *39*
2.7.3 Safety and Handling Considerations *40*
2.8 Implications for Food Structure and Quality *40*
2.8.1 Texture Modification and Structural Integrity *41*
2.8.2 Flavour and Volatile Compound Retention *42*
2.8.3 Nutritional Stability and Enzyme Inactivation *42*
2.8.4 Structural Enhancement in Composite Foods *42*
2.9 Molecular Basis of Bioactive Stabilization *42*
2.9.1 Antioxidants, Polyphenols, and Essential Oils *43*
2.9.2 Encapsulation and Retention During Cryogenic Processing *43*

2.10 Emerging Trends and Future Directions *44*
2.10.1 Nanostructured Cryogenic Interfaces *44*
2.10.2 Integration with Smart Packaging and Sensors *44*
2.10.3 Cryogenic for Functional Foods *45*
2.10.4 Industrial Relevance and Application Potential *45*
2.11 Conclusion and Summary *46*
References *46*

3 Cryogenic Impact on Food Material Properties: Structural and Functional Transformations *51*
Bhupendra M. Ghodki, Krishna Bahadur Chhetri, and Naveen Kumar Mahanti
3.1 Introduction *51*
3.1.1 Historical Perspective and Evolution *52*
3.1.2 Importance in Food Systems *52*
3.2 Fundamentals of Cryogenic Processing in Food Industry *53*
3.2.1 Principles and Methodologies *53*
3.2.2 Equipment and Techniques *54*
3.2.3 Advantages of Cryogenic Processing *55*
3.2.4 Applications in the Food Industry *55*
3.3 Structural Alterations in Cryogenically Treated Food Materials *56*
3.3.1 Molecular and Microstructural Changes *56*
3.3.1.1 Molecular Changes *56*
3.3.1.2 Microstructural Changes *57*
3.3.2 Crystalline Structures and Phase Transitions *58*
3.3.2.1 Ice Crystallization *58*
3.3.2.2 Glass Transition and Vitrification *58*
3.3.2.3 Phase Separation and Recrystallization *59*
3.3.3 Advanced Characterization Techniques *60*
3.3.3.1 Microscopy Techniques *60*
3.3.3.2 Spectroscopic Techniques *61*
3.3.3.3 Thermal Analysis *62*
3.3.3.4 Chromatographic Techniques *62*
3.3.4 Sensory Analysis *62*
3.4 Functional Enhancements and Impacts on Food Quality *63*
3.4.1 Texture and Flavour Improvements *63*
3.4.1.1 Texture Improvements *63*
3.4.1.2 Flavour Improvements *63*
3.4.2 Nutritional Content and Bioavailability *64*
3.4.2.1 Retention of Essential Nutrients *64*
3.4.2.2 Preservation of Bioactive Compounds *64*
3.4.2.3 Improved Bioavailability *64*
3.4.3 Shelf Stability and Safety *65*
3.4.3.1 Shelf Stability *65*
3.4.3.2 Safety *65*
3.5 Applications of Cryogenic Processing *65*

3.5.1 Food Preservation and Shelf-Life Extension *65*
3.5.2 Meat and Poultry Processing *66*
3.5.3 Frozen Desserts and Baked Goods *66*
3.5.4 Functional Foods and Nutraceuticals *67*
3.6 Future Directions and Challenges *67*
3.7 Conclusion *69*
References *70*

4 Industrial Design and Engineering Aspects in Cryogenic Applications *73*
Awani Shrivastav and Tridib Kumar Goswami
4.1 Introduction *73*
4.2 Fundamental Principles of Cryogenics *74*
4.2.1 Heat Transfer Mechanism in Cryogenics *75*
4.2.2 Thermodynamic Applications in Cryogenics *76*
4.3 Industrial Role of Cryogenics in the Food Processing *76*
4.4 Industrial Design and Engineering Considerations in Food Processing Applications *78*
4.4.1 Material Selection for Cryogenics *79*
4.4.2 Thermal Insulation Technique *80*
4.4.3 Safety Measures and Standards *81*
4.5 Role of LN_2 and LCO_2 in Food Processing Applications *82*
4.6 Key Design Elements of Cryogenic Equipment for Food Processing Applications *83*
4.6.1 Cryogenic Freezers *83*
4.6.1.1 Cabinet Freezers 83
4.6.1.2 Cryogenic Freezer Tunnel *83*
4.6.1.3 Cryogenic Fluidized Bed Freezers *84*
4.6.1.4 Cryogenic Spiral Freezers *84*
4.6.1.5 Cryogenic Immersion Freezers *84*
4.6.2 Cryogenic Storage Tanks *84*
4.6.3 Cryogenic Grinding System *85*
4.6.4 Cryogenic Encapsulation by Spray Chilling *86*
4.7 Challenges and Future Trends *86*
4.8 Conclusion and Summary *88*
References *88*

5 Cryogenic Freezing Process for the Preservation of Fruits and Vegetables *95*
Prateek Gururani, Abhirup Mitra, Jesni Manjula Sivaprasad, and Mahipal Singh Tomar
5.1 Introduction *95*
5.2 Overview of Cryogenics in Food Processing *96*
5.3 Conventional Freezing Methods for Fruits and Vegetables *97*
5.3.1 Air-Blast Freezing *97*
5.3.2 Plate-Freezing Systems *98*
5.3.3 Immersion Freezing *98*
5.4 Cryogenic Freezing Techniques *98*

5.4.1 Direct-Contact Cryogenic Freezing *99*
5.4.1.1 Immersion Cryogenic Freezer *99*
5.4.1.2 Spray Cryogenic Freezer *100*
5.4.2 Indirect Contact Cryogenic Freezing *101*
5.5 Case Studies on Cryogenic Freezing of Fruits and Vegetables *101*
5.6 Advantages and Limitations of Cryogenic Freezing *103*
5.6.1 Advantages of Cryogenic Freezing *103*
5.6.2 Economic Feasibility, Challenges, and Limitations *103*
5.7 Conclusion *104*
References *104*

6 Cold Revolution: Cryogenic Innovations in Dairy Production and Preservation *107*
Revathi P. and Sudarshan Ramanathan
6.1 Introduction *107*
6.2 Cryogenics in Dairy Production and Preservation *108*
6.2.1 Cryogenic Technology in Milk Processing and Preservation *109*
6.2.2 Cryogenic Cooling for Milk Quality Preservation *110*
6.2.3 Applications in Milk Processing *110*
6.2.4 Energy Efficiency and Environmental Benefits *111*
6.2.5 Challenges and Future Perspectives *111*
6.3 Cryogenic Processing of Dairy Products *112*
6.3.1 Cryogenic Technology in Cheese Processing *112*
6.3.2 Cryogenic Processing in Starter Culture Preservation *113*
6.3.2.1 Importance of Starter Culture Preservation *113*
6.3.2.2 Cryogenic Processing Technique *113*
6.4 Cryogenic Drying: A Revolution in Milk Powder Production *114*
6.4.1 Cryogenic Drying in Milk Powder Processing *114*
6.4.2 Cryogenic Drying Processing in Dairy Products (Milk Powder) *115*
6.4.3 Preservation of Nutritional and Bioactive Compounds *115*
6.4.4 Energy Efficiency and Environmental Sustainability *115*
6.4.5 Challenges and Innovations in Cryogenic Drying *116*
6.5 Ice Cream and Frozen Desserts *116*
6.5.1 Role of Cryogenic Freezing in Texture and Structure Preservation *117*
6.5.2 Enhancing Product Quality and Flavours *117*
6.5.3 Energy Efficiency and Operational Flexibility *118*
6.5.4 Challenges and Innovations in Cryogenic Freezing *118*
6.6 Cryogenic Technology in Fermented Milk Processing *118*
6.6.1 Preserving Product Quality Through Rapid Freezing *119*
6.6.2 Ensuring Microbial Viability and Probiotic Efficacy *119*
6.6.3 Energy Efficiency, Sustainability, and Operational Benefits *119*
6.6.4 Challenges and Future Prospects *120*
6.7 Cryogenic Preservation of Cultures and Enzymes *120*
6.7.1 Storage and Transportation of Dairy Products *120*
6.7.2 Applications in Cryogenic Storage *121*

6.7.3 Applications in Cryogenic Transportation *121*
6.7.4 Benefits of a Cryogenic System *122*
6.7.5 Sustainability and Future Implications *122*
6.8 Cryogenic Technology in Value-Added Dairy Products *122*
6.8.1 Nitrogen-Infused Milk Beverages *122*
6.8.2 Nutrient Encapsulation *123*
6.8.3 Colostrum Preservation *124*
6.8.4 Environmental and Economic Considerations *124*
6.9 Technological Evolution of Cryogenic Systems *125*
6.10 Challenges and Future Prospects *125*
6.11 Conclusion *126*
References *126*

7 Advanced Cryogenic Applications in Plant-Based Food Preservation *129*
Mahipal Singh Tomar, Rahul Kumar Sharma, and Abhirup Mitra
7.1 Introduction *129*
7.2 Emerging Trend for Plant-Based Foods *130*
7.2.1 Plant-Based Protein Products *131*
7.2.2 Fermented Mycoprotein as a Protein Source *131*
7.2.3 Plant-Based Milk: Process and Stability *132*
7.2.4 Plant-Based Cheese Substitutes *132*
7.2.5 Plant-Based Seafood Mimics *132*
7.2.6 Functional Plant-Based Egg Replacers *132*
7.2.7 Microalgae *132*
7.3 Basic Principles and Cryogenic Techniques for Plant-Based Product Development *133*
7.3.1 Cryopreservation in Plant-Based Systems *134*
7.3.2 Cryogenic Grinding for Plant-Derived Ingredients *135*
7.4 Engineering Challenges in Cryogenic Plant-Based Food Processing *136*
7.5 Conclusion and Future Perspective *138*
References *138*

8 Frosty Enhancements: Cryogenic Approaches in Meat, Poultry, and Marine Product Processing *141*
Sourav Misra, Sanchita Biswas Murmu, Sivashankari Manickam, Rahul Kumar Rout, Ankit Kumar, Sitesh Kumar, and Ritu Bharat Kukde
8.1 Introduction *141*
8.2 Cryobiology: Principles and Methods *141*
8.2.1 Principles of Cryopreservation *142*
8.2.2 Methods of Cryopreservation *143*
8.2.2.1 Slow Freezing *143*
8.2.2.2 Vitrification *143*
8.2.2.3 Subzero Non-Freezing Storage *143*
8.2.2.4 Preservation in the Dry State *144*
8.3 Cryogenic Freezers for Meat, Poultry, and Marine Products *144*
8.3.1 Tunnel Freezers *144*

8.3.2 Rotary Freezers *145*
8.3.3 Immersion Freezers *146*
8.3.4 Cabinet Freezers *146*
8.3.5 Pellet Freezers *146*
8.3.6 Combined Freezers *146*
8.4 Effect of Cryogenic Processing on Quality Changes of Meats *146*
8.4.1 Physical Properties *147*
8.4.1.1 Microstructure *147*
8.4.1.2 Loss of Moisture *147*
8.4.1.3 Tenderness *148*
8.4.1.4 Colour *148*
8.4.2 Biochemical Properties *148*
8.4.2.1 Protein Denaturation *148*
8.4.2.2 Lipid and Protein Oxidation *149*
8.4.3 Microbial Properties *149*
8.5 Effect of Cryogenic Processing on Quality Changes of Marine/Aquatic Products *149*
8.5.1 Freezing Loss *150*
8.5.2 Weight Loss During Storage *150*
8.5.3 Drip Loss *151*
8.5.4 Water Content *151*
8.5.5 Protein Denaturation *152*
8.5.6 Lipid Oxidation *152*
8.5.7 pH and Colour *153*
8.5.8 Morphological Characteristics *153*
8.5.9 Sensory Attributes *154*
8.6 Cryogenic Freezing of Shrimp *154*
8.6.1 Harvesting *154*
8.6.2 Holding *154*
8.6.3 Peeling *155*
8.6.4 Freezing *155*
8.6.5 Glazing *155*
8.6.6 Packaging *155*
8.6.7 Frozen Storage *155*
8.7 Major Considerations for Design and Safety *156*
8.8 Safety Measures in Cryogenics *156*
8.9 Conclusion *156*
References *158*

9 Spice Chills: Cryogenic Technologies in Spice Processing *163*
Gintu Sara George, Akshay D. Sonawane, and Shivanand S. Shirkole
9.1 Introduction *163*
9.2 Overview of Conventional Spice Processing Techniques *164*
9.2.1 Conventional Post-harvest Practices Responsible for Contamination of Spices *164*
9.2.2 Treatment Based on Harvest and the Post-harvesting *167*

9.2.3 Process of Drying the Raw Material *167*
9.2.4 Reduction in Particle Size *167*
9.2.5 Storage and Packaging *168*
9.2.6 Limitations with Conventional Spice Grinding *168*
9.3 Cryogenic Grinding Technology *168*
9.3.1 Benefits of Cryogenic Grinding for Spices *169*
9.3.2 Benefits of Cryogenic Grinding over Conventional Grinding *170*
9.4 Design and Engineering of Cryogenic Grinding Equipment *171*
9.4.1 Construction Methods *172*
9.4.2 Pre-coolers *172*
9.4.3 Distribution System for Liquid Nitrogen *172*
9.4.4 Grinder *172*
9.4.5 Sensors *173*
9.5 Impact on Spice Quality and Preservation *173*
9.5.1 Effects of Cryogenic Processing on Flavour and Aroma Retention *173*
9.5.2 Comparison with Traditional Spice Processing Method *174*
9.6 Challenges and Limitations of Cryogenic Grinding *175*
9.7 Conclusion *175*
References *175*

10 Preserving Nature's Remedies: Cryogenic Applications in Herbal Medicine Processing *181*
K. U. Pavitra Krishna
10.1 Introduction *181*
10.2 Herbal Medicine Processing *183*
10.3 Quality Control for Phytomedicines *184*
10.3.1 Production Controls *187*
10.3.2 Good Manufacturing Practices for Herbal Products *188*
10.3.3 Process Control *189*
10.3.4 Concept of Work-in-Progress Testing *190*
10.3.5 Concept of Purity Management *191*
10.4 Importance of Cryogenic Applications in Herbal Medicine Processing *193*
10.5 Conclusion *194*
References *195*

11 Vegan Chills: Cryogenic Technologies for Vegan-Based Products *205*
Dhiraj Kumar Yadav, Khalid Bashir, Pintu Choudhary, and Shweta Suri
11.1 Introduction *205*
11.2 Concept of Vegan Chills *206*
11.3 Concept of Cryogenic Techniques *207*
11.3.1 Importance of Cryogenics in Food Preservation *208*
11.4 Cryogenic Principle and Mechanism *208*
11.4.1 Definition, Temperature Range, and Principle *208*
11.4.2 Cryogenic Agents *209*
11.4.3 Advantages and Disadvantages of Cryogenic Freezing and Conventional Freezing *210*

11.5 Cryogenic Techniques for Vegan Products *210*
11.5.1 Emerging Cryogenic Technologies *210*
11.5.1.1 Cryogenic Freezing *210*
11.5.1.2 Pulse Freezing *211*
11.5.1.3 Cryo-milling *212*
11.5.1.4 Ultra Cryo-milling *214*
11.5.1.5 Cryogenic Cooling *214*
11.5.1.6 Cryo-concentration *215*
11.6 Effect of Cryogenic Processing on Quality Characteristics of Vegan Products *216*
11.6.1 Packaging Materials for Frozen Foods *217*
11.6.2 Concentration Effects *217*
11.6.3 Freeze Cracking *217*
11.6.4 Moisture Loss and Freezer Burn *217*
11.6.5 Rancid Flavour Development *217*
11.6.6 Ensuring High-Quality Frozen Products *217*
11.6.7 Texture *218*
11.6.8 Drip Loss *218*
11.6.9 Colour *218*
11.6.10 Suppressed Microbial Growth and Reduced Activity of Spoilage Bacteria *218*
11.7 Applications of Cryogenically Treated Vegan Products *220*
11.8 Consumer Attitudes Towards Frozen Vegan Products and Cryogenic Processing *221*
11.9 Future Perspectives and Challenges Associated with Production of Vegan Products *222*
11.9.1 Challenges and Cost Implications *222*
11.9.2 Sustainability and Scalability *223*
11.9.3 Future Perspectives *223*
11.10 Conclusion *223*
References *224*

12 Impact of Cryogenics on Microbial Populations and Food Safety *227*
Pratiksha Wankhede, Jessica Hannah, Sumit Sudhir Pathak, Surendra Babu, and Bhushan Chaudhari
12.1 Introduction *227*
12.2 Overview of Cryogenics in Food Microbiology *229*
12.2.1 Microbial Populations in Food Products *229*
12.3 Role of Temperature in Microbial Survival and Proliferation *231*
12.4 Mechanisms of Microbial Inactivation Through Cryogenic Freezing *232*
12.5 Comparison of Cryogenic Freezing with Conventional Freezing Methods *233*
12.5.1 Advantages in Using LN_2 as Refrigerant *234*
12.5.2 Impact of Cryogenic Processing on Microbial Viability *234*
12.6 Safety Considerations in Cryogenic Processing *234*
12.7 Case Studies on Cryogenic Processing and Microbial Safety *235*
12.8 Challenges, Limitations, and Future Direction *237*
12.9 Conclusion *240*
References *241*

13 Environmental Impact and Sustainability of Cryogenic Processing in Food Industries *243*
Abhishek Kumar
13.1 Introduction *243*
13.1.1 Environmental Impact of Vapour-Compression Refrigeration Systems *243*
13.1.2 Cryogenic Refrigeration as a Sustainable Alternative *244*
13.1.3 Comparative Analysis of Cryogenic and Vapour-Compression TRUs *244*
13.2 Technologies and Mechanisms *245*
13.2.1 Cryogenic Transport Refrigeration Systems *246*
13.2.2 Direct and Indirect Cryogenic Systems *246*
13.3 Environmental Impact Analysis *247*
13.3.1 Greenhouse Gas Emissions of Cryogenic Systems *247*
13.3.2 Emissions from Diesel-Driven Systems *248*
13.3.3 Comparative Analysis of Production and Operational Emissions *248*
13.4 Challenges in Cryogenic Processing *248*
13.4.1 High Mass Intensity and Distribution Range Limitations *249*
13.4.1.1 Comparison with Diesel-Driven Refrigeration Systems *249*
13.4.1.2 Cryogenic Fluid Consumption and Its Impact on Logistics *249*
13.4.2 Infrastructure for Replenishing Cryogens *249*
13.4.2.1 Current Status of Cryogen Replenishment Infrastructure *249*
13.4.2.2 Economic and Logistical Constraints of Expanding Infrastructure *250*
13.4.2.3 Potential Solutions for Infrastructure Development *250*
13.4.3 Operational and Economic Barriers *250*
13.4.3.1 High Capital and Operating Costs *250*
13.4.3.2 Limited Economic Incentives and Regulatory Support *250*
13.4.3.3 Safety and Handling Challenges *251*
13.4.3.4 Strategies for Economic Viability *251*
13.5 Benefits of Cryogenic Processing *251*
13.5.1 Advantages in Food Quality Preservation *251*
13.5.1.1 Faster Cooling and Lower Temperature Impact *251*
13.5.1.2 Extended Shelf Life *251*
13.5.2 Impact on Nutritional Value, Flavour, and Texture *252*
13.5.2.1 Nutrient Retention *252*
13.5.2.2 Flavour and Aroma Conservation *252*
13.5.2.3 Texture Preservation *252*
13.5.3 Reduction of Ice Crystal Formation *253*
13.5.3.1 The Formation of Ice Crystals and Their Effect on Food *253*
13.5.3.2 Cryogenic Freezing for Delicate Food Products *253*
13.5.3.3 Preventing Freezer Burn and Moisture Loss *253*
13.6 Sustainability Considerations *253*
13.6.1 Balancing Energy Requirements and Environmental Benefits *253*
13.6.1.1 Energy Consumption in Cryogenic Refrigeration *253*
13.6.1.2 Well-to-Wheel (WTW) Emissions of Cryogenic vs. Diesel Systems *254*
13.6.1.3 Trade-Off Between Cryogen Production and Usage *254*
13.6.2 Technological Advancements for Improved Sustainability *254*

13.6.2.1 Innovations in Cryogenic Refrigeration Systems *254*
13.6.2.2 Renewable Energy Integration in Cryogen Production *255*
13.6.2.3 Advances in Cryogenic Fuel Infrastructure *255*
13.6.3 Efficiency of Cryogen Production Methods *255*
13.6.3.1 Energy Intensity of Cryogen Production *255*
13.6.3.2 Improving Liquefaction Efficiency *255*
13.6.3.3 Cryogen Storage and Distribution Optimization *256*
13.7 Case Studies and Applications *256*
13.7.1 Examples of Cryogenic Systems in the Food Industry *256*
13.7.1.1 Cryogenic Freezing for Perishable Foods *256*
13.7.1.2 Cryogenic Grinding for Spice and Herb Processing *256*
13.7.1.3 Cryogenic Sample Preparation in Food Safety Analysis *256*
13.7.2 Real-World Performance and Benefits *256*
13.7.2.1 Enhanced Product Quality and Shelf Life *257*
13.7.2.2 Energy Efficiency and Reduced Environmental Footprint *257*
13.7.2.3 Food Waste Reduction and Sustainability *257*
13.8 Future Directions in Cryogenic Processing *257*
13.8.1 Innovations and Emerging Technologies *257*
13.8.1.1 Hybrid Cryogenic Refrigeration Systems *257*
13.8.1.2 AI-Powered Predictive Maintenance and Process Optimization *258*
13.8.1.3 Renewable Energy Integration in Cryogen Production *258*
13.8.1.4 Superconducting Cryogenic Technologies *258*
13.8.2 Potential for Broader Adoption *258*
13.8.2.1 Expanding Cryogenic Processing Beyond High-End Applications *258*
13.8.2.2 Infrastructure Development for Cryogenic Refuelling *258*
13.8.2.3 Policy and Regulatory Incentives *259*
13.8.3 Recommendations for Industry Practices *259*
13.8.3.1 Optimizing Cryogen Consumption for Cost Efficiency *259*
13.8.3.2 Enhancing Cryogenic Food Logistics *259*
13.8.3.3 Sustainable Supply Chain Integration *259*
13.9 Conclusion *260*
References *260*

14 Advancements and Future Avenues in Cryo-Processing Technologies in Food *267*
Keerthivasan Ragupathi, Sankar Manokaran, and Krishnakumar Panneerselvam
14.1 Introduction *267*
14.1.1 Significance of Cryogenic Technology *267*
14.1.2 Overview of Recent Trends and Global Outlook *268*
14.2 Cryogens *268*
14.2.1 Liquid Nitrogen *269*
14.2.2 Carbon Dioxide (CO_2) *269*
14.3 Advanced Cryogenic Processing in the Food Sector *269*
14.3.1 Freezing and Cooling *269*
14.3.1.1 Cryogenic Freezing *270*

14.3.1.2 Cryogenic Immersion Freezers *271*
14.3.2 Fruits and Vegetables *271*
14.3.3 Meat and Seafood *272*
14.3.3.1 Changes to the Physical Characteristics of Frozen Meat Products *272*
14.4 Cryogenic Grinding *272*
14.4.1 Effects of Cryogenic Grinding on Spices *273*
14.5 Advantages of Cryogenic Technology *274*
14.6 Challenges of Cryogenic Processing *275*
14.7 Integration of Artificial Intelligence in Cryogenic Processing *275*
14.8 Hybrid Cryogenic Systems *276*
14.9 Quantum Cryogenics in Food: A Theoretical Approach *276*
14.9.1 Benefits of Quantum Cryogenics *276*
14.10 Precision Cryogenic Processing *277*
14.11 Advances in Cryogenic Equipment and Material Science *277*
14.11.1 Machining Technology *277*
14.11.2 Storage and Transportation *278*
14.11.3 Material Science *279*
14.12 Start-ups, Collaborations, and Commercial Innovations *279*
14.13 Future Goals of Cryogenics *280*
14.14 Technological Gap *281*
14.15 Vision 2030: Strategic Roadmap for Cryogenic Food Processing *281*
14.16 Conclusion *282*
References *282*

Index *287*

List of Contributors

Sakkaravarthy Abishek
Department of Food Technology
School of Bioengineering
SRM Institute of Science and Technology
Kattankulathur
Chennai, Tamil Nadu
India

Surendra Babu
Rudhra Consultancy and Services
Bangalore, Karnataka
India

Khalid Bashir
Department of Food Technology
School of Interdisciplinary Sciences and Technology
Jamia Hamdard, New Delhi
India

Teena Bijo
School of Agriculture, Food and Ecosystem Sciences
The University of Melbourne
Melbourne, Victoria
Australia

Bhushan Chaudhari
Kavayitri Bahinabai Chaudhari North Maharashtra University
Jalgaon, Maharashtra
India

Krishna Bahadur Chhetri
Krishi Vigyan Kendra, Bhagwanpur Hat
Siwan, Dr. RPCAU, Pusa
Samastipur, Bihar
India

Pintu Choudhary
Department of Food Technology
College of Agricultural Engineering and Technology
Dr. Rajendra Prasad Central Agricultural University
Samastipur, Bihar
India

Gintu Sara George
Plant Lipids Private Limited
Ernakulam, Kerala
India

Bhupendra M. Ghodki
Agricultural and Food Engineering Department
IIT Kharagpur
Kharagpur, West Bengal
India

Tridib Kumar Goswami
Department of Agricultural and Food Engineering
IIT
Kharagpur, West Bengal
India

Parameswaran Gurumoorthi
Department of Food Technology
School of Bioengineering
SRM Institute of Science and Technology
Kattankulathur
Chennai, Tamil Nadu
India

Prateek Gururani
Department of Biotechnology
Graphic Era Deemed University
Dehradun, Uttarakhand
India

Jessica Hannah
Department of Food Processing and
Preservation Technology
School of Engineering
Avinashilingam Institute for Home Science
and Higher Education for Women
Coimbatore, Tamil Nadu
India

K. U. Pavitra Krishna
Department of Food Science and
Processing Management
Subbalakshmi Lakshmipathy College
of Science
Madurai, Tamil Nadu
India

Ritu Bharat Kukde
ICAR – Central Institute of Post-Harvest
Engineering and Technology
Ludhiana, Punjab
India

Abhishek Kumar
CSIR-National Metallurgical Laboratory
Jamshedpur, Jharkhand
India

Ankit Kumar
ICAR – National Research Centre on Litchi
Muzaffarpur, Bihar
India

Sitesh Kumar
Department of Agricultural Engineering
Rabindra Nath Tagore Agriculture College
Birsa Agricultural University
Deoghar, Jharkhand
India

Naveen Kumar Mahanti
Post-Harvest Technology Research
Station
Dr. Y.S.R. Horticultural University
West Godavari, Andhra Pradesh
India

Sivashankari Manickam
ICAR – Central Rice Research Institute
Cuttack, Odisha
India

Sankar Manokaran
Department of Food Process
Engineering
National Institute of Food Technology
Entrepreneurship & Management
Thanjavur, Tamil Nadu
India

Sourav Misra
ICAR – National Institute of Natural Fibres
Engineering and Technology
Kolkata, West Bengal
India

Abhirup Mitra
Department of Agriculture
Invertis University
Bareilly, Uttar Pradesh
India

Department of Food Process
Engineering
NIT
Rourkela, Odisha
India

Sanchita Biswas Murmu
ICAR – National Institute of Natural Fibres Engineering and Technology
Kolkata, West Bengal
India

Krishnakumar Panneerselvam
Department of Food Process Engineering
National Institute of Food Technology Entrepreneurship & Management
Thanjavur, Tamil Nadu
India

Sumit Sudhir Pathak
Division of Food Processing Technology
School of Engineering and Technology
Karunya Institute of Technology and Sciences
Coimbatore, Tamil Nadu
India

Keerthivasan Ragupathi
Department of Floriculture and Landscape Architecture
Horticultural College and Research Institute, Tamil Nadu Agricultural University
Coimbatore, Tamil Nadu
India

Sudarshan Ramanathan
Division of Food Processing Technology
School of Engineering and Technology
Karunya Institute of Technology and Sciences
Coimbatore, Tamil Nadu
India

Revathi P.
Department of Food Science and Technology
Arul Anandar College
Madurai, Tamil Nadu
India

Rahul Kumar Rout
ICAR – National Research Centre for Makhana
Darbhanga, Bihar
India

Rahul Kumar Sharma
Department of Agriculture
Invertis University
Bareilly, Uttar Pradesh
India

Shivanand S. Shirkole
D. Y. Patil Agriculture and Technical University, DYP Educational City
Kolhapur, Maharashtra
India

Awani Shrivastav
Department of Agricultural and Food Engineering
IIT
Kharagpur, West Bengal
India

Jesni Manjula Sivaprasad
School of Food Science and Environment Health
Technological University Dublin
Dublin
Ireland

Akshay D. Sonawane
Leibniz Institute for Agricultural Engineering and Bioeconomy (ATB)
Potsdam, Brandenburg
Germany

Shweta Suri
School of Health Sciences & Technology
University of Petroleum and Energy Studies (UPES)
Dehradun, Uttarakhand
India

Mahipal Singh Tomar
Department of Food Science and Technology, Amity University
Mohali, Punjab
India

Pratiksha Wankhede
Kavayitri Bahinabai Chaudhari North Maharashtra University
Jalgaon, Maharashtra
India

Dhiraj Kumar Yadav
Department of Agricultural Engineering
North Eastern Regional Institute of Sciences and Technology
Nirjuli, Arunachal Pradesh
India

Preface

सर्वभूतेषु येनैकं भावमव्ययमीक्षते ।

(sarva-bhūteṣu yenaikaṁ bhāvam avyayam īkṣate)

अविभक्तं विभक्तेषु तज्ज्ञानं विद्धि सात्त्विकम् ॥

(avibhaktaṁ vibhakteṣu taj jñānaṁ viddhi sāttvikam)

'That knowledge by which one sees the one indestructible reality in all beings, undivided in the divided, knows that knowledge to be pure (sattvic)'.

Bhagavad Gita, 18.20

Cryogenic processing signifies a watershed moment in food science and technology. *Cryogenic Processing: Advances in the Food Industry*, published by John Wiley & Sons, has emerged from the combined knowledge of recognized researchers, food technologists, and practitioners from across the globe, each offering a significant contribution based on research and industry experience. It constitutes the most thorough and innovative treatment of cryogenic technology applicable to the food sector, a sector experiencing rapid and transformational change. Cryogenic processing relies on extremely low temperatures and can offer considerable benefits across the preservation, quality improvement, and safety of food products. This application can be applied to a vast spectrum of possible avenues of application: maintaining the structural integrity of fruits and vegetables, nutritional quality in dairy products, breaking through traditional hurdles in spice grinding, and herbal or pharmaceutical processing while also looking to best practices in food safety and sustainability principles. At a time when international food systems are under ever-increasing pressure to be more efficient to be safe, to have an extended shelf life, and to reduce environmental impact, the goal of this book is to extend a channel of scientific thought, with a one-stop shopping for students, scientists, industry, and policy makers. Each chapter captures some of the important knowledge – from the molecular level and processing alternatives to industrial engineering, quality assurance, and environmental considerations – culminating in case studies that are both cutting-edge in research but also applied. Collectively, the authors have not only highlighted important aspects of scientific rigor but also reflected on the ethical and sustainable practices that will establish the future of food science. Through detailed discussion and practical assistance, readers will be made aware of the ability that cryogenic processing may have to challenge conventional limits and encourage real-time solutions for longstanding issues in food manufacture, storage, and innovation. Whether for academics wishing to deepen their theoretical knowledge or practitioners wishing to bring ideas into practice, it is our intention that this volume will serve as a vehicle for knowledge and facilitate the bridging of the gap between basic research and meaningful transfer to the real world. Ultimately, we hope and believe that the collective effort within this book will provide new clarity and enable sound scientific discussion that will be beneficial to the ongoing development of the global food industry.

1

Cryogenics: An Overview and Introduction

Sudarshan Ramanathan[1], Teena Bijo[2], and Sumit Sudhir Pathak[1]

[1] *Division of Food Processing Technology, School of Engineering and Technology, Karunya Institute of Technology and Sciences, Coimbatore, Tamil Nadu, India*
[2] *School of Agriculture, Food and Ecosystem Sciences, The University of Melbourne, Melbourne, Victoria, Australia*

1.1 Introduction

Cryogenics is the science that addresses the production and processing of materials at very low temperatures. The word 'cryogenics' is taken from Greek, where the term '*kryos*' means 'frost or cold' and '*genic*' means 'to generate'. In scientific terms, a low-temperature environment is referred to as a cryogenic environment when the temperature falls below the threshold at which gases begin to liquefy. Till date various definitions have been given by various researchers around the globe based on their research experiences. Some of the important definitions of cryogenics are listed below:

- According to Professor Kamerlingh Onnes (1894), 'Cryogenics is the production of very low temperatures, specifically for the liquefaction of permanent gases like oxygen, nitrogen, hydrogen, and helium' (Zohuri 2018a,b).
- 'Cryogenic treatment is freezing a polymer solution at low temperatures, then storing it in a frozen state and then defrosting it', according to Auriemma et al. (2006).
- Cryogenics is defined by Merriam-Webster (2018) as 'the branch of science that deals with the production of low temperatures and the corresponding effects upon matter'.
- Cryogenics is 'the study and application of materials and systems at low temperatures', according to Qiu et al. (2021). The goal of this wide-ranging field is to reach extremely low temperatures, usually below 120 K (−153 °C, −244 °F).
- According to Font-Palma et al. (2021), 'Cryogenics strictly refers to processes occurring below 120 K, such as the condensation of nitrogen and oxygen, although the term is often used more broadly to imply low-temperature separation processes'.

Cryogenic Processing: Advances in the Food Industry, First Edition. Edited by Sumit Sudhir Pathak, Rama Chandra Pradhan, and Sudarshan Ramanathan.

1.1.1 Simple Concept of Cryogenics

As per these definitions, the utilization of liquefied gases for cold treatments is known as cryogenics. The commonly used gases for cryogenics are liquid oxygen, nitrogen, hydrogen, and helium (Shokrani et al. 2013). Permanent gases are elements that exist in their natural gaseous form and were previously thought to be impossible to liquefy. Defining a temperature criterion for the term cryogenics is difficult since the properties of materials can be quite different at varying temperatures. However, it may be said that it usually pertains to low-temperature phenomena at temperatures below those of chemical freezing; that is, temperatures from −150 °C to absolute zero (0 K), the latter being equivalent to −273.15 °C (Bosch 2021).

Cryogenics is also the study of material characteristics at low temperatures, including heat transmission, thermal design, and refrigeration. The field of cryogenics is not restricted to these uses. It also investigates the impact of low temperatures on biological systems and the feasibility of cryogenic preservation of biological materials (Radha et al. 2022). The technology of cryogenics utilizes very low temperatures, i.e. those below −150 °C, to solidify gases for storage and transportation for subsequent usages. Cryogens are defined as materials having a normal boiling point below about −150 °C; naturally occurring cryogens include helium, hydrogen, nitrogen, oxygen, and methane (Ganesan et al. 2021).

1.1.2 Significance of Cryogenics in General Science and Technology

These elements, i.e. hydrogen, oxygen, nitrogen, and helium, are liquefied for large-scale production and widespread applications; at present, hydrogen is the most widely used material for many industries, including aerospace and metallurgy. However, recent technological development has made large-scale liquefaction feasible for many other low-boiling substances. The main use of cryogenic fluids can be categorized as fuel for rocket propulsion; cooling agent in electronic, medical, and scientific devices; and raw material for gas-producing industry, distribution, and storage. This fundamental area of physics and engineering is practiced by many professionals, including physicists, chemists, engineers, materials scientists, and technicians.

Among the various industries and research institutions that make up the cryogenic working area, the chemical industry is the oldest one. Its activity primarily focuses on economic reasons such as the separation and purification of raw materials, and the storage and transport of products that require cryogenic temperatures during production or application phases.

1.2 Historical Developments of Cryogenics

For centuries, cryogenics has a history that goes back to the sixteenth century and made important contributions to science and technology. Two people who contributed substantially to the liquefaction of gases and therefore helped the field develop in the mid-1800s were James Dewar and Michael Faraday. Cryogenics has progressed over the decades through both theories and applications – from creating superconductors to modern-day medical applications using a magnetic resonance imaging (MRI) machine. Furthermore, cryogenics has been used for hundreds of years in the food processing industry. The chronological incidents of the evolution of cryogenics in the food processing sector are tabulated in Table 1.1.

Cryogenics has played a very significant role for the development of science and technology in the crucial periods from 1848 to 1969 (Flynn 2004). In 1848, John Gorrie constructed the first-ever mechanical refrigeration machine, signifying the advent of artificial refrigeration technology. Followed by this in 1866, the critical points and phase transitions of cryogenic fluids were researched by Van der Waals. Later, in the year 1877, Coleman and Bell produced the commercial version of Gorrie's System for beef processing to produce frozen meat and trading it to leading consumer markets. Significantly in the year 1882, Jules Violle invented the first vacuum-insulated flask, essential for preserving low temperatures in cryogenic applications.

During these developments in 1902, Pieter Zeeman received the Nobel Prize for his work on magnetism upon radiation, with links to cryogenics and fundamental physics, and another Nobel Prize was given to D. A. Glaser in 1960 for his invention of the bubble chamber, which utilized cryogenics for particle detection. In between these years of 1952; The National Bureau of Standards Cryogenic Engineering Laboratory was established to provide engineering data and to produce large quantities of liquid hydrogen (Flynn 2004). Further, the success of cryogenics and cryogenic technology prolonged in various fields of science till date with significant scientific achievements.

1.3 Fundamental Concepts of Cryogenics

Cryogenics, a field focused on studying materials at extremely low temperatures, relies on several fundamental principles that are crucial for its applications across various domains. One key principle in cryogenics is the utilization of cryogenic temperatures, below −150 °C, to induce unique physical and chemical properties in materials (Eghbali et al. 2012). These low temperatures result in reduced thermal motion of particles, allowing for the study of phenomena such as superconductivity and superfluidity (Militzer 2006). Additionally, the cryogenic environment helps to stabilize their unique properties, enabling detailed characterization and analysis (Wang et al. 2018). The fundamentals of cryogenics revolve around some basic principles of sciences and are discussed in the following subsections.

1.3.1 Behaviour of Materials at Room Temperature

The behaviour of materials at ambient temperature is an essential concern in cryogenics, especially concerning the performance and stability of materials under diverse thermal applications. At room temperature, materials are generally stable, but at cryogenic temperatures they may display altered mechanical and thermal characteristics, potentially impacting their employment in diverse technologies (Dasari et al. 2022). Johnson et al. (2014) suggested that all laboratory test devices and spacecraft constructions required mechanical assembly of separate elements. The mechanical characteristics of materials

Table 1.1 Chronological sequences of evolution of cryogenic processing in view of the food processing sector.

Year(s)	Events	References
1643	Evangelista Torricelli and the Barometer – Evangelista Torricelli invented the barometer, leading to the understanding of atmospheric pressure and vacuum, foundational concepts in cryogenics.	Torricelli (1643)
1867	Transportation of Frozen Beef – Henry P Howard used Gorrie's air chilling system to transport frozen beef from Indianola, Texas to the Gulf of Mexico.	Flynn (2004)
1877	Discovery of Liquid Oxygen and Liquid Nitrogen – Louis-Paul Cailletet and Raoul Pictet independently liquefy oxygen, and shortly after, nitrogen. First successful shipment of frozen meat from Marseilles to Buenos Aires	Cailletet (1877)
1877	Coleman and Bell produce a commercial version of Gorrie's system – The frozen meat trade becomes more successful and stems the Malthusian revolution	Flynn (2004)
1895	Carl von Linde and Industrial Liquefaction – Carl von Linde develops a practical method for liquefying large volumes of air, facilitating the production of liquid nitrogen and oxygen.	von Linde (1895)
1908	Heike Kamerlingh Onnes and Liquid Helium – Heike Kamerlingh Onnes liquefies helium, achieving temperatures close to absolute zero.	Onnes (1908)
1920s	Clarence Birdseye and Flash-Freezing – Clarence Birdseye developed the first commercial quick-freezing process for food, especially fruits and vegetables.	Birdseye (1929), Evers (1950)
1930s	Development of Cryopreservation Techniques – Initial exploration of cryopreservation techniques for biological materials.	Polge et al. (1930)
1950s	Advances in Freezing Technologies – Development of improved freezing technologies, such as blast freezing and plate freezing, which enhance the quality of frozen foods.	Smith (1950)
1970s	Cryogenic Freezing in the Food Industry – Adoption of cryogenic freezing using liquid nitrogen and carbon dioxide in the food industry for high-quality freezing of delicate products.	Thurston (1969)
1980s	Introduction of Individual Quick Freezing (IQF) – Introduction of Individual Quick Freezing (IQF) technology, allowing for the rapid freezing of small food items individually.	Andersen (1979)
2000s	Cryogenic Processing for High-Quality Food Preservation – Continuous improvements in cryogenic freezing methods to preserve the texture, flavour, and nutritional value of foods.	Goswami (2010)
2010s	Integration of Cryogenics with Automation – Integration of cryogenic freezing with automated processing lines for efficiency and consistency in food production.	Fluder et al. (2018).
2020s	Sustainable Cryogenic Technologies – Development of more sustainable cryogenic technologies, focusing on energy efficiency and reduced environmental impact.	Popov et al. (2019)

can fluctuate with temperature changes at the component level. The structural integrity and strength of materials under mechanical stresses at elevated temperatures must be evaluated to prevent plastic deformation or fracture of components. At the assembly level, interfaces (and related fastening procedures) need to consider material coefficient of thermal expansion (CTE) effects as well as potential loss of mechanical load while cooling down to temperature.

High-entropy alloys (HEAs) such as chromium (Cr), manganese (Mn), iron (Fe), cobalt (Co), and nickel (Ni) exhibit excellent tensile properties at cryogenic temperatures due to deformation twinning, a process that is challenging to achieve at room temperature due to insufficient stress. However, research has shown that room-temperature deformation twinning, coupled with non-recrystallization, can serve as a strengthening mechanism for ultra-high-strength HEAs. This suggests that while the mechanical properties of these materials are optimized for low temperatures, they can still maintain significant strength at room temperature through specific microstructural manipulations (Jo et al. 2017).

In the case of HEAs such as vanadium (V_{10}), chromium (Cr_{10}), iron (Fe_{45}), and cobalt (Co_{35}), both transformation-induced plasticity (TRIP) and twinning-induced plasticity (TWIP) are observed at room and cryogenic temperatures. Because of the strain-rate-hardening effect, which intensifies as twinning is triggered, this alloy's dynamic compressive strength is noticeably greater than its quasi-static strength. This implies that dynamic loading conditions may enhance room-temperature strength, although phase transformations typically produce an increase in cryogenic-temperature strength (Song et al. 2019).

1.3.2 Phenomena and Temperature Scales Related to Low Temperatures

Using cryogenic fluids to generate and maintain low temperatures, such as liquid nitrogen and helium, is another key principle of cryogenics (Pereira et al. 2010). According to Tang et al. (2019), these fluids are necessary for cooling systems in cryogenic applications because they ensure effective heat transfer and temperature control. Cryogenic fluids are perfect for a variety of cooling processes because of their high specific heat capacities and boiling points (Buzulutskov 2012).

Effective cooling techniques in cryogenics depend on accurate temperature control and measurement. Cryogenics mainly involves the use of very precise temperature scales due to the extremely low temperatures involved. The widely used temperature scales are Kelvin (K), Celsius (°C), and Fahrenheit (°F) scales, as they serve various purposes depending on the specific cryogenic application. The Kelvin scale is the gold standard for scientific temperature measurement in cryogenics. Unlike Celsius and Fahrenheit, Kelvin is an absolute temperature scale that begins at absolute zero (0 K), the theoretical point when all molecular motion ends, signifying the lowest limit of thermodynamic temperature. This is crucial in cryogenics because it exactly conforms to the rules of thermodynamics, where absolute zero (0 K, or −273.15 °C) acts as the standard for all temperature-related phenomena in the quantum mechanical realm of matter.

For many cryogenic processes, notably those involving superconductivity and superfluidity, Kelvin is crucial as these phenomena often appear only at temperatures near to absolute zero. The superconducting transition temperature of various materials, for example, is frequently in the range of 1–10 K. The temperature control in cryogenic applications such as cooling liquid helium (which transitions to a superfluid state below 2.17 K) is only conceivable with the accurate measurements afforded by the Kelvin scale (Chen et al. 2003). While the Celsius scale is not directly connected to thermodynamic rules, it remains widespread in industrial cryogenics, notably in food processing. Since Celsius is defined by the phase transitions of water – 0 °C as the freezing point and 100 °C as the boiling point – it offers practical benefits for defining temperatures in terms that are known to a broad variety of businesses. In cryogenic food processing, where temperatures generally range between −80 °C and −196 °C, the Celsius scale offers an easy framework for industrial operators to evaluate process temperatures.

For instance, liquid nitrogen, often employed in cryogenic freezing, has a boiling point of −196 °C. However, for more scientific explanations, these temperatures are typically adjusted to Kelvin to coincide with the absolute thermodynamic foundation needed for deeper research (Manickam 2013). Though less widespread in scientific circles, the Fahrenheit scale continues to see usage in commercial cryogenic applications in nations like the United States. In cryogenics, the freezing points and boiling points of cryogens, such as liquid nitrogen and liquid helium, are commonly given in Fahrenheit for clarity in commercial contexts. For instance, liquid nitrogen boils at −320.8 °F and liquid helium at −452.2 °F. Despite its practical value in some geographical contexts, the Fahrenheit scale is rarely adopted in the scientific literature because of its less obvious link to thermodynamic concepts (Barron 1985).

The second key pillar in cryogenics is the study of how matter behaves at low temperatures, particularly near absolute zero. At these high settings, classical physics gives way to quantum processes, leading to extraordinary changes in material characteristics. At low temperatures, quantum mechanical phenomena dominate the behaviour of some materials. Superconductivity, for example, is a condition in which a substance may conduct electricity with zero electrical resistance. Discovered in 1911, superconductivity only occurs below a material's critical temperature, which for many elemental superconductors is generally below 10 K. Materials like niobium-titanium, utilized in the manufacture of cryogenic magnets for MRI scanners and particle accelerators, transition into a superconducting state at roughly 9.2 K. This quantum state is vital in technologies requiring the transmission of electrical current without energy loss, which is allowed by the cryogenic freezing of these materials (Eremets and Troyan 2011).

Similarly, superfluidity is another quantum phenomena found in cryogenic settings. Helium-4, when cooled to 2.17 K, transforms to a superfluid state, defined by its ability to flow without viscosity. In this form, superfluid helium defies traditional fluid mechanics, allowing it to travel up walls and through small capillaries without resistance. Although superfluidity is of special interest in low-temperature physics, its applicability is limited in industrial processes due to the extensive infrastructure needed to sustain such low temperatures (Kolatzki et al. 2022). Designing functional cryogenic systems requires an understanding of phase transitions at cryogenic temperatures. For example, at cryogenic temperatures, hydrogen, oxygen, and nitrogen all transition from gas to liquid phases. Because it can quickly freeze food products while maintaining quality and minimizing the formation of ice crystals, liquid nitrogen, which has a boiling point of 77 K (−196 °C), is frequently used in cryogenic food processing.

Helium exhibits unusual behaviour since, at normal pressures, it remains a liquid until absolute zero and only solidifies at very high pressures. Picking the best cryogen for industrial uses, where specific temperature ranges and phase stability are needed, like cryogenic freezing or long-term storage of biological materials, requires an understanding of these phase behaviours. Cryogenics also demands a detailed grasp of thermodynamic concepts as they relate to low temperatures. As temperature falls, so does the kinetic energy of molecules, altering the specific heat, thermal conductivity, and vapour pressure of materials. For example, at cryogenic temperatures, the specific heat capacity of solids reduces drastically, meaning they require less energy to chill further. This is why materials in cold conditions frequently display different thermal characteristics, necessitating particular insulation to reduce heat transmission.

The phenomena of heat transmission in cryogenic systems, particularly radiative heat transfer, becomes insignificant at extremely low temperatures. Instead, conduction and convection dominate, with good cryogenic designs taking use of these principles to create stable, low-temperature conditions. For example, in cryogenic food processing, the quick heat transfer from the food product to the cryogenic fluid (e.g. liquid nitrogen) allows fast and uniform freezing, retaining the texture and quality of the food product. Understanding temperature scales and the behaviour of materials at low temperatures is vital for both research and industrial cryogenics. The Kelvin scale gives the absolute measure needed for precision in low-temperature research, whereas Celsius and Fahrenheit scales offer practical value in commercial applications. Low-temperature superconductivity and superfluidity are two quantum phenomena revealed by physics that redefine material properties and energy efficiency at cryogenic temperatures. These ideas form the basis of process optimization for cryogenic food processing, where rapid freezing and product quality preservation are essential.

1.3.3 Behaviour of Solids, Liquids, and Gases at Extremely Low Temperatures

Decreased thermal energy and quantum mechanical processes cause fundamental changes in the behaviour of gases, liquids, and solids at ultra-low temperatures. These changes are significant in cryogenics, which operates below 120 K, and have an immediate effect on a variety of industrial applications, including cryogenic food processing. Particularly, gases exhibit significant differences from how they behave in ambient conditions. Intermolecular interactions become more dominant as gases cool because molecules' kinetic energy drastically decreases. Gases transform into liquid phases during condensation as the process culminates. The widely used cryogen nitrogen, which condenses into a liquid at 77 K (−196 °C), is a notable example. When liquid nitrogen comes into contact with heated food, it vaporizes quickly, allowing for efficient heat evacuation and even freezing without the formation of large ice crystals that could detract from the texture of the food. Liquid nitrogen is an essential ingredient in cryogenic food processing because of its potent heat-absorbing capacity, which produces excellent preservation results by quickly and evenly freezing food (Johnson et al. 2014).

Additionally, gases can reach a supercritical state at extremely low temperatures, where they exhibit characteristics of both liquids and gases. Supercritical fluids are especially useful in specialized processes like supercritical CO_2 extraction because they are thick like liquids but can flow freely like gases. Although supercritical fluids are not commonly used in conventional cryogenic freezing, they offer potential for future developments in food preservation techniques. At cryogenic temperatures, liquid hydrogen, helium, and nitrogen exhibit very different behaviours. All of these liquids quickly freeze food, which lowers its temperature over a substantially shorter amount of time and produces smaller ice crystals that keep food, such as expensive prepared meals and seafood, intact, as opposed to freezing with larger ice crystals. While helium is not frequently used in food processing as nitrogen is, one unique aspect of helium is its superfluidity below 2.17 K. Due to being free of viscosity and frictionless, more is understood about ultra-low temperature physics and how any liquid drops of hydrogen and nitrogen behave in unusual temperature conditions.

The mechanical and thermal properties in solid-state record significant changes at cryogenic temperatures. As heat energy is lowered, atomic vibrations within solids are greatly lowered causing the material to contract. The need for construction of very durable materials that resist these temperatures for extended periods of time is essential within cryogenic food processing equipment, because the reduction in physical dimensions is clearly visible in cryogenic temperatures. Furthermore, the thermal conductivity of solids tends to increase with cooling, facilitating more efficient heat transfer from the frozen food items. Another crucial characteristic of solid behaviour is the inhibition of crystalline ice formation during fast freezing. In cryogenic freezing, foods adopt an amorphous or glassy state where ice crystals are either missing or substantially smaller than in ordinary freezing. This helps preserve the structural integrity of the food, avoiding cell wall breakage and ensuring that the product keeps its original texture, moisture, and nutritional content. Through such quick freezing, cryogenics also helps seal in nutrients, tastes, and colours, therefore keeping the freshness of food retained long after thawing.

1.3.4 Low-Temperature Thermodynamics

Cryogenic temperatures are generally achieved utilizing cryogenic fluids in either the liquid or solid phase. The fluid is used to acquire the sensible heat of the item (or system) in order to chill it down to the necessary operational temperatures. For a specific cooling application, the cryogenic fluid used has either a saturation temperature at 101 kPa (1 atm) or a triple point that roughly resembles the specified operational temperature for the item of interest (Johnson et al. 2014). In order to function efficiently at low temperatures, cryogenics also depends on cutting-edge technologies such as cryogenic detectors and cryogenic control systems (Hornibrook et al. 2015). Applications in physics and material research depend heavily on cryogenic detectors, like Cryogenic Avalanche Detectors (CRADs), which are specifically designed to function at cryogenic temperatures. Particularly in fields like quantum computing, cryogenic control systems are essential for managing and regulating operations in cryogenic environments.

1.3.5 Various Cryogenic Fluid Phase Transitions

The phase transitions of cryogenic fluids are necessary for their effective use in various scientific fields, such as food processing. The condition-dependent phase behaviour is different for important cryogenic fluids such as hydrogen, nitrogen, and helium. Because of their peculiar thermodynamic properties relative to other fluids, cryogenic fluids represent good cryogenics, high-pressure physics, and geophysical applications starting points, especially at low temperatures and high pressures, as shown in Figure 1.1.

Properties of liquid helium, a cryogenic fluid with unique low-temperature properties have been subjected to significant study. Below 2.2 K, liquid helium undergoes a phase change into the superfluid state and becomes a fluid with zero viscosity, as represented by the λ-line in Figure 1.1a. The properties of helium at temperatures below 2.2 K, in which it occupies a superfluid phase, as opposed to a normal liquid state, cannot be easily discussed aside from this transition. The graphic also emphasizes the critical point, beyond which helium transitions into a supercritical fluid phase, eliminating the boundary between liquid and gas phases (Kolatzki et al. 2022). Moreover, it is seen that at pressures below 1 bar, liquid helium changes immediately to a gas, circumventing the solid phase. Kolatzki et al. (2022) majorly in his research focused on the formation of micrometer-sized droplets from

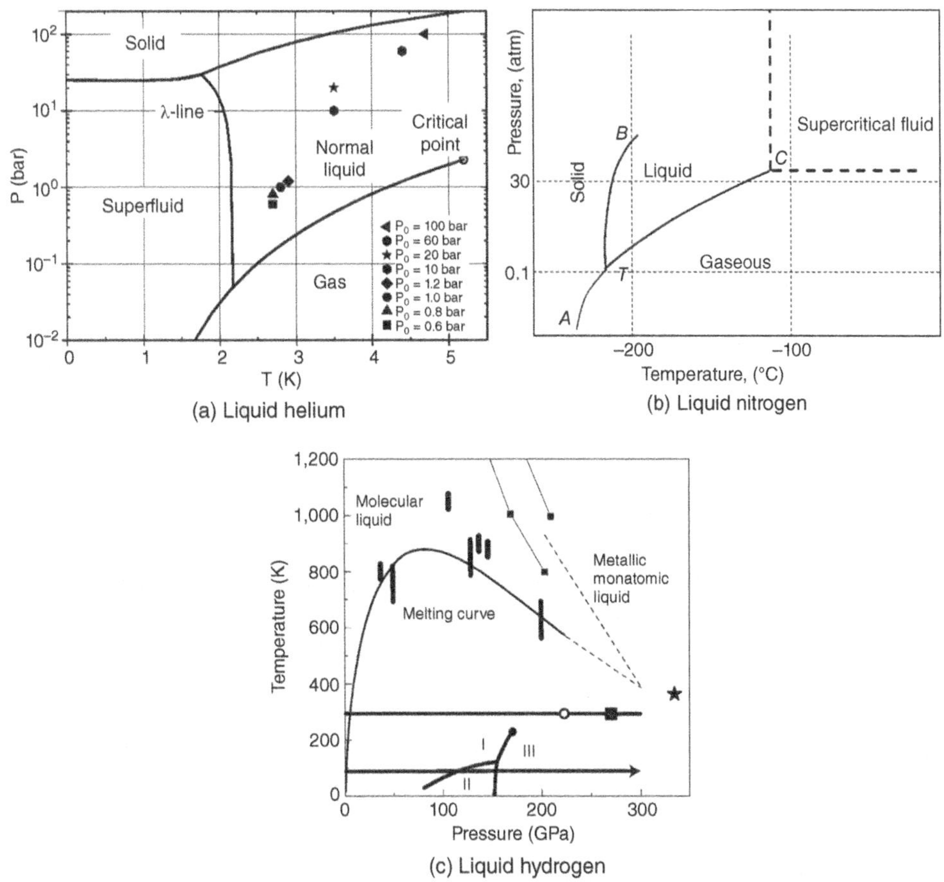

Figure 1.1 Phase transition graphs of (a) liquid helium (b) Liquid nitrogen and (c) liquid hydrogen. *Source:* Adapted from Kolatzki et al. (2022), Khuzin et al. (2019) and Eremets and Troyan (2011).

liquid helium, a phenomenon visible in the low-pressure superfluid area of helium's phase transition.

Another most widely used cryogenic fluid is liquid nitrogen, frequently utilized in cooling and food processing. The phase transitions and thermodynamic activity of liquid nitrogen are represented in Figure 1.1b. The graphic figure displays the boundaries between nitrogen's solid, liquid, and gaseous forms under varied pressures and temperatures. As temperature changes, nitrogen undergoes phase transitions between solid, liquid, and gaseous states at atmospheric pressure (roughly 0.1 atm). The supercritical fluid phase of nitrogen, which, like that of helium, denotes a state in which the fluid is neither clearly liquid nor clearly gas, begins at the critical point (C). For applications that depend on nitrogen's low boiling point and its behaviour in a variety of atmospheric conditions, this change is especially important (Khuzin et al. 2019).

Cryogenic fluid with potentially complex phase behaviour shows both metallic and molecular phases at very high pressures. The phase behaviour of hydrogen is timely, in particular, to understanding the potential for its use with respect to superconductivity and potential energy storage. In Figure 1.1, the boundary between the liquid metallic monatomic phase, which occurs only at extreme pressures, and the liquid molecular phase is represented in black. The melting curve indicates the transition of hydrogen from solid phase to liquid phase, with pressure facilitating a transition to the metallic phase. The dotted lines indicate a path to the metallic monatomic liquid, a rare form of hydrogen that exists at pressures greater than several hundreds of gigapascals (Eremets and Troyan 2011).

1.4 Cryogenic Equipment and Systems

In the field of cryogenics, cryostats play a critical role in maintaining very low temperatures for extended durations. These devices are commonly utilized in food processing for operations such as flash-freezing and preserving product freshness over long storage periods that require precise temperature regulation. Utilizing various cooling methods, including liquid nitrogen and helium, cryostats can achieve temperatures as low as −196 °C. Their ability to preserve samples at these extremely low temperatures makes them vital in cryogenic food applications.

Dewar flasks are another key component within cryogenic systems. These specialized containers are designed to store and transport cryogenic substances like liquid nitrogen and liquid oxygen. In food processing, Dewar flasks serve the purpose of temporarily holding cryogens during freezing processes. Typically constructed with double walls and vacuum insulation, these flasks minimize evaporation of the stored materials while reducing heat transfer. The vacuum layer effectively prevents heat from escaping, allowing the cryogenic liquids to maintain their low temperatures for prolonged periods.

Cryogenic refrigeration systems are necessary for food processing operations, especially on a large, industrial scale. These systems are used continuously and below freezing temperatures to allow for the rapid and effective preservation of food commodities. Industrial cryogenic freeze systems, like tunnel or spiral freezers, use liquid nitrogen or carbon dioxide as refrigerants. Quick cooling prevents ice crystal formation, which negatively impacts texture and flavour, and the quick freeze process not only preserves food but also benefits the quality of the product.

The performance of cryogenic processes is affected by vacuum insulation technology. Vacuum insulation uses a vacuum barrier that prevents conduction and convection (two important heat transfer processes), and this vacuum barrier limits the heat loss into cryogenic processes. Vacuum insulation as insulation for pipe, storage tank, and refrigeration systems ensure food products are kept at the necessary low temperature with little heat loss. Furthermore, superinsulation technology uses multiple layers of reflective material to also improve insulation performance. Superinsulation provides an additional level of protection by reflecting radiant heat in the vacuum layer, which is important for transportation systems and storage tanks where thermal leakage needs to be minimized. Superinsulation works particularly well when used in association with vacuum insulation, offering an excellent option for maintaining ultra-low temperatures in food processing applications.

1.4.1 Cryogenic Fluids and Their Properties

At very low temperatures, usually below 120 K, compounds known as cryogenic fluids, or cryogens, exist in a liquid state. These fluids – such as hydrogen, oxygen, nitrogen, and helium – are essential in many technical, scientific, and industrial applications because of their special qualities at cryogenic temperatures. Each cryogen has different thermodynamic properties; some of the most common cryogens are tabulated in Table 1.2. These properties completely influence the processing parameters.

For instance, argon's efficiency in cryogenic applications is demonstrated by its 87.3 K boiling point, 150.9 K critical temperature, 1,402 g/L liquid density, and 860 significant expansion ratio. Helium, on the other hand, is appropriate for ultra-low temperature applications due to its significantly lower liquid density of 125 g/L and expansion ratio of 780, as well as its noticeably low boiling point of 4.2 K. Additionally, carbon dioxide has a lower expansion ratio of 790 and a greater liquid density of 1,560 g/L, which may affect its choice for particular industrial processes; it also has a boiling point of 194.7 K (Karabuga et al. 2019).

Table 1.2 Thermodynamic properties of various common cryogens.

Common cryogens	Boiling point (K)	Critical temperature (K)	Critical pressure (atm)	Liquid density (g/L)	Gas specific density	Liquid to gas expansion ratio
Argon	87.3	150.9	48.3	1,402	1.39	860
Carbon dioxide	194.7	304.2	72.8	1,560	1.70	790
Helium	4.2	5.2	2.2	125	0.14	780
Hydrogen	20.3	33.0	12.8	71	0.07	865
Nitrogen	77.3	126.3	33.5	808	0.7	710
Oxygen	90.2	154.8	50.1	1,410	1.11	875

Source: Adapted from Zohuri (2018a,b).

1.4.2 Challenges in Cryogenic Fluid and Material Selection

The implementation of these systems involves major hurdles, notably in the selection of acceptable materials and the design of equipment capable of working under very low temperatures. The choice of materials that can withstand the heat stresses connected to cryogenic environments is one of the most important considerations. Since many metals weaken at cryogenic temperatures, materials must be extremely durable and resistant to brittleness. For example, due to its strength and ductility, stainless steel is used extensively; however, it may not be appropriate for all cryogenic applications, necessitating a careful assessment of substitutes such as copper or nickel for higher temperature ranges (Peterson and Weisend 2019).

Furthermore, in order to minimize heat transfer and sustain low temperatures, cryogenic systems need to be constructed with appropriate thermal insulation management. Adopting better insulating materials and making sure pipe systems are built to withstand the specific pressures caused by thermal expansion and contraction are necessary to achieve this. Another major issue is safety; cryogenic system failures can result in disastrous events like explosions or severe thermal burn injuries. Therefore, during the design and installation process, strict adherence to safety standards and norms (like ASME B31.3) is required. Furthermore, the financial aspect cannot be overlooked; although contemporary technologies may increase productivity, they are often more expensive, which presents a challenge for businesses seeking to implement cryogenic solutions in a sustainable manner (Taylor et al. 2015).

1.4.3 Safety Measures – Working with Cryogenic Systems

Safety precautions in cryogenic systems are vital to prevent accidents and protect the well-being of those involved. Key dangers include cold burns, asphyxiation from vaporized cryogens, and pressure-related events. The fast expansion of cryogenic liquids, such as liquid nitrogen and helium, can displace oxygen, causing an asphyxiation danger in poorly ventilated environments, necessitating the use of portable oxygen monitors (Bostock and Scurlock 2019).

Cryogenic liquids, with their extremely low temperatures, offer severe dangers of cold burns or frostbite upon direct contact with skin or tissue. These injuries occur when bodily components come into touch with surfaces or liquids at temperatures well below freezing, causing rapid cell destruction. Personal protective equipment (PPE), such as insulated gloves, aprons, and face shields made to withstand cryogenic temperatures, is one of the safety precautions. To prevent unintentional skin exposure, handling cryogenic materials should always be done with specialized tools like clamps or tongs. In order to prevent further tissue damage, operators must also be trained in cold injury first aid, which includes techniques like gradually warming the affected areas.

In cryogenic systems, asphyxiation poses a major risk, particularly in small or poorly ventilated spaces. Colourless, odourless, and heavier-than-air vaporized cryogens like nitrogen, helium, or argon displace oxygen and create an atmosphere that is oxygen-deficient. If inhaled in excess, this could cause unconsciousness or even death. In facilities that handle cryogens, safety protocols include installing adequate ventilation systems, such as oxygen monitors and exhaust fans. Sensors for oxygen shortage should be placed in strategic locations to provide early warnings of hazardous air conditions. Employees operating in such situations should also receive training on emergency evacuation procedures and be equipped with a portable oxygen supply if needed.

Liquids that can quickly turn into gases when heated and cause a significant pressure build-up are stored and transported in cryogenic systems. If there are not enough pressure relief devices, this growth may result in explosions or equipment failure. All cryogenic systems must be equipped with rupture discs, venting devices that can safely control pressure spikes, and pressure relief valves in order to get around this. To ensure operation, these parts must be regularly inspected and repaired. Additionally, because of the associated pressure, operators should not seal cryogenic liquids in closed containers, as this could lead to their rupture. Alarms and emergency shutdown procedures should be set up to address any pressure irregularities right away.

Beyond particular issues, rigorous monitoring and procedural discipline are essential to a facility's overall safety when handling cryogenic systems. Temperature, pressure, and gas composition sensors in automated systems allow for constant real-time monitoring and alert operators to anomalies. To stop unauthorized individuals from handling cryogenic equipment, facilities should have strict access controls. All employees should receive training on the characteristics and dangers of cryogens, emergency response protocols, and the importance of reporting incidents or equipment malfunctions. Ensuring operational safety and efficiency requires regular audits of safety procedures and compliance with international standards, such as those established by the Occupational Safety and Health Administration (OSHA) or comparable authorities. In cryogenic systems, safety measures are essential to avert accidents and safeguard human health. Cold burns, asphyxiation from vaporized cryogens, and pressure-related incidents are the main risks. The fast expansion of cryogenic liquids, such as liquid nitrogen and helium, can displace oxygen, causing an asphyxiation danger in poorly ventilated spaces, necessitating the use of portable oxygen monitors (Bostock and Scurlock 2019).

1.5 Impacts and Application of Cryogenics

Cryogenics has evolved as a transformative technology around various fields of science, which significantly enhances the properties of various materials at low temperatures. The major fields where cryogenics has shown its broad impacts are visualized in Figure 1.2.

1.5.1 Cryogenics in Space Science

Among these, cryogenics in space science has evolved over the last 25 years, enhancing reliability and simpler operations, matching the needs of advanced focal-plane detectors and complex science instrumentation (Rando 2013). Aerogel-based insulation systems improve performance for space launch applications by providing thermal insulating ability and not cryopumping beyond the initial cooldown and thermal stabilization phase (Fesmire 2006). Cryogenic propellants are critical to mission success in the thermal and microgravity environment of space (Meyer et al. 2013).

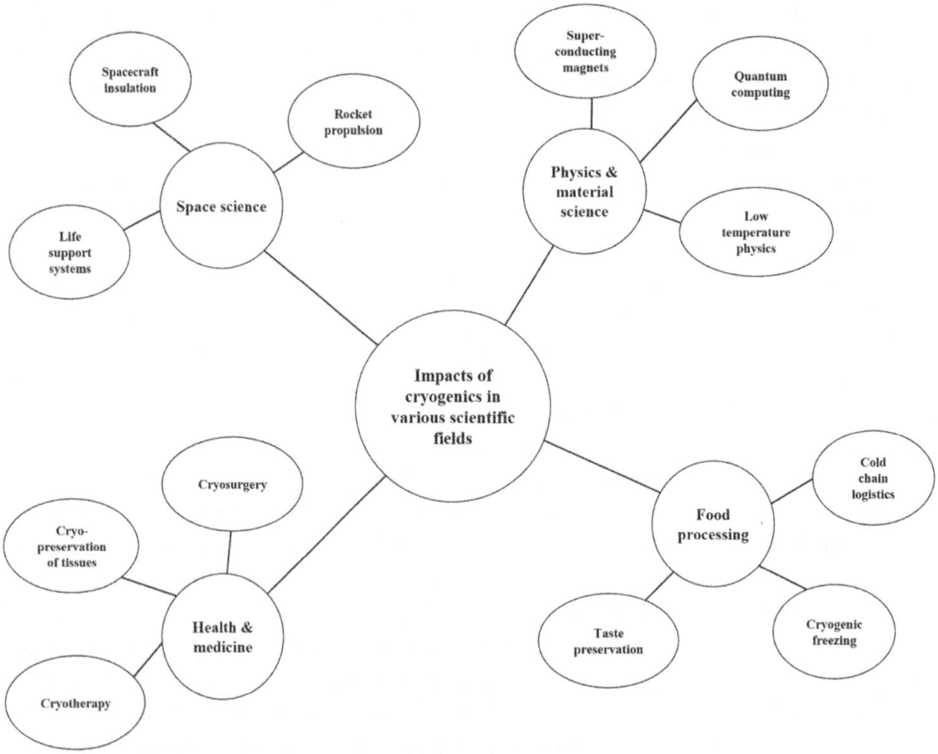

Figure 1.2 Impacts of cryogenics in various fields of science.

1.5.2 Cryogenics in Physics and Material Science

Cryogenic techniques have been crucial in solid-state physics, particularly in the study and application of superconductors. These techniques have enabled the development of superconductive devices used in large-scale scientific projects like bubble chambers and linear accelerators, as well as in cryoelectronics and cryogenic power systems (Erdmann 1969). Cryogenic electron microscopy can offer access to inaccessible length scales for materials characterization and address central questions in materials science, such as batteries, soft polymers, metal–organic frameworks, perovskite solar cells, electrocatalysts, and quantum materials (Li et al. 2020). Cryogenic electron microscopy enables high-resolution characterization of reactive and beam-sensitive materials, such as lithium in rechargeable batteries, and reduces the critical dose for damage. Cryogenics in quantum computing preserves quantum coherence using superconducting or high-electron-mobility topological materials, requiring deep cryogenic cooling for high-performance computing (Yazawa and Shakouri 2019).

1.5.3 Cryogenics in Health and Medicine

Cryogenics in advanced medical technologies, clinical treatments for life-threatening conditions, and high-quality biospecimen preservation in biomedicine (Mokbel et al. 2023). Especially in biology and medicine include cryosublimation fractionation, lipid extraction, and human whole body extreme cryotherapy, with potential applications in pharmacy, cosmetics, and the food industry. Cryosurgery, which resides at the interface of medicine and engineering, has attracted the interest of engineers, scientists, and medical doctors where cryosurgery is impacted by solution concentration, cooling rate, and embryo number, requiring thermophysical studies to solve problems during freezing and thawing. In the end of 1980s, technological advancements in cryotherapy equipment and enormous advances in imaging techniques, such as computed tomography and ultrasonography, have enabled surgeons and interventional radiologists to precisely guide cryogenic probes into tumours while minimizing damage to surrounding tissues.

1.5.4 Cryogenics in Food Processing

Cryogenics has made significant impacts in various scientific and industrial fields. One such field is the food processing industry, where cryogenics has found widespread use. Among the cryogenic fluids, liquid nitrogen (LN_2) stands out as a sustainable and safe choice in the food industry due to its versatility. Its adoption in food and beverage processing is driven by its superiority over mechanical refrigeration techniques in terms of food chilling and freezing. The demand for cryogenic processes in the food industry is staggering. According to a survey conducted in the United Kingdom alone, the food industry consumes approximately 100,000 tons of LN_2 per year. This cryogenic fluid is used to freeze a wide array of food products, including seafood, meat, vegetables, poultry, fruit, desserts, and bakery products. Cryogenic freezing and cooling equipment play a significant role in this sector, accounting for over 10% of the frozen food market and contributing to the rapid growth of both the frozen food and chilling markets.

The discovery of cryogenics has paved the way for advancements in food processing and preservation techniques. Cryogenics has been instrumental in areas such as cryopreservation, materials science, and cryogenic grinding (Balasubramanian et al. 2012; Saxena et al. 2018). Cryogenic freezing technology has come and changed the preservation process by which any living cells, tissues, and food products are protected from spoilage or decay by preserving them at very low temperature. In this technology, all types of moisture are removed from the food product and made solid. When liquid N_2 or CO_2 (cryogenic liquid) comes into contact with a product, it converts the phase from liquid to vapour with the help of heat transfer (Kumar 2018). Cryogenic techniques in ensuring food security including long-term conservation, maintenance of the regenerative ability of embryogenic tissues, enhancement of genetic transformation, and efficient production of pathogen-free plants (Wang et al. 2014). The application of low temperature shows a promising pathway to produce higher-quality end products with higher flavour and volatile oil retention. Application of cryogenics in the food industry has grown from its inception 12 years ago to nearly 100,000 tons per annum; most of this involves liquid nitrogen.

1.6 Influence of Cryogenics in Food

Cryogenics has a role in sustaining food quality at the molecular and particle levels. The use of cryogenic techniques such as cryogenic grinding and cryogenic freezing will help preserve bioactive components and phytochemicals necessary for nutrition and flavour (Aradwad et al. 2021). When looking to preserve both the chemical and physical structures of materials that may denature at higher temperatures, cryogenic procedures will optimize particle sizes through processing (Bhatta et al. 2020). The finer particle sizes produced through cryogenic grinding allow better extraction of phytochemicals with improved nutritional qualities and health benefits (Saxena et al. 2018). Pavlyuk et al. (2016) referred other advantages of a submicron particle size to hold phenolic and antioxidant compounds which are important for food products; shelf stability and health.

Cryogenic freezing improves the organoleptic properties of food products (e.g. taste, colour, texture or aroma) while still preserving the chemical structure of the product by reducing the size of the ice crystals that form during the freezing response. Cryogenic freezing reduces cellular damage to tissue cells in fruits and vegetables, improving their visual and sensory qualities. Studies have shown that volatile compounds and chemical essential oils are retained better with cryogenic treatments than traditional freezing methods, demonstrating the potential of cryogenically processed products for food quality, which is distinctively linked to their phytochemical content. Furthermore, cryogenic conditions preserve organoleptic features and strengthen the physical structure; so, for customers, the loss of food quality or nutritional value is practically nonexistent, making this a clear and continuously growing practice in the current food service industry.

1.7 Safety in Cryogenics

Since cryogenics means working with very low-temperature stuff, safety is one of the most important things to take care of. If not done correctly, it might lead to injury or equipment failure, or other accidents. Cryogenic fluids, which usually consist of liquid nitrogen, liquid oxygen or it may be one more like a liquid helium have properties that makes them very risky to work with. It is thus crucial to make sure the right handling and protection gear alongside system design are accounted for before operating in a cryogenic environment.

Specialized containment vessels such as Dewar flasks are used to minimize heat transfer and maintain low temperatures. This training should include the characteristics of cryogens, e.g. low temperatures in some cases (with boiling points below −150 °C), or how they expand and vaporize when coming into contact with an atmosphere at conventional temperature. This training also teaches how to minimize the chances of coming into contact with these fluids, which have been known to cause cases of severe frostbite and cold burns. Many heavily kerogen-rich sediments were believed to be under- or unsampled, particularly those from the Cretaceous and Eocene time intervals, where interest may focus on source potential of shale oil.

1.7.1 Hazards of Cryogenic Fluids

Cryogenic fluids, due to extremely low temperatures and rapid vaporization, are little unsafe. Direct contact may cause extensive burning of tissue whilst gases such as nitrogen and helium enrich air in confined spaces (leading to asphyxiation). Upon increasing the temperature of cryogenic liquids, they can rapidly expand to form gases possessing substantially increased pressure in the process – more than enough force is produced that will surely burst or blow up any closed container. The problem with the application of cryogenic fluids is that, in certain cases, it leads to embrittlement of materials, i.e. they become brittle and may fail under critical loads. Additionally, oxygen-rich atmospheres resulting from the spillage of liquid oxygen can increase fire hazards.

1.7.2 Safe Practices for Transporting

Handling cryogenic fluids requires correct PPE. This includes insulated cryogenic gloves, face shields or goggles, and full body apron. Therefore, cryogenic gloves are developed using materials that can reduce extreme cold but at the same time provide flexibility. Face protection like full-face shield protects against splashes or vapours. Workers must also wear coveralls with long sleeves to minimize skin exposure, and shoes that are closed top and insulated foot. The workplace shall be constructed to reduce exposure, and with a ventilation system and emergency shut-off systems close by, the workplace must be designed to minimize exposure.

1.7.3 Cryogenic System Safety

The safety risk in cryogenic systems is that against the mechanical effort of the temperature cycle, applied to each equipment from storage tanks through piping. Cryogenic materials have to be carefully selected as they may fail due to the fact that they become very brittle at these low temperatures. Leaks or structural failures that will affect the system integrity should be discovered during general maintenance and pressure testing. Vaporizing cryogenic liquids can lead to a dangerous pressure build-up within a tank, so emergency pressure relief systems are necessary to prevent imminent explosion.

1.7.4 Safety Aspects: Designing Cryogenic Systems to Prevent Failure and Accidents

Designing a safety cryogenic system requires multiple redundancy in the case of all devices, using appropriate materials that resist thermal stress under low temperatures and installing automatic pressure or temperature control systems. The heat transfer in the storage vessels for cryogenic liquids can be very fast, so attention needs to be paid while designing these vessels that sufficient insulation is given; otherwise, it could lead to quick vaporization of such a liquid. Most design features that prevent accidents such as double walled vessels, vacuum insulation, and pressure relief valves are part of the glycol dispensing system. In addition to this, system designs should be carefully engineered for thermal contraction and expansion during cryogenic operations.

Liquid nitrogen gas is non-hazardous, inflammable, and inert in nature, but several other risks are linked with it. Since these cryogens, such as liquid nitrogen, are held at a very low temperature of −196 °C, they are exceedingly cold. Hence, they may inflict a severe burn on the skin when spilled or exposed on the naked skin. If exposed to eyes in extreme conditions, it may cause loss of eyesight. It should always be kept in mind that the room should be adequately ventilated when the cryogenic equipment is operated. If it generates saturation in the room of more than 40%, it may cause respiratory issues. To avoid the instants of accidents, several safety precautions are enumerated below:

- The oxygen levels should be checked in the room where cryogenic activities are going on to avoid respiratory failure which is caused by the saturation of liquid nitrogen (LN_2) in the air.
- Avoid the exposure of liquid nitrogen on the exposed skin and other body parts, as it may lead to tissue embrittlement. Cryo-hand gloves with long sleeves, cryo-non-absorbent shoes, cryo-non-absorbent apron, face mask, and a face shield with glasses to protect the eyes. Exposure to liquid nitrogen for a prolonged time may induce tissue embrittlement.
- Careful handling, storage, transportation, and operations should be carried out under professional supervision to prevent the possibilities of mishaps.
- The pipes or the Teflon tubes transporting cryogens (LN_2) must be examined and inspected at regular intervals to avoid the leakage problems and loss of cryogen (LN_2).

1.7.5 Pressure Curtailment and Self-Explosion Management

One of the major potential hazards in a cryogenic system is pressure build-up – cryogens produce large quantities of gas from relatively small amounts of liquid. Pressure relief devices systems are also designed to include devices that will automatically relieve pressure by allowing excess gas to vent. Efforts must also be made not to create dead spaces where gas might pool, so cryogenic systems need venting as well. Regular inspection and maintenance are essential for valves, as well as pressure relief systems, to enable them to perform efficiently to contain overpressure. In addition to these safeguards, explosion-proof containment and relief systems also must be included in order facilitate the safe handling of pressure surges.

1.8 Limitations and Challenges in Cryogenics

Different hurdles and constraints in the employment of cryogenics have been linked with its application for food processing. A specific difficulty is the enormous overheads paid by cryogenic systems – not least of which are volumes upon copious expenditure on consumables such as liquid nitrogen (LN_2) and liquid carbon dioxide (CO_2). Some of these cryogens are pretty expensive, and if you were flash-freezing tons of food at a time, it may wind up costing too much money to be economically practical (Kumar 2018). On the other hand, while cryogenic freezing procedures can alleviate these concerns and increase product quality many times over, it is typically too expensive to justify its usage in commercial contexts unless high-value items are being employed. Another problem is the technological complexity of cryogenic systems. And these are systems in which you require particularly trained individuals with specific equipment to make them run safe and effective (Joy and Vanapalli 2023).

Due to the presence of cryogens, it is necessary for this sort of work, which demands rigorous safety precautions, since direct exposure might lead to serious burns produced by low temperatures. This also makes cryogenic technology less expensive for smaller food processors to employ, as they may not have the current utilities or expertise base. Fast-freezing techniques but product quality problems. Although cryogenic freezing decreases crystallization and assures a more natural texture after defrosting, there are possible difficulties linked with the rapid rates of temperature change, which might lead to cracking stress fractures in some types of food products. This phenomenon arises when the exterior layers freeze fast while the interior layers stay unfrozen longer, causing internal tensions that might damage product integrity (Kumar 2018).

The limited choice of appropriate food items for cryogenic processing is a difficulty. Not all foods respond positively to ultra-low temperatures; some may lose flavour or texture when subjected to quick freezing. This constraint needs careful selection and testing of materials intended for cryogenic treatment. Regulatory and environmental factors also play a role in the implementation of cryogenic technology. Compliance with safety standards regulating the use and storage of cryogens can complicate operations, particularly in locations with severe environmental legislation. The possible environmental impact of greater cryogen use must also be addressed, since it might lead to higher greenhouse gas emissions if not regulated appropriately (Joy and Vanapalli 2023).

1.9 Summary and Structure of the Book

The book *Cryogenic Processing: Advances in the Food Industry* gives a thorough analysis of how cryogenic technology is emerging in the field of food processing. This cutting-edge technology, which includes the utilization of extremely low temperatures, is being employed to increase the quality, shelf life, and performance of many food products. The book elaborates on the application of cryogenics not just in common food processing but also in specific fields such as major food groups, spices, herbs, and even medicinal items. Each chapter gives insights into how cryogenic methods improve the structural and functional characteristics of food components, changing the preservation, texture, and nutritional profile of processed meals. The book also dives into the technical and design elements of cryogenic systems, giving it a comprehensive reference for industry specialists.

The first chapter, 'Cryogenics: An Overview and Introduction', sets the tone by establishing the basic concepts of cryogenics, emphasizing its significance and expanding importance in food sectors. 'Chemical Foundations of Cryogenic Processes: A Chemical Perspective', the second chapter, takes a thorough dive into the chemical alterations that occur during cryogenic processing. This chapter discusses the chemical interactions at low temperatures that impact food quality and shelf life. The major changes in food texture, structure, and functional qualities generated by cryogenic treatment are discussed in the third chapter, i.e. 'Cryogenic Impact on Food Material qualities: Structural and Functional Transformations'. This chapter focuses on how cryogenic freezing assists in keeping moisture, avoiding spoilage, and maintaining the natural texture of food goods. Chapter 4, i.e. 'Industrial Designing and Engineering Aspects in Cryogenic Applications', examines the industrial-scale deployment of cryogenic systems. It examines the engineering issues involved in constructing effective cryogenic equipment for large-scale food processing.

The cryogenic applications with respect to particular food groups such as fruits, vegetables, plant based, dairy products, meat, poultry, marine products, spices and condiments are deeply discussed in the chapter from 5 to 9. Chapter 5 explains how cryogenic freezing aids in retaining the inherent features of these items without altering their flavour or texture. Similarly, Chapter 6 investigates the importance of cryogenics in dairy processing, notably in maintaining the quality and increasing the shelf life of milk, cheese, and other dairy products, and how the technique keeps the texture and flavour of these items, preventing quality losses. Chapter 7 focuses on the importance on cryogenics in plant-based food products and their texture properties. Chapter 8 completely focuses on the overall impacts of cryogenics in the meat and fish processing industries, while Chapter 9 discusses how cryogenic processing helps maintain important oils and volatile chemicals that are commonly lost during traditional processing methods.

The book also investigates uses outside food, such as in Chapter 10, 'Preserving Nature's Remedies: Cryogenic Uses in Herbal Medicine Processing', where cryogenic techniques are used for herbal medicine processing. This chapter shows how ultra-low temperatures retain the potency and usefulness of plants used in therapeutic goods. With these, the advanced emerging food products that vegan foods are also focused on in this book in Chapter 11, 'Vegan Chills: Cryogenic Technologies for Vegan-Based Products'. It addresses the growing market for vegan food products and explains how plant-based products' texture and stability can be enhanced through the use of cryogenics.

A significant chapter, Chapter 12, studies how cryogenic technology can impact microbial populations, resulting in improved food safety and decreased risk of bacterial contamination. Chapter 13 continues the discussion on the environmental implications of cryogenics, stressing the sustainability and energy efficiency of cryogenic systems in current food businesses. Finally, the last chapter of the book provides a glimpse into the future of cryogenic processing and shares information on developments and trends that may have an influence on food processing and preservation in the future.

1.10 Conclusion

Cryogenics, an important multidisciplinary subject, involves law enforcement in the physics, engineering, and applied sciences to manipulate materials and manipulate materials at ultra-low temperatures. This chapter has covered the fundamental features of cryogenics, beginning with its history from early studies of refrigeration to sophisticated processes like producing liquid helium and flash-freezing foods. The salient features of cryogenics include behaviour of gases, liquids, and solids at cryogenic temperatures, phase transitions, and thermodynamic concepts, especially those related to heat transfer and energy transfer. Understanding these features is important for proper application of cryogenic procedures in both research and industrial applications. The material also describes how cryogenic fluids (liquefied nitrogen, liquefied helium, and liquefied hydrogen) each have their own thermodynamic properties, which dictate the area of application.

Cryogenic systems (including cryostats, Dewar flasks, and vacuum-insulated technologies) are intended to be precise, efficient, and safe. However, challenges arise from material brittleness, pressure build-ups, and cost limitations. As such, using careful design considerations, ensuring inspections are undertaken, and properly training the methods used is crucial. Crucially, cryogenics has changed the food processing industry with methods like individual quick freezing (IQF), cryogenic grinding, and cryopreservation. These processes are beneficial in promoting quality, flavour retention, and nutrient retention. In addition, cryogenics is prevalent in a variety of fields, including space exploration, quantum computing, medicine, and materials science – showing its large impact and versatility.

The high operating costs, complex systems, and safety issues continue to be barriers, mostly for small companies, and overcoming them with innovation, increased integration of systems, and sustainable practices will shape the future of cryogenics. To sum up, cryogenics is not only important for the present technology but has many potential benefits in transforming typical processes within industries. The amount of knowledge and creativity discussed here allows for possibilities to advanced uses, which are discussed in the following chapters. Advanced uses in food technology such as cryofreezing and grinding make it possible to preserve shelf life, texture, and nutrients of the food. The chapter discusses the various aspects of the design of cryogenic systems, temperature scales, properties of materials, fluid dynamics, and the basics of cryogenics. It highlights challenges with safety, proper use of equipment, and industry standards. In conclusion, this chapter gives a solid grounding in understanding cryogenics, its significance in industry, and potential uses in advancing technology and food preservation.

References

Andersen, E. (1979). Forming of frozen liquids and pastes into convenient briquettes. *International Journal of Refrigeration* 2: 237–241. https://doi.org/10.1016/0140-7007(79)90015-X.

Aradwad, P.P., Kumar, A., Sahoo, P.K., and Mani, I. (2021). Key issues and challenges in spice grinding. *Cleaner Engineering and Technology* 5 (100347): 100347. https://doi.org/10.1016/j.clet.2021.100347.

Auriemma, F., Rosa, C., and Triolo, R. (2006). Slow crystallization kinetics of poly(vinyl alcohol) in confined environment during cryotropic gelation of aqueous solutions. *Macromolecules* 39 (26): 9429–9434. https://doi.org/10.1021/ma061955q.

Balasubramanian, S., Gupta, M.K., and Singh, K. (2012). Cryogenics and its application with reference to spice grinding: a review. *Critical Reviews in Food Science and Nutrition* 52 (9): 781–794. https://doi.org/10.1080/10408398.2010.509552.

Barron, R.F. (1985). *Cryogenic Systems*. US: McGraw-Hill Inc.

Bhatta, S., Stevanovic Janezic, T., and Ratti, C. (2020). Freeze-drying of plant-based foods. *Foods (Basel, Switzerland)* 9 (1): 87. https://doi.org/10.3390/foods9010087.

Birdseye, C. (1929). Some scientific aspects of packaging and quick-freezing perishable flesh products I—more rapid freezing means better preservation. *Industrial and Engineering Chemistry* 21 (5): 414–417.

Bosch, O.M.N. (2021). Thermodynamic and economic assessment of cryogenic energy storage (CES) systems operating in cogeneration regime.

Bostock, T.D. and Scurlock, R.G. (2019). Chapter 1. Safe handling and storage of cryogenic liquids. In: *Low-Loss Storage and Handling of Cryogenic Liquids: The Application of Cryogenic Fluid Dynamics* (ed. T.D. Bostock and R.G. Scurlock). Cham: Springer https://doi.org/10.1007/978-3-030-10641-6_8.

Buzulutskov, A. (2012). Advances in cryogenic avalanche detectors. *Journal of Instrumentation* 7 (02): –C02025. https://doi.org/10.1088/1748-0221/7/02/c02025.

Cailletet, L. (1877). Sur la liquéfaction de l'oxygène. *Annales De Chimie Et De Physique* 45–54.

Chen, J., Wang, J., Rao, R. et al. (2003). Research on the temperature control of thermal properties experiment of superconducting material. *Physica C: Superconductivity and Its Applications* 386: 544–546. https://doi.org/10.1016/S0921-4534(02)02164-0.

Dasari, S., Patnaik, S., Ray, B., and Prusty, R. (2022). Effects of in-situ cryogenic testing temperature and ex-situ cryogenic aging on the mechanical performance of glass fiber reinforced polymer composites with waste short carbon fibers as secondary reinforcements. *Polymer Composites* 44 (1): 294–304. https://doi.org/10.1002/pc.27045.

Eghbali, H., Sandra, K., Tienpont, B. et al. (2012). Exploring the possibilities of cryogenic cooling in liquid chromatography for biological applications: a proof of principle. *Analytical Chemistry* 84 (4): 2031–2037. https://doi.org/10.1021/ac203252u.

Erdmann, J. (1969). Applications of cryogenic techniques to materials research. *Cryogenics* 9: 411–421. https://doi.org/10.1016/0011-2275(69)90278-1.

Eremets, M.I. and Troyan, I.A. (2011). Conductive dense hydrogen. *Nature Materials* 10 (12): 927–931. https://doi.org/10.1038/nmat3175.

Evers, C. (1950). Sanitary problems in the processing and distribution of frozen foods. *Journal of Milk and Food Technology* 13: 35–39. https://doi.org/10.4315/0022-2747-13.1.35.

Fesmire, J. (2006). Aerogel insulation systems for space launch applications. *Cryogenics* 46: 111–117. https://doi.org/10.1016/J.CRYOGENICS.2005.11.007.

Fluder, C., Lefebvre, V., Wolak, T. et al. (2018). JACoW: Automation of the software production process for multiple cryogenic control applications. https://doi.org/10.18429/JACOW-ICALEPCS2017-TUPHA006.

Flynn, T. (2004). *Cryogenic Engineering, Revised and Expanded*. CRC Press.

Font-Palma, C., Cann, D., and Udemu, C. (2021). Review of cryogenic carbon capture innovations and their potential applications. *C – Journal of Carbon Research* 7 (3): 58. https://doi.org/10.3390/c7030058.

Ganesan, V., Patel, R., Hartwig, J., and Mudawar, I. (2021). Review of databases and correlations for saturated flow boiling heat transfer coefficient for cryogens in uniformly heated tubes, and development of new consolidated database and universal correlations. *International Journal of Heat and Mass Transfer* 179: 121656.

Goswami, T. (2010). Role of cryogenics in food processing and preservation. *International Journal of Food Engineering* 6: https://doi.org/10.2202/1556-3758.1771.

Hornibrook, J., Colless, J., Lamb, I., Pauka, S., Lü, H., Gossard, A., ... & Reilly, D. (2015). Cryogenic control architecture for large-scale quantum computing. *Physical Review Applied*, 3(2), 024010. https://doi.org/10.1103/physrevapplied.3.024010

Jo, Y., Jung, S., Choi, W. et al. (2017). Cryogenic strength improvement by utilizing room-temperature deformation twinning in a partially recrystallized VCrMnFeCoNi high-entropy alloy. *Nature Communications* 8: 15719. https://doi.org/10.1038/ncomms15719.

Johnson, W., Tomsik, T., and Moder, J. (2014). *Fundamentals of Cryogenics*. Cambridge University Press https://doi.org/10.1017/978110.149914.007.

Joy, M. and Vanapalli, S. (2023). Cryogenic nitrogen in the food and beverage industry. *iifiir.org*. https://iifiir.org/en/encyclopedia-of-refrigeration/cryogenic-nitrogen-in-the-food-and-beverage-industry. Accessed date: 18-02-2024

Arif Karabuga, Zafer Utlu, Reşat Selbaş. (2019). Examination of the liquefaction system for the use of different cryogenics in terms of thermodynamic analysis. *International Journal of Exergy*, 29, 1https://doi.org/10.1504/IJEX.2019.10021239.

Khuzin, R., Gazpromneft Badra, B.V., Yushchenko, T. et al. (2019). Changes in the properties of fluids and chemical reactions products caused by acid treatment of carbonate reservoirs. *Вестник Пермского национального исследовательского политехнического университета. Геология. Нефтегазовое и горное дело* 19 (3): https://doi.org/10.15593/2224-9923/2019.3.7.

Kolatzki, K., Schubert, M.L., Ulmer, A. et al. (2022). Micrometer-sized droplets from liquid helium jets at low stagnation pressures. *Physics of Fluids (Woodbury, N.Y.: 1994) 34* (1): 012002. https://doi.org/10.1063/5.0074026.

Kumar, Y. (2018). Cryogenic freezing technology. *International Journal of Pure & Applied Bioscience* 6 (2): 1343–1346. https://doi.org/10.18782/2320-7051.6458.

Li, Y., Huang, W., Li, Y. et al. (2020). Opportunities for cryogenic electron microscopy in materials science and nanoscience. *ACS Nano* 14 (8): 9263–9276. https://doi.org/10.1021/acsnano.0c05020.

Manickam, P. (2013). Cryogenics –an engineering tool for textiles and apparels. *Journal of Textile and Apparel Technology and Management* 8: 1–7.

Meyer, M., Chato, D., Plachta, D. et al. (2013). Mastering cryogenic propellants. *Journal of Aerospace Engineering* 26: 343–351. https://doi.org/10.1061/(ASCE)AS.1943-5525.0000297.

Militzer, B. (2006). First principles calculations of shock compressed fluid helium. *Physical Review Letters*, 97(17), 175501. https://doi.org/10.1103/physrevlett.97.175501

Mokbel, K., Kodresko, A., Ghazal, H. et al. (2023). Cryogenic media in biomedical applications: current advances, challenges, and future perspectives. *In Vivo* 38: 1–39. https://doi.org/10.21873/invivo.13407.

Onnes, H. (1908). *The Liquefaction of helium*. Communications from the Physical Laboratory of the University of Leiden.

Pavlyuk, R., Pogarska, V., Pavlyuk, V. et al. (2016). The development of cryogenic method of deep treatment of inulin-containing vegetables (topinambour) and obtaining of prebiotics in the nanopowders form. *Eureka: Life Sciences* 3: 36–43.

Pereira, H., Haug, F., Silva, P. et al. (2010). Cryogenic loop heat pipes for the cooling of small particle detectors at cern. https://doi.org/10.1063/1.3422264

Peterson, T.J. and Weisend, J. (2019). Properties of fluids and materials at cryogenic temperatures. In: *Cryogenic Safety: A Guide to Best Practice in the Lab and Workplace*, 1. Cham: Springer International Publishing https://doi.org/10.1007/978-3-030-16508-6_1.

Polge, C., Smith, A.U., and Parkes, A.S. (1930). Revival of spermatozoa after vitrification and dehydration at low temperatures. *Nature* 164: 666.

Popov, D., Fikiin, K., Stankov, B. et al. (2019). Cryogenic heat exchangers for process cooling and renewable energy storage: a review. *Applied Thermal Engineering* 153: 275–290.

Qiu, Y., Yang, H., Tong, L., and Wang, L. (2021). Research progress of cryogenic materials for storage and transportation of liquid hydrogen. *Meta* 11.

Radha, K., Decruse, W., and Krishnan, P. (2022). Plant cryopreservation. https://doi.org/10.5772/35919.

Rando, N. (2013). Cryogenics in space. In: *Observing Photons in Space*, vol. vol. 9 (ed. M.C.E. Huber, A. Pauluhn, J.L. Culhane, et al.), 639–655. https://doi.org/10.1007/978-1-4614-7804-1_37.

Saxena, S.N., Barnwal, P., Balasubramanian, S. et al. (2018). Cryogenic grinding for better aroma retention and improved quality of Indian spices and herbs: a review. *Journal of Food Process Engineering* 41 (6): https://doi.org/10.1111/jfpe.12826.

Shokrani, A., Dhokia, V., Muñoz-Escalona, P., and Newman, S. (2013). State-of-the-art cryogenic machining and processing. *International Journal of Computer Integrated Manufacturing* 26: 616–648. https://doi.org/10.1080/0951192X.2012.749531.

Smith, M. (1950). Technological advances in food freezing. In: *Food Engineering*, 1–50.

Song, H., Kim, D.G., Kim, D.W. et al. (2019). Effects of strain rate on room- and cryogenic-temperature compressive properties in metastable V10Cr10Fe45Co35 high-entropy alloy. *Scientific Reports* 9: 6163. https://doi.org/10.1038/s41598-019-42704-x.

Tang, Y., Mortimer, R., and Pan, G. (2019). Cryogenic circulation for indoor air pollution control. *The Science of the Total Environment* 651: 1451–1456. https://doi.org/10.1016/j.scitotenv.2018.09.220.

Brian Taylor, Jarvis Caffrey, Ali Hedayat, Jonathan, Stephens, Robert, Polsgrove. (2015). Cryogenic Fluid Management Technology Development for Nuclear Thermal Propulsion. 51st AIAA/SAE/ASEE Joint Propulsion Conference. https://doi.org/10.2514/6.2015-3957.

The Merriam-Webster Dictionary (2018). *Cryogenics*. Springfield, MA: Merriam-Webster, Inc.

Thurston, R.S. (1969). Fifteenth cryogenic engineering conference. *Cryogenics* 9 (5): 393–397.

Torricelli, E. (1643). The invention of the barometer. *Journal of Historical Science* 5–10.

Von Linde, C. (1895). The Linde technique for liquefying air. *Journal of Industrial Chemistry* 15–35.

Wang, B., Wang, R., Cui, Z. et al. (2014). Potential applications of cryogenic technologies to plant genetic improvement and pathogen eradication. *Biotechnology Advances* 32 (3): 583–595. https://doi.org/10.1016/j.biotechadv.2014.03.003.

Wang, X., Li, Y., and Meng, Y. (2018). Cryogenic electron microscopy for characterizing and diagnosing batteries. *Joule* 2 (11): 2225–2234. https://doi.org/10.1016/j.joule.2018.10.00.

Yazawa, K. and Shakouri, A. (2019). Integrated cryogenic quantum refrigeration of electrons ... https://doi.org/10.18462/IIR.ICR.2019.1653

Zohuri, B. (2018a). Cryogenics and liquid hydrogen storage. In: *Hydrogen Energy*. Cham: Springer https://doi.org/10.1007/978-3-319-93461-7_4.

Zohuri, B. (2018b). The beginning and concept of cryogenics, basic principles. In: *Physics of Cryogenics*, 387–473. https://doi.org/10.1016/b978-0-12-814519-7.00014-8.

2

Molecular Foundations of Cryogenic Processes: A Chemical Perspective

Sakkaravarthy Abishek and Parameswaran Gurumoorthi

Department of Food Technology, School of Bioengineering, SRM Institute of Science and Technology, Kattankulathur, Chennai, Tamil Nadu, India

2.1 Introduction

2.1.1 Overview of Cryogenic Processing in the Food Industry

Cryogenic processing refers to the application of extremely low temperatures, primarily using liquid nitrogen (LN_2, −196 °C) or solid or liquid carbon dioxide (CO_2, −78.5 °C), to rapidly freeze or stabilize food products. This technique leverages the high latent heat absorption capacity of cryogens and direct-contact heat transfer to achieve ultrafast freezing rates. The principal objective is to minimize the formation of large extracellular ice crystals by driving water within the food matrix into a vitrified or microcrystalline state, thereby preserving the cellular architecture and maintaining organoleptic properties, such as texture, flavour, and colour (Goswami 2017). Cryogenic freezing is commonly executed through immersion freezing, where products are submerged in a liquid cryogen, and spray or tunnel freezing, where atomized cryogen is directed over products moving along a conveyor system. In some systems, fluidized-bed technology is employed to suspend and evenly freeze products in a cold gas stream, particularly for individual quick freezing (IQF) applications (Zohuri 2017).

The effectiveness of the process relies on the optimization of equipment parameters, i.e. nozzle geometry, cryogen flow rate, belt speed, bed depth, and residence time. These parameters regulate the velocity of freezing front, degree of super cooling, and nucleation site density, which collectively determine the morphology and distribution of ice crystals, with direct implications for the quality of the final product (Yazdani-Asrami et al. 2022). Cryogenic processing also offers significant technological advantages, including the formation of smaller ice crystals, leading to improved preservation of food texture, colour, aroma, and nutritional value. The reduced dehydration and drip loss are important for high-moisture foods such as seafood, fruits, and muscle-based products, and lower oxidative degradation and protein denaturation compared to traditional methods (Goswami 2010).

Cryogenic Processing: Advances in the Food Industry, First Edition. Edited by Sumit Sudhir Pathak, Rama Chandra Pradhan, and Sudarshan Ramanathan.

Cryogenic grinding also improves the performance of heat-sensitive materials, such as spices, by preserving volatile compounds and yielding finer, more potent powders (Junghare et al. 2017). In addition, it plays a key role in advanced food processes, such as microencapsulation, freeze-drying pretreatment, and enzymatic inactivation during minimal processing. However, this technology is associated with high operating costs owing to cryogen consumption and the need for special maintenance. It also offers the advantages of lower capital investment and compact equipment footprint, thus making it suitable for small-batch, high-value applications. From a regulatory and safety perspective, strict compliance with the Food and Drug Administration (FDA), and European Food Safety Authority (EFSA) and Codex standards is essential to ensure food safety, and appropriate precautions must be taken to manage risks related to asphyxiation, frostbite, and equipment thermal stress (Balasubramanian et al. 2012).

A chronological overview of the development of cryogenic technology, tracing its progression from foundational discoveries to modern advancements, along with the associated implications for food preservation, processing techniques, and product innovation, is presented in Table 2.1. Recent innovations have enabled the integration of artificial intelligence (AI)-based monitoring systems, real-time sensor technologies, and hybrid freezing systems that combine the use of cryogenic freezing with ultrasound and high-pressure processes. Additionally, there is growing interest in developing environmentally friendly cryogens with reduced global warming potential (GWP) (Boehm 2013). Despite these advancements, challenges such as energy inefficiency, high operating costs, and incomplete life-cycle assessments persist in the field. Future directions are expected to prioritize modular design, energy recovery, process intensification, and synergy with other preservation methods, such as super cooling, isochoric freezing, and modified atmosphere storage. Altogether, cryogenic processing is a highly precise, scientifically grounded method that merges the principles of advanced thermal engineering, food microstructure protection, and sustainable manufacturing, positioning it as a key enabler of the next generation of high-quality food preservation technologies.

Table 2.1 Evolution of cryogenic technology and its impact on the food industry.

Era	Key milestones	Food industry implications	References
1877–1900s	Liquefaction of gases (oxygen, nitrogen, CO_2)	Foundation of cryogenic science	Sloane (1900)
1940–1950s	Military rations, experimental uses	Long-term storage of food in defence and space sectors	Cecil and Woodruff (1962)
1960–1970s	Commercial cryogenic freezing systems introduced	Rapid freezing, IQF technology, reduced food spoilage	Cleland (2020)
1980–1990s	Automation, cryogenic grinding, encapsulation	Improved preservation of spices, dairy, bioactives	Singh and Meghwal (2021)
2000s–Now	Smart sensors, hybrid freezing, sustainability focus	Functional foods, frozen ready meals, clean label tech	Alam et al. (2025)

2.2 Thermodynamic Principles of Cryogenic Systems

Cryogenic systems in food processing are governed by the foundational thermodynamic principles that control the interactions between cryogenic agents and food matrices. The success of cryogenic freezing depends not only on reaching ultra-low temperatures but also on the rapid and uniform removal of thermal energy from the product. This involves complex heat and mass transfer mechanisms, phase equilibria behaviour, and energy exchange phenomena, all of which impact the physical and quality attributes of frozen foods.

2.2.1 Heat Transfer and Temperature Gradients

In cryogenic freezing, heat transfer occurs primarily via convection and conduction as the cryogen (usually liquid nitrogen or carbon dioxide) comes into direct or near-direct contact with the food surface. The large temperature differential often exceeding 200 °C between the cryogen and the product drives a steep thermal gradient. This gradient is essential for rapid nucleation and high freezing front velocity, leading to the formation of fine and uniformly distributed ice crystals. The three stages of freezing are pre-cooling, phase transition, and subcooling, which are governed by transient heat transfer phenomena. During pre-cooling, the sensible heat is removed until the product reaches its initial freezing point. The latent heat of fusion is then extracted to initiate and complete the phase change from water to ice. Finally, subcooling is continued as the product temperature drops below the freezing point, stabilizing the structure. These stages are governed by transient heat transfer phenomena. The Fourier's law of heat conduction (Eq. 2.1) describes the rate at which heat is conducted within the food:

Governing Equation: Fourier's Law of Heat Conduction

$$q = -k \times A \times \frac{dT}{dx} \tag{2.1}$$

where q = heat transfer rate (W), k = thermal conductivity of food (W/m·K), A = surface area (m^2), and $\frac{dT}{dx}$ = temperature gradient (K/m).

Biot and Fourier Numbers: Transient Heat Transfer
To evaluate the effectiveness of heat transfer within a food body, two dimensionless numbers are commonly used:

Biot Number (B_i):
The Biot number defines whether surface convection resistance or internal conduction resistance dominates. *A* small B_i ($\ll$1) indicates uniform internal temperature; a large B_i ($\gg$1) implies a steep internal gradient common in cryogenic processes, as shown in Eq. (2.2):

$$B_i = \frac{hL_c}{k} \tag{2.2}$$

where h is convective heat transfer coefficient (W/m^2·k); L_c is characteristic length (m); and k is thermal conductivity of food (W/m·k).

Fourier Number (F_o):
The Fourier number represents the extent of internal temperature homogenization during freezing. A higher F_0 indicates greater penetration of cold into the product overtime, as shown in Eq. (2.3).

$$F_0 = \frac{\propto t}{L_c^2} \tag{2.3}$$

where $\propto$ is thermal diffusivity (m^2/s); t is time (s); and L_c is characteristics length (m).

By applying Eqs. (2.1)–(2.3), engineers can predict freezing times based on food composition, size, geometry, and process conditions, ensuring optimal quality preservation during cryogenic freezing.

2.2.2 Phase Diagrams and Freezing Point Depression

Phase diagrams of water and multicomponent food systems provide critical insights into cryogenic freezing behaviour. In pure water, the freezing point is 0 °C under 1 atm, but in complex food systems, freezing point depression occurs because of the presence of solutes such as salts, sugars, and proteins. This colligative property of ice formation begins at a lower temperature, and not all water freezes at once, leading to the creation of a freeze-concentrated matrix. The progressive concentration of solutes during freezing further lowers the local freezing point and increases the viscosity, thereby affecting the ice morphology and glass transition behaviour. Understanding the solid–liquid equilibrium and associated eutectic points is essential for optimizing cryogenic conditions, especially in products where textural or biochemical stability is sensitive to freezing dynamics.

Freezing Point Depression Equation (Ideal Case):
As ice forms, the concentration of unfrozen solutes increases further, lowering local freezing temperature. Cryogenic systems often reach below the glass transition temperature (Tg) of the freeze-concentrated phase, stabilizing sensitive food components. The extent of freezing point depression can be calculated using Eq. (2.4):

$$\Delta T_f = i \times K_f \times m \tag{2.4}$$

where ΔT_f is depression in freezing point (°C), i is van't Hoff factor (number of particles), K_f is cryoscopic constant of water (1.86 °C·kg/mol), and m is molality of solutes.

In food systems, i varies depending on whether the solute is a non-electrolyte (e.g. sugars, $i \approx 1$) or an electrolyte (e.g. salts, $i > 1i > 1i > 1$). A higher solute molality (mmm) increases ΔT_f, meaning the remaining unfrozen water stays liquid at lower temperatures. This is a critical factor in cryogenic freezing because it contributes to super-cooling and enhances structural preservation during storage.

2.2.3 Energy Changes During Cryogenic Transitions

The energy transformations during cryogenic freezing include both sensible heat removal and latent heat extraction (Figure 2.1). Sensible heat is the energy required to cool the product from its initial temperature to the onset of ice formation. Latent heat, typically

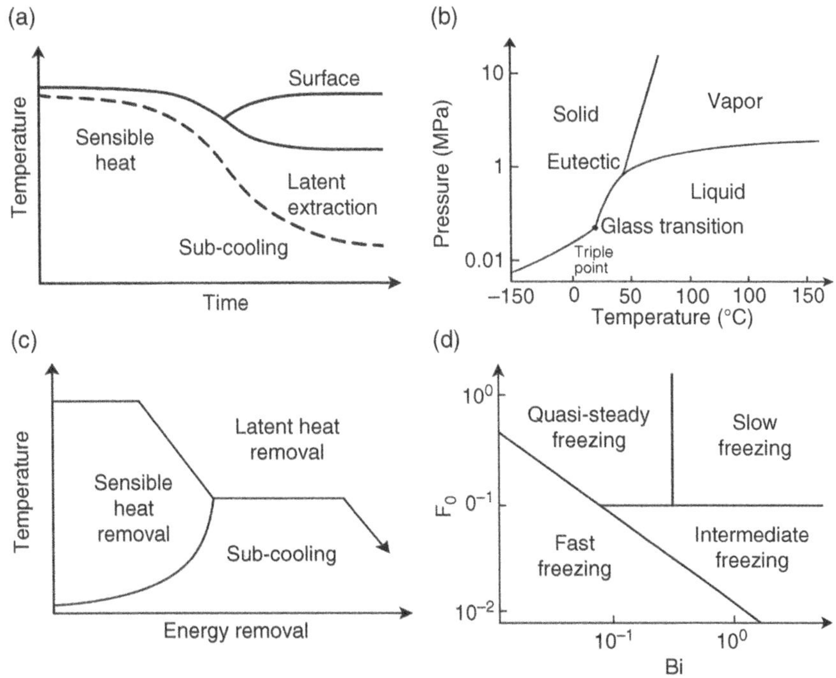

Figure 2.1 (a) Transient temperature profile in a cryogenically frozen food (core vs. surface), (b) water phase diagram with eutectic and glass transition zones marked, (c) energy pathway showing sensible + latent heat removal and subcooling, (d) Biot and Fourier number map demonstrating different freezing regimes.

around 334 kJ/kg for water, is the energy absorbed or released during the phase change without a change in temperature. Cryogenic agents themselves undergo significant energy transformations. For instance, when liquid nitrogen contacts the food, it absorbs heat and undergoes a liquid-to-gas phase change (boiling/evaporation), absorbing latent heat of vaporization (~199 kJ/kg for LN_2). This endothermic process is extremely efficient for heat extraction, enabling rapid surface cooling. In CO_2 systems, both sublimation (solid to gas) and vaporization (liquid to gas) transitions may occur, depending on the delivery form. Each transition consumes energy, which is drawn from the food product, contributing to rapid thermal reduction. The overall enthalpy change experienced by the food system must be carefully managed to avoid thermal shock, surface cracking, or non-uniform freezing.

Total Energy Equation

The total energy removed to bring the product from its initial temperature to final subcooled state is shown in Eq. (2.5):

$$Q_{\text{total}} = m \times c_p \times \left(T_i - T_f\right) + m_w \times L_f \tag{2.5}$$

where Q_{total} is the total energy removed (kJ), m = mass of food (kg), c_p is specific heat capacity (kJ/kg· K), T_i, T_f are initial and final temperatures (°C), m_w is mass of freezable water, and L_f is latent heat of fusion (~334 kJ/kg for water).

The first term represents sensible heat removal before and after phase change, while the second term represents latent heat extracted during ice formation.

Cryogen-Specific Energy Consumption

The energy absorbed by the cryogen during the freezing process depends on its phase-change properties, as shown in Eq. (2.6):

Liquid nitrogen (LN_2):

- Boiling point: −196 °C
- Latent heat of vaporization: ~199 kJ/kg
- 1 kg of LN_2 can absorb ~199 kJ as it vaporizes:

$$Q_{LN_2} = m_{LN_2} \times \Delta H_{vap} \tag{2.6}$$

Carbon dioxide (CO_2):

- Sublimation point: −78.5 °C
- Latent heat of sublimation: ~574 kJ/kg
- Higher energy absorption per unit mass than LN_2 but operates at higher temperature.

2.3 Kinetic and Molecular Motion in Cryogenic Conditions

Cryogenic processing of food induces profound changes in molecular dynamics owing to the extreme reduction in temperature. This section explores the influence of cryogenic temperatures on key molecular behaviours, including vibrational, rotational, and translational motion. The discussion includes the impact on diffusion rates, reaction kinetics, and the glass transition phenomenon, all of which underpin the stability and functionality of cryogenically preserved food systems.

2.3.1 Molecular Vibrations and Rotations at Low Temperatures

Nucleation Dynamics: Molecular dynamics simulations reveal that nucleation rates and cluster formation in cryogenic gases (e.g. N_2 and O_2 in H_2 and He) are highly sensitive to temperature. At 30–80 K, the energy released during cluster formation can significantly increase local temperature, but overall, the critical cluster size grows as temperature decreases, and nucleation rates can deviate from classical predictions (Song et al. 2023).

Non-Arrhenius Behaviour: Some processes, such as grain boundary migration, can exhibit unexpected kinetics at cryogenic temperatures. When activation barriers are low, certain motions may proceed more rapidly at low temperatures than predicted by Arrhenius behaviour, due to the unique interplay of thermal energy and barrier heights (Homer et al. 2022). Molecular vibrations and rotations are thermally driven and thus significantly constrained at cryogenic temperatures. According to the equipartition theorem, each degree of freedom contributes energy proportional to temperature (2.7):

$$E = \frac{1}{2} \times kT \tag{2.7}$$

where E is the average energy per degree of freedom, k is the Boltzmann constant, and T is the absolute temperature (in Kelvin).

At cryogenic temperatures (typically <100 K), the available thermal energy is insufficient to maintain higher vibrational and rotational states:

Vibrational Modes: At cryogenic temperatures, molecular vibrations are significantly dampened. According to quantum mechanical models, vibrational energy levels are quantized, and the population of molecules in excited vibrational states follows the Boltzmann distribution. As thermal energy decreases, fewer molecules occupy higher vibrational states, reducing the amplitude of bond oscillations. This phenomenon contributes to the suppression of enzymatic and degradation reactions in foods, as molecular rearrangement become energetically unfavourable.

Rotational Motion: Similar to vibrations, rotational motions are also quantized. In low-temperature environments, the energy required for molecular rotation surpasses the thermal energy available. This leads to a near-complete 'freezing' of rotational degrees of freedom. As a result, polar molecules that typically reorient in response to electric fields, such as those in dielectric materials become rigid, affecting dipole relaxation behaviours and dielectric constants.

Anharmonic Contributions: At room temperature, molecular vibrations often exhibit anharmonicity, leading to thermal expansion and non-linear reaction pathways. In cryogenic environments, harmonic behaviour dominates due to minimal vibrational excitation, reducing entropy-driven conformational transitions and providing structural stabilization in biological tissues and food macromolecules (e.g. proteins, polysaccharides).

2.3.2 Diffusion and Reaction Rate Alterations

Diffusion and chemical reaction rates are strongly temperature-dependent. The diffusion coefficient D follows an Arrhenius-type relationship (2.8) (Peleg et al. 2012).

$$D = D_0 \times e^{-E_a/RT} \tag{2.8}$$

where D is diffusion coefficient, D_0 is pre-exponential factor, E_a is activation energy for diffusion, R is universal gas constant, and T is absolute temperature (K).

As T decreases:

- Molecular diffusion slows dramatically, impacting mass transport in the food matrix
- Enzymatic reactions are inhibited both by enzyme immobilization and reduced substrate diffusion.
- Microbial metabolism ceases due to blocked nutrient transport and enzyme inactivity.

These effects explain the high preservation potential of cryogenic systems: enzymatic browning, lipid oxidation, and microbial spoilage are nearly halted.

2.3.2.1 Effects on Diffusion

Translational Mobility Reduction: Molecular diffusion in both liquids and amorphous solids becomes negligible under cryogenic conditions. The diffusion coefficient (described by the Einstein–Stokes relation) is directly proportional to temperature and inversely proportional to viscosity. As water transitions to ice or enters a glassy state, viscosity becomes effectively infinite, and diffusion ceases.

Implications in Food Preservation: This immobilization drastically inhibits lipid oxidation, Maillard reactions, enzymatic browning, and microbial activity in cryopreserved foods. Molecules such as oxygen, pro-oxidants, and flavour-active volatiles become trapped in place, preserving the original quality and aroma of the food matrix.

2.3.2.2 Effects on Chemical Reaction Rates

Inhibition of Autocatalytic Processes: Self-propagating reactions such as lipid peroxidation or polyphenol oxidase-driven reactions are halted due to kinetic suppression. The energy required to break molecular bonds is not available, even in the presence of catalytic enzymes.

Quantum Tunnelling at Ultra-Low Temperatures: In some extremely low-temperature systems (e.g. $<10\,K$), certain reactions may proceed via quantum tunnelling rather than classical activation. Although rare in food systems, this underscores the fundamental shift in reaction mechanism in cryogenic regimes.

2.3.3 Glass Transition and Molecular Immobilization

The glass transition temperature (Tg) is a critical concept when considering amorphous food systems at cryogenic temperatures. Unlike crystalline solids, amorphous materials (such as sugars, starches, and proteins in freeze-dried or frozen foods) do not exhibit a sharp melting point but instead undergo a gradual transition from a viscous or rubbery state to a brittle glass-like state.

2.3.3.1 Glass Transition in Cryogenic Food Systems

Thermodynamic Arrest: Below Tg, molecular motion becomes so limited that the system is considered to be in a non-equilibrium solid-like state. Water, for example, when trapped in an amorphous matrix, exhibits a glass transition around −135 °C, forming a 'vitreous ice' phase that lacks crystalline structure.

Stabilization of Biological Structures: The vitrification of cytoplasm and cellular components in plant and animal tissues is essential for maintaining integrity during cryopreservation. The immobilization of water and solutes prevents ice crystal formation, which would otherwise cause mechanical damage to membranes and proteins.

Food Processing Applications: Understanding and manipulating Tg is key to designing stable cryogenic processes. For instance, sugars such as trehalose and sucrose are often used as cryoprotectants due to their high Tg values, which contribute to vitrification and molecular immobilization at higher subzero temperatures.

2.3.3.2 Kinetic Consequences of Glassy States

Cessation of Diffusion and Relaxation: Below Tg, molecular diffusion coefficients drop by several orders of magnitude, and structural relaxation timescales increase to the point where no macroscopic changes occur within practical time frames. This effectively 'locks in' the molecular configuration, offering long-term stability for bioactive compounds, aromas, and pigments in foods.

Impact on Freeze-Drying: The freeze-drying process often exploits the glass transition to maintain structural integrity. During sublimation, if the food remains below its Tg, collapse and shrinkage are prevented, resulting in better rehydration properties and texture retention.

2.4 Intermolecular Forces and Structural Stabilization

In cryogenic conditions, the drastic slowdown of molecular motion greatly amplifies the effects of intermolecular forces. The non-covalent interactions such as hydrogen bonding, van der Waals forces, and hydrophobic effects play a pivotal role in preserving the structural and functional integrity of food systems, especially during freezing and long-term low-temperature storage. As thermal energy diminishes, these forces largely dictate the physical state, molecular organization, and stability of both biological and synthetic food components. This section highlights how such interactions stabilize frozen matrices and prevent degradation during cryogenic processing.

2.4.1 Hydrogen Bonding and Cryoprotection

Hydrogen bonding is the most dominant and structurally significant interaction in aqueous food systems. It involves the attraction between a hydrogen atom covalently bonded to a highly electronegative atom, typically oxygen or nitrogen, and another electronegative atom nearby. This interaction underlies the organization of water molecules and is essential in maintaining the native conformation of proteins, nucleic acids, and polysaccharides (Grabowski 2020). At ambient conditions, water forms a dynamic and flexible network of transient hydrogen bonds. However, as temperatures drop, especially during cryogenic cooling, these bonds become more ordered, leading to the formation of hexagonal ice crystals. The growth of such ice structures can damage cellular membranes and disrupt the texture of foods (Tanaka 2001). Cryoprotectants like trehalose, sucrose, glycerol, and sorbitol counteract this by forming extensive hydrogen bonds with water molecules and biomacromolecules. These agents interfere with ice nucleation and growth by competing for hydrogen-bonding sites, lowering the freezing point through colligative effects, and encouraging vitrification – an amorphous solid state that prevents crystallization (Šesták et al. 2011). Additionally, they stabilize hydration shells around proteins and membranes, reducing structural denaturation (Murray et al. 2024).

In proteins and polysaccharides, hydrogen bonding preserves secondary and tertiary structures by maintaining intramolecular and intermolecular cohesion (Babu et al. 2024). At cryogenic temperatures, the reduction in thermal motion prevents hydrogen bond breakage, which helps retain native folding. For example, in dairy systems, disaccharide cryoprotectants protect whey proteins, enhancing their solubility and emulsifying properties post-thaw. Similarly, in fruits and vegetables, the integrity of pectin and cellulose networks is preserved through hydrogen bonding, which helps maintain firmness and reduce mushiness after thawing by limiting ice-induced cell wall rupture.

2.4.2 Hydrophobic and van der Waals Interactions in Frozen Matrices

While hydrogen bonding is central to aqueous systems, hydrophobic and van der Waals interactions also significantly influence the stability of complex food matrices, especially those containing lipids and proteins. Hydrophobic interactions are entropy-driven associations where non-polar molecules cluster together to minimize contact with water, thereby releasing structured water molecules and increasing system entropy. At low temperatures, the rigidification of water structure reduces this entropic drive, which may diminish the strength of hydrophobic interactions. However, the reduced molecular mobility can stabilize pre-existing hydrophobic domains, such as lipid bilayers, protein cores, and micellar structures, thereby preserving their integrity during freezing. In lipid-containing systems, cryogenic processing induces phase transitions in membrane lipids, typically from a fluid, liquid-crystalline phase to a more rigid gel phase. These transitions can compromise membrane functionality and integrity. Cryoprotectants help mitigate such damage by modulating water-lipid interactions, maintaining hydrophobic core stability, and preserving bilayer cohesion. It is particularly relevant in frozen emulsions and cell-based foods, where lipid membranes must retain selective permeability and structural fluidity (Kharakoz 2000).

Van der Waals forces, although individually weak, are ever-present and contribute to the physical cohesion of molecules. These forces include London dispersion and dipole-induced dipole interactions, and they are particularly important in amorphous systems. In vitrified food matrices, van der Waals interactions help stabilize the molecular arrangement by contributing to mechanical rigidity and preventing collapse or deformation. This is particularly critical during freeze-drying, where amorphous structures can easily collapse without sufficient internal bonding. In frozen and dried foods like confections, emulsions, and aerated products, van der Waals interactions helps to maintain the spatial arrangement of fat globules, protein networks, and air spaces. Their stabilizing effect ensures microstructural integrity during both storage and rehydration or thawing. Together with hydrogen bonding and hydrophobic effects, these interactions form a comprehensive framework for molecular stabilization, ensuring high-quality preservation of food structure and functionality under cryogenic conditions.

2.5 Cryogenic Effects on Biomolecules

Biomolecules such as proteins, lipids, and carbohydrates play essential roles in defining the structure, functionality, and sensory characteristics of food (Figure 2.2). Their physico-chemical stability during cryogenic processing governs the success of freezing, freeze-drying, and ultra-low temperature storage. At cryogenic temperatures, molecular motion is significantly suppressed, but the associated changes in hydration, phase state, and intermolecular interactions can induce both beneficial and adverse transformations (Kimijima et al. 2024). This section outlines the impact of cryogenic conditions on the structure and behaviour of key biomolecular components in food.

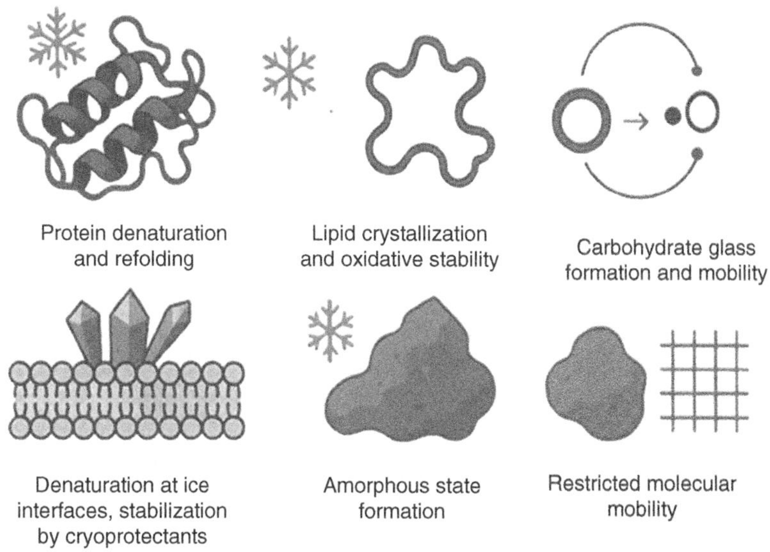

Figure 2.2 Cryogenic effect on biomolecules.

2.5.1 Protein Denaturation and Refolding

Proteins are intricate macromolecules whose biological and functional roles are tightly linked to their native three-dimensional structures. However, freezing imposes several stresses that can disrupt protein conformation. One of the primary causes of protein denaturation is interaction with ice crystal surfaces, where adsorption can lead to distortion of hydration shells and conformational changes, particularly for surface-active or membrane-bound proteins (Brytan and Padrela 2023). Additionally, as water crystallizes, the remaining unfrozen solution becomes concentrated with solutes, which can shift pH and ionic strength, destabilizing electrostatic interactions and hydrogen bonds. Slow freezing amplifies these effects, allowing time for unfolding and aggregation. In contrast, ultra-rapid cryogenic freezing, such as immersion in liquid nitrogen, can arrest proteins in their native or near-native states by preventing transitional unfolding.

Cryoprotectants play a vital role in mitigating protein damage during freezing. Compounds such as trehalose, sorbitol, and proline help stabilize protein structures by replacing hydration water through hydrogen bonding, promoting vitrification of the surrounding medium, and sterically hindering aggregation (Rudolph and Crowe 1985). The extent to which proteins can refold after thawing is highly variable. While some globular proteins regain their native structure spontaneously, others, like certain enzymes, require additional stabilizing agents to retain activity. For instance, enzymes such as lactate dehydrogenase and catalase often lose catalytic function during freezing unless adequately protected. Functionally, denaturation can impair solubility, emulsifying capacity, and enzymatic performance in food matrices. However, in some cases, partial unfolding may enhance properties like foaming or gelation, beneficial in dairy or meat applications.

2.5.2 Lipid Crystallization and Oxidative Stability

Lipids, particularly those rich in unsaturated fatty acids, are prone to both structural and oxidative changes during cryogenic treatment (Domínguez et al. 2019). Crystallization behaviour is influenced by cooling rates, leading to the formation of various polymorphic forms. Rapid cooling tends to favour the formation of less stable α-crystals, while controlled cooling or post-tempering can shift the balance towards more stable β′ and β forms, which affect textural properties and mouthfeel. In emulsified systems such as frozen desserts or salad dressings, the lipid crystallization affects the morphology of the droplets. Although partial coalescence may improve the structure of aerated food products such as ice cream, excessive crystallization can result in undesirable graininess or phase separation. Furthermore, membrane lipids in biological tissues undergo a phase change from fluid to gel phases, which compromises membrane permeability and induces cell lysis unless cryoprotective procedures are observed.

Oxidative stability is another challenge in the freezing process. Although cryogenic temperatures significantly reduce the molecular mobility and oxygen diffusion, thereby slowing lipid peroxidation, oxidation is not entirely halted. Residual enzymatic activity or free radical pathways may remain, particularly in microenvironments where oxygen becomes concentrated on freezing water. These hotspots have the potential to induce oxidation unless counteracted by vitrification or the use of antioxidants. Protective strategies include encapsulating lipids within cryoprotective matrices (such as polysaccharide gels), adding antioxidants like tocopherols or ascorbic acid, and minimizing oxygen exposure through vacuum packaging or inert gas flushing. All these measures contribute to extending lipid stability and sensory quality under prolonged cryogenic storage.

2.5.3 Carbohydrate Glass Formation and Mobility

Carbohydrates have a unique cryogenic character due to their inherent hydrogen bonding capacity and their tendency to form glassy, amorphous states. Many mono-, di-, and polysaccharides, such as trehalose, maltodextrin, and pectin, do not crystallize upon rapid cooling; rather, they produce a vitreous phase. The glassy state is defined by the glass transition temperature (Tg), below which molecular mobility is greatly reduced. When maintained below Tg, carbohydrates inhibit diffusion and reaction kinetics, thus stabilizing labile ingredients such as enzymes, probiotics, and volatile aromas during frozen or freeze-dried storage (Shirke and Ludescher 2005).

The formation of a glassy matrix leads to diffusional arrest, significantly limiting the mobility of water, oxygen, and small metabolites. This effect contributes to the enhanced stability of carbohydrate-rich systems such as fruit purées, jams, and bakery fillings. In frozen desserts, glassy sugar matrices are particularly effective in suppressing ice recrystallization, which helps maintain smooth texture over time. Moreover, certain polysaccharides like starch, cellulose, and hydrocolloids interact with water to form tightly bound hydration layers that are resistant to freezing. This interaction preserves water availability in a non-freezable form, contributing to the textural integrity of frozen plant-based foods.

Carbohydrates also serve critical roles as carriers and stabilizers in cryogenic food applications. In spray drying and freeze-drying, they act as protective encapsulation matrices for sensitive bioactive, including flavours, vitamins, and lipids (Fang and Bhandari 2012). These glassy matrices offer a physical barrier against oxidative degradation and volatilization. Carbohydrate cryoprotectants in frozen foods such as dairy, meat, and bakery products are utilized to reduce ice crystal damage, control water migration, and enhance product stability. Due to their multifunctional behaviour, they are indispensable for both structural and chemical preservation in cryogenic processes.

2.6 Water Behaviour and Ice Crystallization in Cryogenic Processing in the Food Industry

Water is the most abundant and structurally influential component in biological food systems, and its behaviour during freezing determines the efficacy of cryogenic processing. The phase transformation of water into ice characterized by gentle physical process that directly impact the texture, stability, and integrity of food (Li et al. 2018). Understanding the molecular and structural dynamic behaviour of water under cryogenic temperatures is critical for optimizing food preservation technology.

2.6.1 Nucleation and Growth of Ice Crystals

The transformation of water into ice begins with nucleation, the initial step where a small cluster of water molecules organizes into a crystal lattice (Zheng et al. 2024). In food systems, this process is usually heterogeneous, caused by the presence of interfaces or particles that lower the energy barrier to nucleation. Homogeneous nucleation, though theoretically possible, only takes place under conditions of extreme supercooling (below −40 °C) and is rare in practical applications. Once nucleated, ice crystals continue to grow by the progressive addition of water molecules onto the lattice (Safari et al. 2017). The rate of freezing profoundly influences this process: slower cooling leads to the formation of fewer but larger ice crystals, which can damage cellular structures and contribute to texture degradation. Conversely, cryogenic freezing, with its ultra-rapid temperature drop (often using liquid nitrogen at −196 °C or carbon dioxide at −78.5 °C), promotes rapid nucleation and limits the time available for crystal growth. This leads to production of many small crystals, which have less devastation for food microstructures and help in retaining sensory attributes.

2.6.2 Amorphous vs. Crystalline Ice

Depending on the rate of cooling and the physicochemical environment, water can solidify into either crystalline or amorphous (vitreous) ice. Crystalline ice, particularly ice with a hexagonal structure, is the most commonly formed under conventional freezing. Its ordered lattice expands upon freezing, potentially causing rupture in cell membranes and mechanical damage to tissues. On the other hand, amorphous ice lacks a defined crystalline structure and forms only under extremely rapid-cooling rates exceeding 10^5–10^6 K/s. This vitrified state avoids volumetric expansion and internal stresses, making it far less damaging to biological structures. While full-scale formation of amorphous ice in bulk food systems is uncommon due to technical constraints, localized vitrification may occur within confined regions such as cellular compartments or nanoemulsions. The presence of solutes, cryoprotectants, and glass-forming compounds also influences the likelihood of amorphous ice formation by increasing viscosity and suppressing crystallization kinetics (Al-Amoudi et al. 2002).

2.6.3 Effects on Cell and Tissue Integrity

The structural integrity of food tissues during cryogenic processing is largely determined by the location, size, and morphology of ice crystals. Large ice crystals, particularly those forming extracellularly during slow freezing, can impose mechanical pressure that dehydrates cells through osmosis and compacts membranes during their expansion (Pérez-Bermúdez et al. 2023). Under extreme conditions, this results in irreversible cell rupture, disrupted texture, and leakage of nutrients during thawing. Cryogenic freezing reduces such effects through the ability to achieve quick intracellular vitrification or the formation of uniformly small crystals that pose minimal risk of structural compromise. In fruits and vegetables, this maintains firmness and reduces thaw-induced drip loss. In muscle-based foods such as meat and seafood, cryogenic freezing helps to retain the organization muscle fibre, thereby preserving juiciness and minimizing textural deterioration. In addition, dairy-based products and emulsions also benefit from this process through improved emulsion stability and reduced phase separation.

Furthermore, cryogenic storage conditions significantly reduce the possibility of ice recrystallization, small ice crystals merge into larger ones during long-term storage. Recrystallization not only disrupts texture but also enhances physical degradation. Through ultra-low temperatures (often below −80 °C), cryogenic systems inhibit molecular mobility, hence stabilizing the ice structure and prolonging product shelf life with minimal structural loss.

2.7 Role of Cryogens: Liquid Nitrogen and Carbon Dioxide

Cryogens such as liquid nitrogen (LN_2) and solid or liquid carbon dioxide (CO_2) are the backbone of cryogenic technologies in the food industry. Their selection is based on a combination of thermal efficiency, availability, inertness, and compatibility with different food systems (Figure 2.3). Each cryogen exhibits distinct physical and chemical properties, interacts uniquely with food constituents, and requires specific safety and handling measures. Understanding these aspects is crucial for optimizing cryopreservation processes while ensuring food safety and operator protection (Murray and Gibson 2022).

2.7.1 Physical and Chemical Properties

Liquid nitrogen and carbon dioxide differ fundamentally in their thermodynamic behaviour and phase transitions. Liquid nitrogen boils at −196 °C (77 K) at atmospheric pressure and transitions directly from liquid to gas (boiling point) with no liquid phase above this

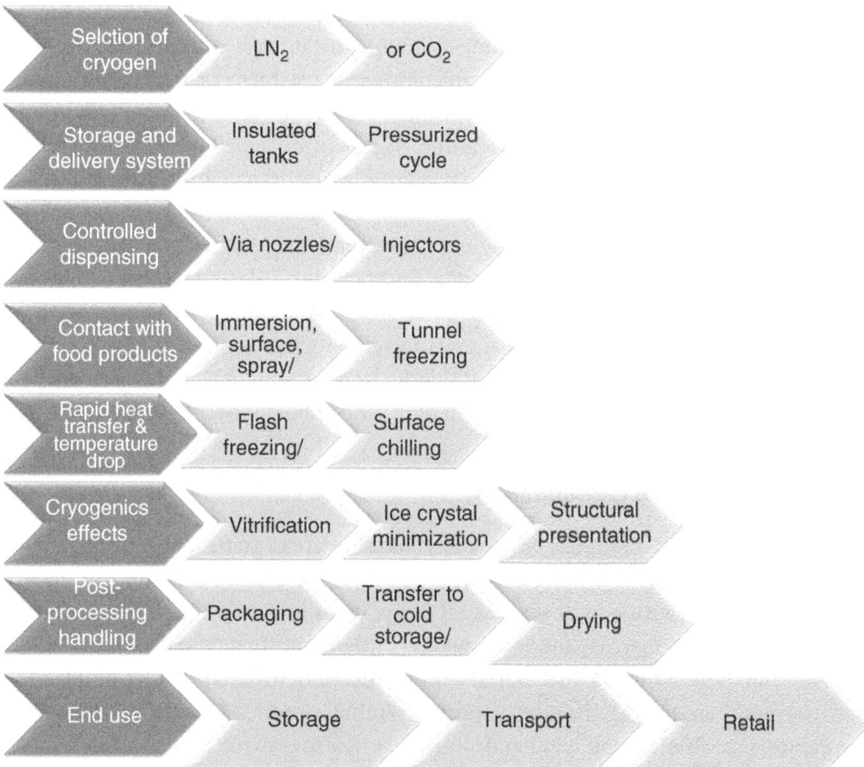

Figure 2.3 Cryogenic food processing using liquid nitrogen and carbon dioxide process flow chart.

temperature, making it highly efficient for rapid heat extraction. It is inert, non-toxic, odourless, and colourless. It's extremely low boiling point provides rapid freezing rates, facilitating vitrification and minimal ice crystal formation, which is vital for preserving microstructure and cellular integrity in delicate foods like berries, seafood, and bakery fillings (Zohuri 2017).

Carbon dioxide, on the other hand, sublimes at −78.5 °C (194.7 K) at atmospheric pressure, meaning it transitions from solid to gas without a liquid phase under normal conditions. In the form of dry ice or liquid CO_2 under pressure, it is also effective for cryogenic freezing but offers lower cooling intensity compared to LN_2. CO_2 is slightly acidic in nature and can dissolve into water to form carbonic acid (H_2CO_3), potentially influencing pH in moisture-rich systems. Its gaseous expansion provides mechanical scouring in surface chilling or snow formation applications, particularly useful in meat and dough processing (Purandare et al. 2023).

2.7.2 Reaction with Food Constituents

Both LN_2 and CO_2 are considered chemically inert under cryogenic conditions, meaning they do not engage in significant covalent or reactive chemical changes with food molecules. However, physical interactions and localized effects can influence the behaviour of food components (Sanjeev and Ramesh 2006). In the case of liquid nitrogen, its inertness makes it ideal for preserving biochemical integrity. It does not alter protein structure or react with lipids, carbohydrates, or vitamins. However, rapid freezing can induce physical stress, including thermal shock, leading to cell rupture in high-moisture foods. This effect is generally beneficial when the goal is to enhance extractability or texture (e.g. freeze-grinding of herbs or spices).

Carbon dioxide, due to its mild acidity, may slightly lower the pH of moist foods, which can affect the stability of pH-sensitive compounds like anthocyanin, chlorophyll, or enzyme systems. This is particularly relevant in beverages, dairy products, and high-water-content fruits (Mattioli et al. 2020). In some applications, the pH-lowering effect can be beneficial, offering antimicrobial or antioxidant properties. Nevertheless, overexposure may induce protein precipitation or carbonate formation, particularly in dairy and confectionery systems.

2.7.3 Safety and Handling Considerations

Despite the fact that both cryogens are widely used in food-grade applications, their extremely low temperatures and unique properties demand careful handling to ensure workplace safety. Liquid nitrogen poses several risks, including cold burns, asphyxiation hazards in poorly ventilated areas, and material embrittlement upon contact with containers or tools not rated for cryogenic use. Rapid vaporization can lead to pressure build-up in sealed environments, requiring the use of pressure-relief systems. Operators must wear cryogenic gloves, face shields, and insulated clothing and also provide adequate ventilation to prevent oxygen displacement.

Carbon dioxide, particularly in the form of dry ice, is associated with pressure and sublimation hazards. Its gaseous state can displace oxygen, causing a risk of hypoxia hazard in enclosed spaces. CO_2 can also accumulate in process areas if ventilation is inadequate, potentially reaching toxic concentrations. Protective equipment is required as with LN_2, and containers must be compatible with handling of pressurized or subliming materials. Notably, CO_2 systems often require monitoring of ambient gas levels using oxygen and CO_2 sensors to ensure compliance with safety regulations. From a regulatory perspective, both LN_2 and CO_2 are approved for direct contact with food and are classified as generally recognized as safe (GRAS) by food safety authorities such as the FDA and EFSA. Traceability, documentation, and material compatibility remain critical aspects in maintaining food safety standards and operational efficiency.

2.8 Implications for Food Structure and Quality

Cryogenic processing, particularly through the application of liquid nitrogen (LN_2) and carbon dioxide (CO_2), plays a transformative role in preserving and enhancing food quality (Table 2.2). Its ultrafast freezing capabilities allow for precise control over ice crystal formation, moisture migration, and biochemical activity that fundamentally affect the texture, flavour, and nutritional value of food. Unlike conventional freezing, which may induce structural breakdown and loss of sensory attributes, cryogenic freezing enables superior

Table 2.2 Influence of cryogenic processing on key food quality parameters.

Quality aspect	Mechanism in cryogenic processing	Impact on food quality	Supporting details/ examples	References
Texture modification and preservation	Rapid freezing forms small ice crystals, limits protein denaturation and aggregation	Maintains cell structure, reduces drip loss, preserves firmness	Cryoprotectants (e.g. oligosaccharides, polyphenols) further stabilize proteins and prevent texture loss in muscle foods	Meghwal et al. (2021)
Flavour retention and volatile compounds	Quick temperature drop limits enzymatic and oxidative reactions	Retains natural flavours and aromas, prevents off-flavour formation	Volatile compounds are better preserved due to minimal molecular motion and enzyme inactivation	Shang et al. (2023)
Nutritional stability	Minimizes nutrient degradation by halting enzyme activity and oxidation	Preserves vitamins, bioactive compounds, and overall nutritional value	Slower freezing or heat can degrade nutrients; cryogenic methods avoid this	Shang et al. (2023), Meghwal et al. (2021)
Enzyme inactivation	Rapid cooling inactivates spoilage and quality-degrading enzymes	Extends shelf life, maintains colour, texture, and flavour	Enzymes like polyphenol oxidase and lipoxygenase are quickly inactivated, reducing spoilage and quality loss	Xu et al. (2024)

retention of food characteristics across various categories, from fresh produce and seafood to bakery and dairy products (Pesce et al. 2025).

2.8.1 Texture Modification and Structural Integrity

Texture is one of the most critical quality attributes in food, influencing both consumer acceptance and functionality. In cryogenic freezing, the rapid rate of heat transfer minimizes the formation of large extracellular ice crystals. Instead, it favours the creation of small, uniform crystals that are less disruptive to cellular structures. This is particularly beneficial for high-moisture foods such as fruits, vegetables, meats, and seafood (Zhang et al. 2023). In conventional freezing, slow cooling often leads to ice recrystallization and expansion, which ruptures cell membranes and results in drip loss, mushy textures, and reduced juiciness upon thawing. On the other hand, the cell-wall matrix is preserved through cryogenic freezing by enhancing firmness, chewiness, and the original mouthfeel of the food. For example, cryogenic methods prevent tissue collapse and surface pitting in leafy vegetables and berries. The preservation of myofibrillar integrity helps retain tenderness and water-holding capacity of muscle-based food such as fish fillets or poultry. Furthermore, in frozen baked goods, rapid freezing maintains the aerated structure of doughs and batters, resulting in a fresher texture after thawing or baking.

2.8.2 Flavour and Volatile Compound Retention

Flavour perception in foods is a combination of taste and aroma, heavily influenced by volatile organic compounds (VOCs). These compounds are sensitive to temperature changes, oxidation, and enzymatic degradation (Porrello et al. 2024). During traditional freezing, slow temperature decline allows for the loss or transformation of these volatiles, diminishing the overall sensory profile of the product. However, cryogenic freezing immediately immobilizes these molecules by arresting molecular motion and enzymatic activity, thereby preserving the original aroma and taste. This preservation is especially important in aromatic foods such as herbs, spices, brewed beverages, and fermented products, where even minor changes in volatiles can significantly alter perceived quality. Cryogenic processing preserves flavour in frozen food and desserts ensures that consumers experience a more authentic and fresh taste after reheating.

2.8.3 Nutritional Stability and Enzyme Inactivation

Food's nutritional quality is susceptible to degradation due to enzymatic reactions, oxidation, and moisture loss. Vitamins, particularly water-soluble ones like vitamin C and B-complex are prone to degradation under conventional freezing due to prolonged exposure to residual oxygen and thermal fluctuation (Giannakourou and Taoukis 2021). Cryogenic freezing dramatically lowers the temperature below the threshold for enzymatic activity. Furthermore, enzyme inactivation through cryogenic processing helps prevent undesirable post-harvest changes. Enzymes like lipase, peroxidase, and polyphenol oxidase responsible for rancidity, nutrient loss, and browning are effectively immobilized. This results in better preservation of nutritional constituents and sensory appeal during long-term storage.

2.8.4 Structural Enhancement in Composite Foods

In complex food systems such as emulsions, foams, or layered desserts, cryogenic processing can improve structural stability by quickly solidifying components and preventing phase separation. This is essential in premium products like mousse cakes, cream-based fillings, or multilayered snacks, where distinct textures and visual layers need to be preserved (Dickinson 2010). In conclusion, cryogenic technologies offer comprehensive benefits for maintaining food structure, flavour, nutrition, and quality. By minimizing damage, preserving volatile and nutrient profiles, and enhancing the mechanical property and qualities of various food systems, cryogenic processing sets a new standard in quality preservation, especially in premium, sensitive, and minimally processed foods.

2.9 Molecular Basis of Bioactive Stabilization

Bioactive compounds such as antioxidants, polyphenols, and essential oils play pivotal roles in food functionality and health promotion. However, these molecules are inherently sensitive to environmental factors such as temperature, light, oxygen, and pH. During conventional processing and storage, many bioactives degrade rapidly, losing their functional and nutritional efficacy (Toydemir et al. 2022). Cryogenic processing offers a unique advantage by reducing molecular motion and enzymatic degradation, which helps to stabilize sensitive compounds. This part explores how cryogenic techniques help stabilize bioactives at a molecular level, with a particular focus on antioxidants, polyphenols, and essential oils, as well as strategies for encapsulation and retention.

2.9.1 Antioxidants, Polyphenols, and Essential Oils

Antioxidants and polyphenols play major role in preventing oxidative damage in both food and biological systems by neutralizing reactive oxygen species (ROS). These compounds include flavonoids, tannins, phenolic acids, and vitamins like ascorbic acid and tocopherols. Essential oils, which are mostly composed of volatile terpenes and phenolics, not only contribute to flavour and aroma but also exhibit antimicrobial and antioxidant properties (Siddiqui et al. 2024). Cryogenic temperatures significantly reduce the rates of oxidation and enzymatic degradation by limiting oxygen diffusion and inactivating polyphenol oxidase, peroxidase, and other degrading enzymes. As a result, phenolic profiles remain intact, and antioxidant potency is preserved over extended periods. For fresh produce, this translates into better colour retention, reduced browning, and maintenance of phytochemical potency. In essential oils, cryogenic processing minimizes evaporation and oxidation, stabilizing volatile fractions and protecting their aroma and bioactive properties, thereby enhancing their effectiveness in functional food and nutraceutical applications.

2.9.2 Encapsulation and Retention During Cryogenic Processing

To further enhance bioactive stability, encapsulation techniques are often employed in conjunction with cryogenic processing. Encapsulation involves trapping active molecules within a protective matrix, typically composed of biopolymers such as maltodextrin, gum Arabic, starches, or proteins (Zabot et al. 2022). Cryogenic freezing supports this by rapidly solidifying the encapsulating matrix, locking the bioactives in a non-reactive and thermodynamically stable state. During cryogenic spray freezing or freeze-drying, bioactive-laden emulsions or suspensions are subjected to ultra-low temperatures, forming a solid matrix with minimal diffusion and migration. This results in high encapsulation efficiency and retention of volatile and heat-labile compounds (Singh et al. 2024).

For instance, encapsulated essential oils in a cryogenically frozen powder maintain their integrity and release characteristics better than those processed through thermal drying. Additionally, glass-transition phenomena at cryogenic temperatures play a key role in molecular immobilization. When the encapsulating matrix forms an amorphous glass, molecular mobility is drastically reduced, which halts oxidation, hydrolysis, and polymer degradation. This vitrified state serves as an ideal medium for the long-term storage of bioactives, especially when followed by freeze-drying to ensure shelf stability. Moreover, in food systems enriched with encapsulated bioactives (such as fortified beverages, baked goods, or dairy products), cryogenic processing ensures minimal loss of functionality during incorporation and storage. The stability of flavonoids in berry extracts, carotenoids in oil emulsions, or vitamin E in microcapsules is significantly improved when cryogenic conditions are employed during or immediately after encapsulation.

2.10 Emerging Trends and Future Directions

Cryogenic processing has traditionally been associated with preservation and storage; however, its application is rapidly expanding with the incorporation of advanced materials science, nanotechnology, and smart food systems. As the demands for high-performance, functional, and sustainable food products increase, emerging trends in cryogenics are changing the future of food processing. This section explores recent advancements and future directions, including the development of nanostructured cryogenic interfaces, integration with smart packaging and sensor technology, and the use of cryogenic biochemistry in functional foods.

2.10.1 Nanostructured Cryogenic Interfaces

Nanostructured cryogenic interfaces are one of the most exciting new developments in cryogenics food science. These interfaces are nanoscale changes to the surfaces of food or encapsulating materials that work with cryogenics environment in a controlled and useful way. Nanostructures can be engineered to modulate the conduct heat, control ice nucleation, or enhance the interaction of bioactive-loaded nanoparticles with surrounding matrices. For example, incorporating nanolayers of hydrophobic or hydrophilic coatings can influence the rate of ice formation, leading to finer ice crystals and minimal structural damage. Nanostructured cryogels and nanofiber mats composed of biopolymers, including cellulose nanocrystals or chitosan, have demonstrated potential in enhancing cryoprotective effects by forming tailored diffusion barriers or controlled-release platforms. In the future, these interfaces may be further refined using stimuli-responsive materials such as temperature- or pH-sensitive nanogels that adapt to changes during freezing and thawing, providing real-time control over molecular interactions and stability.

2.10.2 Integration with Smart Packaging and Sensors

Another emerging trend is the integration of cryogenic technologies in combination with smart packaging and sensor systems to enhance food safety, quality, and traceability issues are addressed in the cold chains. For example, smart cryogenic packaging may involve the incorporation of embedded nano sensors to detect changes in temperature, humidity, gas composition (e.g. oxygen or CO_2), or microbial activity. These sensors can alert handlers to temperature abuse or cryogen loss occurs during transport and storage, ensuring product integrity. Radio-frequency identification (RFID) tags, thermochromic labels, and time–temperature indicators (TTIs) are already being adopted with cryogenic systems.

Further integration may lead to responsive packaging materials that alter permeability or protective behaviour in real-time based on external conditions. For example, cryogenically compatible films with nanocarriers can release antioxidants, antimicrobials, or flavour compounds only upon triggering by thawing or a change in pH which could allow extended shelf-life at ambient conditions and sensory attributes. This convergence of cryogenics and smart systems is particularly important in global logistics for high-value, perishable products including seafood, meat, functional beverages, and ready-to-eat frozen meals.

2.10.3 Cryogenic for Functional Foods

The intersection of cryogenic science and functional food biochemistry offers notable advancement. The use of cryogenic temperatures increasingly affords not merely preservation, but modulation or enhancement of the biochemical profile of food ingredients (Van der Sman 2020). Many new and emerging research focus on how subzero enzymatic reactions conducted exactly or below freezing temperatures can be selectively activated to synthesize or stabilize functional compounds. For example, certain cold-adapted enzymes are active in cryogenic temperatures and can be used to bio-transform precursors into health-promoting molecules, such as converting glucosinolates to isothiocyanates or producing bioactive peptides from frozen protein matrices.

Moreover, the evolving field of cryogenic fractionation could allow for heat-sensitive isolate heat-sensitive compounds to be isolated from specific foods, such as polyunsaturated fatty acids, flavonoids, or probiotics in an active and pure form, and could be delivered in designer foods with specific health benefits, especially for immune modulation, cognitive support, and gut health (Xiang et al. 2023). In addition, combining cryogenic processing with synthetic biology and/or precision fermentation could transform how various functional food constituents are generated and preserved, resulting in customized nutritional product with minimal degradation and enhanced stability. The cryogenic processing of food is not only about preservation, but about more intelligent, interactive, and biochemically enriched food systems. The evolution of nanostructuring, sensor integration, and biochemical processing creates the potential for cryogenics to become a primary approach to developing next-generation foods that support sustainability, health, and global food security.

2.10.4 Industrial Relevance and Application Potential

From an industrial perspective, cryogenic processing offers several clear advantages that align with the goals of food safety, quality retention, and innovation. Its rapid-cooling capabilities make it particularly suitable for high-value, sensitive, or perishable food products such as seafood, berries, herbs, dairy emulsions, and specialty baked goods. By preserving structure, nutrients, and sensory quality, cryogenic methods ensure minimal post-thaw degradation, reduced drip loss, and enhanced shelf life. Industrially, cryogenic systems are employed not only for freezing but also for cryomilling, freeze-drying pretreatment, bioactive encapsulation, and enzymatic inactivation. The integration of cryogenic processing with micro- and nanotechnology is opening pathways towards smart materials, controlled release systems, and tailored food functionalities. For instance, nanostructured cryoprotective interfaces and cryo-compatible packaging with sensors enhance traceability, safety, and process efficiency in cold chains.

Furthermore, cryogenic biochemistry represents an emerging frontier. Subzero enzymatic reactions, cryo-induced metabolite stabilization, and bioengineering of frozen matrices for functional foods are likely to become core components of next-generation food manufacturing. As the demand for nutrient-dense, minimally processed, and high-performance foods increases, the ability of cryogenic technology to maintain biochemical integrity and structural fidelity becomes indispensable. Despite high operating costs and energy demands, cryogenic processing remains competitive due to its precision, small equipment footprint, and compatibility with modular, on-demand systems. As sustainability initiatives advance, innovations in cryogen recycling, hybrid freezing systems, and eco-friendly refrigerants will help make cryogenics a more viable solution across scales. In conclusion, the molecular foundations of cryogenic processing provide not only a robust framework for understanding food preservation but also a platform for innovation in quality control, food design, and future-ready food systems. Its ability to combine thermodynamic precision with biochemical preservation positions cryogenic processing as a cornerstone of advanced food engineering.

2.11 Conclusion and Summary

Cryogenic processing stands at the confluence of molecular science and advanced food engineering, offering unparalleled precision in preserving the structural, nutritional, and sensory attributes of foods. At the molecular level, the suppression of thermal motion, control of phase transitions, and stabilization through intermolecular forces form the scientific basis for cryogenic efficacy. The chapter elucidates how key biomolecules such as proteins, lipids, and carbohydrates respond to ultra-low temperatures and cryoprotectants, and how vitrification and glass transition phenomena contribute to food stabilization. Beyond preservation, cryogenic technology plays a transformative role in enhancing functional food design – enabling encapsulation of bioactives, extending shelf life, and facilitating novel applications such as cryogenic biochemistry and nanostructured interfaces. As the food industry moves towards clean-label, minimally processed, and high-value products, cryogenics offers sustainable and smart solutions tailored to these demands. While challenges remain in energy efficiency and scalability, ongoing innovations such as hybrid systems, smart cryo-packaging, and enzyme-assisted cryo processes promise to redefine the future landscape of food preservation and functionalization. Cryogenics is no longer just a preservation method but a strategic tool for engineering next-generation food systems rooted in molecular precision and sustainability.

References

Alam, M.W., Kumar, J.V., Awad, M. et al. (2025). Emerging trends in food process engineering: integrating sensing technologies for health, sustainability, and consumer preferences. *Journal of Food Process Engineering* 48 (1): e70035.

Al-Amoudi, A., Dubochet, J., and Studer, D. (2002). Amorphous solid water produced by cryosectioning of crystalline ice at 113 K. *Journal of Microscopy* 207 (2): 146–153.

Babu, A., Shams, R., Dash, K.K. et al. (2024). Protein-polysaccharide complexes and conjugates: structural modifications and interactions under diverse treatments. *Journal of Agriculture and Food Research* 18: 101510.

Balasubramanian, S., Gupta, M.K., and Singh, K.K. (2012). Cryogenics and its application with reference to spice grinding: a review. *Critical Reviews in Food Science and Nutrition* 52 (9): 781–794.

Boehm, F.J. (2013). Nanomedical device and systems design in space applications. In: *Nanomedical Device and Systems Design: Challenges, Possibilities, Visions*, 585–652. Boca Raton, FL: CRC Press, Taylor and Francis Group.

Brytan, W. and Padrela, L. (2023). Structural modifications for the conversion of proteins and peptides into stable dried powder formulations: a review. *Journal of Drug Delivery Science and Technology* 89: 104992.

Cecil, S.R. and Woodruff, J.G. (1962). *Long-Term Storage of Military Rations*, vol. 2. Department of the Army, Quartermaster Research and Engineering Command, Quartermaster Food and Container Institute for the Armed Forces.

Cleland, D.J., 2020. The history of food freezing. *ASHRAE Transactions*, 126(1). pg. 616

Dickinson, E. (2010). Food emulsions and foams: stabilization by particles. *Current Opinion in Colloid & Interface Science* 15 (1–2): 40–49.

Domínguez, R., Pateiro, M., Gagaoua, M. et al. (2019). A comprehensive review on lipid oxidation in meat and meat products. *Antioxidants* 8 (10): 429.

Fang, Z. and Bhandari, B. (2012). Spray drying, freeze drying and related processes for food ingredient and nutraceutical encapsulation. In: *Encapsulation Technologies and Delivery Systems for Food Ingredients and Nutraceuticals* (ed. N. Garti and D.J. Mcclements), 73–109. Wood Head Publishing.

Giannakourou, M.C. and Taoukis, P.S. (2021). Effect of alternative preservation steps and storage on vitamin C stability in fruit and vegetable products: critical review and kinetic modelling approaches. *Food* 10 (11): 2630.

Goswami, T.K. (2010). Role of cryogenics in food processing and preservation. *International Journal of Food Engineering* 6 (1): 1–29.

Goswami, T.K. (2017). Recent trends of application of cryogenics in food processing and preservation. *Journal of Food, Nutrition and Population Health* 1 (3): 27.

Grabowski, S.J. (2020). *Understanding Hydrogen Bonds: Theoretical and Experimental Views*, vol. 19. Royal Society of Chemistry.

Homer, E.R., Johnson, O.K., Britton, D. et al. (2022). A classical equation that accounts for observations of non-Arrhenius and cryogenic grain boundary migration. *npj Computational Materials* 8 (1): 157.

Junghare, H., Hamjade, M., Patil, C.K. et al. (2017). A review on cryogenic grinding. *International Journal of Current Engineering and Technology* 7 (7): 420–423.

Kharakoz, D.P., 2000. Liquid-crystalline to gel-state phase transition in lipid membranes. *Russian Journal of Physical Chemistry A*, 74. pg. 177–191.

Kimijima, J., Inagawa, A., Miyagawa, A. et al. (2024). Probing the interaction between biomolecules under sub-zero temperature conditions by electrophoresis in ice grain boundaries. *Analytica Chimica Acta* 1311: 342713.

Li, D., Zhu, Z., and Sun, D.W. (2018). Effects of freezing on cell structure of fresh cellular food materials: a review. *Trends in Food Science & Technology* 75: 46–55.

Mattioli, R., Francioso, A., Mosca, L., and Silva, P. (2020). Anthocyanins: a comprehensive review of their chemical properties and health effects on cardiovascular and neurodegenerative diseases. *Molecules* 25 (17): 3809.

Meghwal, M., Singh, H., and Prabhakar, P. (2021). Cryogenic freezing. In: *Food Processing* (ed. K.K. Dash and S. Chakraborty). CRC Press https://doi.org/10.1201/9781003163213-8.

Murray, K.A. and Gibson, M.I. (2022). Chemical approaches to cryopreservation. *Nature Reviews Chemistry* 6 (8): 579–593.

Murray, A., Kilbride, P., and Gibson, M.I. (2024). Trehalose in cryopreservation. Applications, mechanisms and intracellular delivery opportunities. *RSC Medicinal Chemistry* 15 (9): 2980–2995.

Peleg, M., Normand, M.D., and Corradini, M.G. (2012). The Arrhenius equation revisited. *Critical Reviews in Food Science and Nutrition* 52 (9): 830–851.

Pérez-Bermúdez, I., Castillo-Suero, A., Cortés-Inostroza, A. et al. (2023). Observation and measurement of ice morphology in foods: a review. *Food* 12 (21): 3987.

Pesce, F., Parafati, L., Fallico, B., and Palmeri, R. (2025). Use of liquid nitrogen in food products: a review. *Food Frontiers* 6: 1617–1644.

Porrello, A., Orecchio, S., and Maggio, A. (2024). Matrix-matched quantification of volatile organic compounds (VOCs) in gluten free flours and bakery products. *Food Chemistry: X* 22: 101399.

Purandare, A.S., Verbruggen, W.M., and Vanapalli, S. (2023). Experimental and theoretical investigation of the dry ice sublimation temperature for varying far-field pressure and CO_2 concentration. *International Communications in Heat and Mass Transfer* 148: 107042.

Rudolph, A.S. and Crowe, J.H. (1985). Membrane stabilization during freezing: the role of two natural cryoprotectants, trehalose and proline. *Cryobiology* 22 (4): 367–377.

Safari, A., Saidur, R., Sulaiman, F.A. et al. (2017). A review on supercooling of phase change materials in thermal energy storage systems. *Renewable and Sustainable Energy Reviews* 70: 905–919.

Sanjeev, K. and Ramesh, M.N. (2006). Low oxygen and inert gas processing of foods. *Critical Reviews in Food Science and Nutrition* 46 (5): 423–451.

Šesták, J., Queiroz, C.A., Mareš, J.J., and Holeček, M. (2011). Some aspects of vitrification, amorphisation and disordering and the generated extent of nano-crystallinity. In: *Glassy, Amorphous and Nano-Crystalline Materials: Thermal Physics, Analysis, Structure and Properties* (ed. J. Šesták, J.J. Mareš, and P. Hubík), 59–75. Dordrecht: Springer Netherlands.

Shang, S., Wang, Y., Dong, X. et al. (2023). Progress in the application of novel cryoprotectants for the stabilization of myofibrillar proteins. *Critical Reviews in Food Science and Nutrition* 64: 9756–9770. https://doi.org/10.1080/10408398.2023.2215874.

Shirke, S. and Ludescher, R.D. (2005). Molecular mobility and the glass transition in amorphous glucose, maltose, and maltotriose. *Carbohydrate Research* 340 (17): 2654–2660.

Siddiqui, T., Khan, M.U., Sharma, V., and Gupta, K. (2024). Terpenoids in essential oils: chemistry, classification, and potential impact on human health and industry. *Phytomedicine Plus* 4 (2): 100549.

Singh, H. and Meghwal, M. (2021). Advancement in the processing of condiments and spices. In: *Advances in Processing Technology* (ed. G.K. Sharma, A.D. Semwal, and D.K. Yadav), 143–168. London: CRC Press.

Singh, P., Pandey, V.K., Singh, R., and Dar, A.H. (2024). Spray-freeze-drying as emerging and substantial quality enhancement technique in food industry. *Food Science and Biotechnology* 33 (2): 231–243.

Sloane, T.O.C. (1900). *Liquid Air, and the Liquefaction of Gases: Theory, History, Biography, Practical Applications, Manufacture*. Henley.

Song, J., Berry, J.D., and Goudeli, E. (2023). Nucleation Rate of N_2 and O_2 in Cryogenic H_2 and He. *The Journal of Physical Chemistry B* 127 (46): 9976–9984. https://doi.org/10.1021/acs.jpcb.3c03364.

Tanaka, H. (2001). Hydrogen bonds between water molecules: thermal expansivity of ice and water. *Journal of Molecular Liquids* 90 (1): 323–332.

Toydemir, G., Subasi, B.G., Hall, R.D. et al. (2022). Effect of food processing on antioxidants, their bioavailability and potential relevance to human health. *Food Chemistry: X* 14: 100334.

Van der Sman, R.G.M. (2020). Impact of processing factors on quality of frozen vegetables and fruits. *Food Engineering Reviews* 12: 399–420.

Xiang, J., Mlambo, R., Shaw, I. et al. (2023). Cryopreservation of bioflavonoid-rich plant sources and bioflavonoid-microcapsules: emerging technologies for preserving bioactivity and enhancing nutraceutical applications. *Frontiers in Nutrition* 10: 1232129.

Xu, X., Jin, Y., Zheng, Z. et al. (2024). Inactivation of microorganisms in foods by electric field processing: a review. *Journal of Agriculture and Food Research* 16: 101109. https://doi.org/10.1016/j.jafr.2024.101109.

Yazdani-Asrami, M., Sadeghi, A., and Atrey, M.D. (2022). Selecting a cryogenic cooling system for superconducting machines: general considerations for electric machine designers and engineers. *International Journal of Refrigeration* 140: 70–81.

Zabot, G.L., Schaefer Rodrigues, F., Polano Ody, L. et al. (2022). Encapsulation of bioactive compounds for food and agricultural applications. *Polymers* 14 (19): 4194.

Zhang, R., Realini, C.E., Kim, Y.H.B., and Farouk, M.M. (2023). Challenges and processing strategies to produce high quality frozen meat. *Meat Science* 205: 109311.

Zheng, O., Zhang, L., Sun, Q., and Liu, S. (2024). Basic theory of ice crystallization based on water molecular structure and ice structure. *Food* 13 (17): 2773.

Zohuri, B. (2017). *Physics of Cryogenics: An Ultralow Temperature Phenomenon*. Elsevier.

3

Cryogenic Impact on Food Material Properties: Structural and Functional Transformations

Bhupendra M. Ghodki[1], *Krishna Bahadur Chhetri*[2], *and Naveen Kumar Mahanti*[3]

[1] *Agricultural and Food Engineering Department, IIT Kharagpur, Kharagpur, West Bengal, India*
[2] *Krishi Vigyan Kendra, Bhagwanpur Hat, Siwan, Dr. RPCAU, Pusa, Samastipur, Bihar, India*
[3] *Post-Harvest Technology Research Station, Dr. Y.S.R. Horticultural University, West Godavari, Andhra Pradesh, India*

3.1 Introduction

Cryogenic processing is the treatment of foods in low temperature for preservation and quality enhancement of food. During cryogenic processing, materials are typically exposed to temperature lower than −80 °C. Cryogenic processing has emerged as an alternative method that improves the quality and safety of food items in addition to making them more functional (Goswami 2017). The basic principle on which cryogenic processing operates relies on the rapid cooling of the food mass, causing specific structural and functional changes. However, it does so with minimal possible losses of volatile compounds, essential nutrients, and bioactive components, therefore becoming an attractive vehicle in modern food science.

The cryogenics technology sector is increasingly relevant to the food industry. This emerges from the desire of consumers for high-quality, mildly processed food, which has an extended shelf life. Traditional methods have tended to destroy the sensory and nutritional attributes of food because they have been overexposed to either heat or chemical treatment. Cryogenic treatment maintains the natural texture, flavour, and nutritional quality of foods. New technologies in combination with freezing thus have a positive and promising effect on process optimization since the application of these technologies reduces the negative impacts of traditional freezing methods. For example, because cryogenic methods freeze very quickly, they do not allow for large ice crystal formation, as happens with normal freezing, which destroys the cellular structure and, consequently, product quality.

Cryogenic Processing: Advances in the Food Industry, First Edition. Edited by Sumit Sudhir Pathak, Rama Chandra Pradhan, and Sudarshan Ramanathan.

3.1.1 Historical Perspective and Evolution

Cryogenics originated from early industrial applications. Cryogenic agents such as liquid nitrogen (LN_2) and carbon dioxide (CO_2) were used for quick cooling in various industries. At first, these methods were mainly used in fields like medicine, metallurgy, and aerospace to reach extremely low temperatures for specific industrial tasks. In the food industry, the idea started gaining popularity in the mid-twentieth century, with an emphasis on maintaining the freshness and quality of food items. Foods are frozen at very low temperatures using cryogenic freezing, preventing food from spoiling when it is stored or transported (Dhanya et al. 2023). The cryogenics has advanced to cope with the specific needs of the food industry, since protecting items such as fruits, vegetables, and seafood is crucial. Large ice crystals formed using traditional freezing methods would damage cellular structure in food, making it lose its texture, flavour, and become drier (Köprüalan Aydın et al. 2023). Food is frozen in small, even ice crystals in cryogenic processing, so its texture and nutrients are preserved. As equipment for cryogenics advanced, it became convenient for food industry to make good use of cryogenic materials. Advances in technology now allow cryogenic processes to cover tasks such as cryogenic grinding and cryogenic encapsulation to protect sensitive parts of foods. Cryogenics is used in many parts of the food industry, such as for frozen foods, grinding, freeze-drying, and encapsulation. They are designed to boost the texture, taste, and shelf life of food, allowing people to enjoy their meals, no matter how long it takes them to eat it.

3.1.2 Importance in Food Systems

Cryogenic processing is vital for helping the food industry to achieve sustainability, preserve the quality of foods, and experiment with new ideas. By preserving food materials, cryogenic processing prevents waste and allows for the production of foods high in antioxidants, vitamins, and minerals (Loayza-Salazar et al. 2024). As a result, these foods meet the need of many consumers for natural, health-improving goods made without chemical additives or preservatives. Cryogenic processing plays a vital role in preserving the food industry by maintaining the quality and nutritional value of foods. Cryogenic processing is key to enhancing the qualities of perishable foods and helping them last longer. Additionally, cryogenic processing allows for new food textures and qualities, giving more chances for creating unique food products. Rapid freezing and grinding of food with cryogenic methods helps produce materials that are more easily soluble and bioavailable. For this reason, cryogenic processes are often used for the development of nutraceuticals, fortified foods, and dietary supplements, where nutrients are absorbed efficiently (Ramachandran et al. 2024). In this way, cryogenic encapsulation protects fragile ingredients such as probiotics, omega-3 fatty acids, and other bioactive substances from damage while they are being processed or stored (Wang et al. 2023).

As a result, functional foods become available to consumers. It aids the consumer desire to improve health by taking foods with plenty of nutrients. Despite all the advantages and benefits of cryogenic processing, there are still difficulties and disadvantages that need to be solved in the food industry. The major challenge is the high operational cost, mainly because of the cryogenic agents used, such as liquid nitrogen (LN_2) and carbon dioxide (CO_2), which may be expensive and heavily energy intensive. However, the major drawback associated with cryogenic freezing and grinding methods includes high-energy consumption throughout the entire process of freezing and grinding tissue samples, translating to high costs that producers with limited capital may find it hard to embrace. Another hurdle is the public's attitude towards cryogenic methods, stating safety and environmental concerns. Certain consumer groups could have certain perceptions or constraints about the use of agents like cryogenic in food processing, since they may think those agents pose certain health risks or unusual processing means.

In addition, there are also concerns about the global-warming potential due to the usage of CO_2 gases and cryogenic agents during production and transportation, which are transported globally, primarily in areas where renewable energy is still scarce. However, these challenges also hold potential for the development of new forms of IT organization. Another practical example of more environmentally friendly cryogenic agents is the use of CO_2 obtained from industrial emissions for cryogenic processing or the further development of new cryogenic solutions. Furthermore, current studies and links with the production sector ensure that cryogenic techniques are adopted in food manufacturing. To secure the safety and efficiency of cryogenic food processing, the regulatory environment is evolving; misinformation about this technology is being addressed via educational campaigns. This chapter found that, as these challenges are dealt with, cryogenic processing has the potential to improve food preservation methods, food quality, and food sustainability within the food processing industry.

In this chapter, an attempt has been made to present the state of the art in cryogenic processing in the food processing industry. It explores the essential aspects of cryogenic treatment, which comprise the theoretical concepts, processes, and tools, as well as the structural and functional changes resulting from the treatment. The chapter also reveals the advantages of the cryogenic processing technique over other techniques, the different areas of its usage, and the issues and perspectives of the method. Cryogenic processing is a technique that holds promise for food processing and calls for a better understanding of the mechanisms that support innovation and sustainability in the food industry. Thus, this chapter is a reference for adopting cryogenic technology to respond to consumers' constantly developing demands, increasing the final product's positive characteristics and ensuring food safety.

3.2 Fundamentals of Cryogenic Processing in Food Industry

Cryogenic processing is changing the food industry by offering new ways to preserve, improve, and change food products. This section explores the main ideas, methods, and tools used in cryogenic processing, emphasizing its many benefits and various uses in today's food systems.

3.2.1 Principles and Methodologies

Cryogenic processing works on the basic idea of quickly reaching very low temperatures to create specific changes in food materials. It uses cryogens such as liquid nitrogen (LN_2) and carbon dioxide (CO_2) to achieve temperatures as low as −196 °C. The methods used in cryogenic processing are designed to be efficient and maintain the quality of food products.

1) **Rapid Freezing:** The key aspect of cryogenic processing is rapid freezing. This involves quickly lowering the temperature of food below its freezing point. Unlike traditional freezing methods that create large ice crystals, which can harm cell structures, cryogenic freezing produces small ice crystals uniformly. This helps to reduce physical stress on food tissues, keeping the texture and appearance intact. Its applications include frozen vegetables, fruits, seafood, and ready-to-eat meals (Köprüalan Aydın et al. 2023).
2) **Cryo-Grinding:** It involves reducing food materials to fine powders under very low-temperature conditions. This is particularly useful for spices, herbs, and things that heat can damage, because it helps keep in the oils and important nutrients (Ghodki et al. 2020). The applications are things like powdered spices, health supplements, and herbal remedies. The applications include powdered flavouring agents, nutraceuticals, and herbal supplements.
3) **Cryo-Encapsulation:** This method encapsulates sensitive bioactive ingredients, such as probiotics, omega-3 fatty acids, and antioxidants, in a protective cryogenic shell. The low temperatures prevent thermal degradation and oxidative damage (Benson et al. 2018). Applications include functional foods, beverages, and pharmaceutical-grade nutraceuticals.
4) **Cryo-Assisted Drying:** This method used for preserving the food's structural and nutritional integrity (Ramachandran et al. 2024). This method is ideal for high-value products like instant coffee, astronaut food, and specialty snacks.
5) **Cryo-Mechanochemistry:** This emerging field focuses on using really low-temperature freezing to change the way food materials are broken down and how useful compounds can be released (Pavlyuk et al. 2019). Applications are being studied in areas like making better food additives and nutritional supplements. Applications are being tried out in nanotechnology to make better food additives and dietary supplements.

3.2.2 Equipment and Techniques

The implementation of cryogenic processing relies on advanced equipment tailored to specific food processing needs. Key components include cryogenic chambers, freezers, grinders, and safety mechanisms.

1) **Cryogenic Freezers**
   a) **Tunnel Freezers:** Tunnel freezers freeze products as they pass by using liquid nitrogen or CO_2, and are popular for freezing meat, seafood, and foods from the bakery. These are popular in freezing meat, seafood, and bakery items.
   b) **Batch Freezers:** Suitable for smaller production scales, batch freezers allow precise temperature control for artisanal or specialty products.
2) **Cryogenic Grinders:** Specifically used for grinding particles, cryogenic grinders are machines that possess liquid nitrogen cooling to minimize temperatures during grinding. This averts heat degeneration and provides for standard particle sizes, especially where spices and pharmaceutical-grade powder are involved.

3) **Cryogenic Encapsulation Systems:** These systems use freeze-and-coat techniques to encapsulate the bioactive molecules. This is because liquid nitrogen is employed as the species coolant in order to preserve functional attributes of the susceptible materials in the process.
4) **Auxiliary Systems:** Contemporary cryogenic facilities are equipped with temperature, pressure, and flow-rate control parameters. An engineering draft enables accurate positioning and safe operation of the cryogenic equipment while reducing consumption of cryogenic agents.

3.2.3 Advantages of Cryogenic Processing

Cryogenic processing is exceptional in this respect because it influences many of the basic parameters of food items, such as quality, safety, and functionality. This allows cells in food to remain unsubmerged, since rapid freezing will not allow the formation of big ice crystals. This matters a lot so that it preserves the texture, colour, and taste of things such as seafood, fruits, and vegetables. In particular, low-temperature processing affords an opportunity to minimize nutrient degradation, such as vitamins, antioxidants, and other bioactive compounds. The cryogenic processes normally retain more nutrition than conventional methods.

Cryogenic conditions help to limit oxidation in foods that often leads to them becoming rancid and discoloured. Therefore, cryogenic methods help preserve foods with oil and fats, such as nuts and fish, a lot better than normal methods. In these methods cannot survive in low temperatures, so using them does not require the addition of chemicals. As a result, perishable foods can be kept for a longer time, saving on wastage and making it easier to transport and distribute them.

3.2.4 Applications in the Food Industry

Cryogenic processing has a broad spectrum of applications, making it indispensable across various food industry sectors.

1) **Food Preservation**: Extending the shelf life of perishable items, such as meat, poultry, seafood, and dairy, is one of the most common applications of cryogenic technology. Products retain their freshness, texture, and nutritional profile during storage.
2) **Frozen Foods**: Cryogenic freezing plays a vital role in the production of frozen fruits, vegetables, and ready-to-eat meals. It ensures uniform freezing, reduces crystallization issues, and delivers a consistent product to consumers.
3) **Functional and Nutraceutical Foods**: The ability to preserve and enhance bioactive compounds has led to cryogenic techniques in producing functional foods and supplements. Examples include antioxidant-rich powders, probiotic-enhanced dairy products, and omega-3-enriched snacks.
4) **Confectionery and Desserts**: Cryogenic processing enables the creation of innovative frozen desserts, such as nitrogen-frozen ice cream, which has a smooth texture and retains flavour intensity. Additionally, cryogenic grinding is used to produce high-quality chocolate powders.

5) **Bakery and Snacks**: Freezing and grinding processes are utilized to create bakery products with extended shelf life and improved texture. Cryogenic techniques are also employed to develop powdered flavourings for snacks.
6) **High-Value Ingredients (Spices, Oleoresins, and Essential Oils):** Cryogenic encapsulation protects sensitive ingredients like flavours, enzymes, and essential oils, allowing their controlled release in food products.
7) **Meat and Poultry:** Cryogenic freezing ensures minimal drip loss and colour retention in meat and poultry products. It is widely used in the export market, where product quality is of paramount importance.

The growing need for top-notch processed foods has made cryogenic processing a key part of food breakthroughs. Progress in machines' use of green energy and new ways to apply cryogenics will boost its use in the food world. Also, studies on cutting running costs and improving how people see it will be vital for its widespread use. By grasping the basics of cryogenic processing, the food industry can keep tapping its power to make better products while tackling green and safety issues. The story of cryogenics in food processing isn't over yet – promising more new ideas and growth in the coming years.

3.3 Structural Alterations in Cryogenically Treated Food Materials

Cryogenic processing induces significant structural changes in food materials at both molecular and microstructural levels. These alterations, driven by the extreme low-temperature conditions, directly influence the food's texture, stability, and functionality (Giannakourou and Taoukis 2021). Understanding these changes is crucial for optimizing cryogenic techniques to produce high-quality food products. This section explores the molecular and microstructural changes, the evolution of crystalline structures and phase transitions, and the advanced characterization techniques employed to study these transformations.

3.3.1 Molecular and Microstructural Changes

Cryogenic processing induces dramatic modifications at food materials' molecular and microstructural levels, which are attributed to rapid freezing rates and extremely low temperatures.

3.3.1.1 Molecular Changes

1) **Protein Stabilization and Denaturation Prevention**: Cryogenic freezing in meat and seafood protects the proteins so the food stays tender and juicy, which makes customers satisfied. This quick freezing in dairy foods such as cheese or ice cream keeps the framework of proteins unharmed, avoiding bad flavour. This rapid freezing in dairy products, such as ice cream or cheese, ensures that the delicate protein networks remain intact, preventing textural defects and enhancing the sensory experience (Liu and Zhou 2015).
2) **Preservation of Lipids**: If unsaturated fats in food lipids oxidize, it will make the product rancid and reduce its quality. The process keeps lipids fresh by quickly freezing them

to a low temperature. Foods like nuts, seeds, and oily fish are perfect examples of food that need this technique to protect lipids. Cryogenic freezing helps in maintaining the natural goodness of fat in food and prevents oil from becoming rancid or spoiled, improving its shelf life (Buzan et al. 2014). By freezing quickly, it reduces the effects of oxidative enzymes and oxygen, so the food's lipids are conserved. By rapidly reducing the temperature, the process inhibits the activity of oxidative enzymes and reduces the exposure to oxygen, which are primary factors are contributing to lipid oxidation.

3) **Retention of Bioactive Compounds**: Cryogenic processing keeps vitamins and other items in good condition, which helps in fighting ageing and other healthy parts. When you make food cold fast, it stops the heat from breaking down these good parts. It also slows down the tiny things in food that can make the healthy bits go away. This works great for fruits, vegetables, and spices. It helps them stay healthy and fresh for longer. The cold conditions in cryogenic processing stop bioactive compounds from breaking down, which means food keeps its health benefits (Jafari and Capanoglu 2022). This is especially important for products marketed for their nutritional benefits, as it allows consumers to enjoy the full spectrum of nutrients without compromise.

3.3.1.2 Microstructural Changes

1) **Cellular Integrity**: Another advantage of cryogenic processing is the maintenance of the tissue's ultrastructure in plant and animal cells. For a long time, simple freezing methods have been used, resulting in the formation of large ice crystals inside the cell structure of food, which severely affects the food. This leads to the deterioration of texture, which also affects the surface and appearance of food (Khaydukova et al. 2024). Cryogenic freezing as a method involves very rapid cooling, usually with the use of liquid nitrogen (LN_2) or carbon dioxide (CO_2). This type of freezing forms small and orderly ice crystals whose formation exerts little pressure on the cells of food and serves as an added advantage due to its ability to freeze rapidly. Therefore, cell integrity retains the shape of the food with the right texture and colour. This is especially so where physical attributes like the appearance and texture of the product form the basis on which many consumers are willing to purchase the product.
2) **Reduction of Drip Loss**: Drip loss is the phenomenon experienced when frozen food products melt, and the amount of juice that comes out causes a change in the juiciness of the food product. This is always a drawback of normal freezing techniques because very large extracellular ice crystals that form are destructive, bursting the cell walls and releasing intracellular fluids. Cryogenic freezing helps prevent very large ice crystals from forming. The main advantage of the process is that due to high-speed freezing, small ice crystals are formed inside the cell structures (Goswami 2010). This assists in preserving the structure of the cell walls and minimizes moisture loss during a thawing process. Therefore, cryogenically frozen meat and seafood products have less percent of their moisture loss, which makes them juicy and tastier. This preservation of juiciness is important for choice portions of meats and fragile fish, for juiciness is an attribute of quality.
3) **Micronization**: Micronization is the process of converting food materials into micro- or nanoparticles, increasing the surface area and solubility of food materials. Micronization by cryogenic is done at low temperatures to minimize the disintegration

of precarious elements. This method has a major advantage when handling materials sensitive to heat, such as spices, herbs, and nutraceuticals.

4) In cryogenic grinding, first the material is cooled by cryogenic agents, which include liquid nitrogen. This makes the material brittle and easier reducing them into fine particles without producing heat, which can harm the essential oils, vitamins, and bioactive compounds in health supplements. The final ultrafine particles possess higher solubility and bio assimilation; they are suitable for flavouring agents, nutraceuticals, and functional food. This process is used throughout the spice and nutraceutical industries for applications in which the stability of active compounds is crucial.

3.3.2 Crystalline Structures and Phase Transitions

The cryogenic environment induces specific changes in the crystalline structure of food materials, which affect their physical properties, such as texture and stability. The freezing rate and the composition of the food matrix govern these transitions.

3.3.2.1 Ice Crystallization

1) **Small and Uniform Ice Crystals**: Cryogenic freezing is associated with small and uniform ice crystal formation. This is mainly because of the high rate of nucleation and short phase of crystal growth, which is common in these types of steel. Once food tends to undergo cryogenic temperature, for instance, with the help of LN_2 or CO_2, the change in temperature forces many ice nuclei to develop at once. This fast nucleation ensures that any of the ice crystals do not grow big, and instead, there will be many small-sized, equal-distance ice crystals form (Zheng et al. 2024). This contrasts with conventional freezing methods where the cooling rate is slower and fewer nuclei of ice form. Therefore, the fewer nuclei have more time to develop considerably larger crystals that are not favourable for the structure of cells in foods.
2) **Impact on Texture**: Freezing temperature and the resulting size and distribution of ice crystals play important roles in the texture of frozen products. The smaller proportion of the ice crystals and their homogenous size when using cryogenic freezing allow minimal physical destruction on different cells of the food. In the context of fruits, vegetables, and some special delicate seafood products, the maintenance of this cellular integrity is specifically significant in texture retention. For instance, in freezing small fruits and vegetables, the formation of small ice crystals in a product prevents the destruction of the cell walls, and hence, the firmness and normal texture of the fruits and vegetables are preserved. Likewise, in seafood, no loss of cell integrity would help the texture maintain almost the quality of the fresh one. In products such as ice cream, the formation of small ice crystals is crucial for achieving a smooth and creamy texture. Larger ice crystals, which are more likely to form during conventional freezing, can result in a coarse and gritty texture, which is undesirable. Cryogenic freezing, by ensuring the formation of small ice crystals, helps to maintain the desired smoothness and creaminess of the final product.

3.3.2.2 Glass Transition and Vitrification

1) **Glass Transition Temperature**: In cryogenic processing, one of the critical factors influencing food stability is the glass transition temperature (Tg). The moisture and

solutes present in the food matrix, when they are cooled below Tg, transform into a vitreous state. This state considerably decreases molecular diffusivity; water and solutes practically cease to move about or 'flow' in the material. The reduction of molecular mobility decreases chemical reactions that cause deterioration, like enzymatic action and microbial presence (Grzybowski 2023). The method keeps the food in this glassy state, reduces the shelf life, and also enhances the quality of the product through cryogenic processing. This mechanism is especially beneficial when the foods require special handling, as their nutrient and organoleptic properties should be retained.

2) **Vitrification**: A process through which the food matrix solidifies in the process of cryopreservation is called vitrification. This transformation, on its part, improves the stability of the food during storage by avoiding crystallization and phase separation. Vitrification is particularly applied in freeze-dried products and powders, as it guarantees the protection of the reactions and bioactive compounds and nutrients for a long time (Amini and Benson 2023). The glass-like state imposed by vitrification offers food safety from physicochemical transformation, making the method suitable for high-value foods where there is a need to retain quality and functionality for extended periods.

These processes are integral to optimizing cryogenic techniques, ensuring that food products remain stable and retain their desirable attributes during storage and distribution.

3.3.2.3 Phase Separation and Recrystallization

1) **Cryo-Induced Phase Separation**: Cryogenic operation is particularly useful in the demulsification of multiphases such as in foods. For instance, in ice cream, which is not a stable emulsion, freezing the aqueous phase stabilizes fat globules rapidly. This stabilization allows the fat not to come together into bigger globules that would further affect the smoothness of the ice cream. It also ensures that after freezing, the fat particles remain well distributed so that the texture and mouthfeel characteristics of such products are achieved.
2) **Recrystallization During Storage**: One of the main problems that occur during frozen storage is ice recrystallization: when temperatures change, small ice crystals melt and then refreeze, enlarging and detrimental to the texture and quality of the product. Cryogenic freezing that forms smaller and more uniform ice crystals minimizes the risk of recrystallizations that tend to scatter cells. This stability is highly desirable when preserving texture, particularly in ice cream and other frozen fruits and vegetables for long retail periods of storage.
3) **Impact on Starch and Polysaccharides**: It has been discovered that cryogenic treatment may cause considerable alterations to the crystalline structure of the starch, which directly affects gelatinization and retrogradation of the starch (Lu et al. 2023). In the case of starch, when cooled to cryogenic temperatures, the relatively fast rate of freezing destroys the crystal lattice portions. This change influences the moisture absorption capacity of the starch, which is a significant parameter during gelatinization. During gelatinization, the crystalline structure is upset by water molecules, hence improvement in the ability of water to swell the starchy particles at a uniform and faster rate. This can refine the mouth feel of bakery and processed foods and can make the products

softer in texture and taste. Also, it has been suggested that putting starch through the cryogenic treatment can slow down how fast it changes back to a hard shape after being cooked and heated, which is what can make it firm and tough to eat. In baked goods, this means that using the cryogenic method with starch makes the bread softer and helps to stop it from getting dry and stale too quickly. For processed foods, it can help them stay the same in texture and not change too much over time while they are in storage. The precise control over the freezing rate and temperature in cryogenic processing changes the starch while cooking, which is helpful for making foods better and keeping the quality the same. The precise control over the freezing rate and temperature in cryogenic processing allows for tailored modifications to the starch properties, making it a valuable tool in food formulation and quality improvement (Lu et al. 2023).

3.3.3 Advanced Characterization Techniques

Advanced analytical techniques are used to find out how the structure changes during cryogenic freezing. These methods help researchers to see the changes how molecules are organized, check for ice crystals in food, and see how different parts of the food structure changes the shape

3.3.3.1 Microscopy Techniques

3.3.3.1.1 Scanning Electron Microscopy (SEM)

It is used to study specific features of food structure, such as the size and location of ice crystals. It is especially helpful for researchers who need to investigate the effects of 'cryogenic freezing' on the cells' organization. Investigating the diameter, size distribution, and microstructure of ice crystals in foods under cryogenic conditions. Evaluating the thawing effect on fruits, vegetables, and meat cell damage caused by freezing. Control of coalescence and emulsion structure in products such as dairy products and frozen desserts. Assessing cellular damage in fruits, vegetables, and meats caused by freezing and monitoring emulsions' uniformity and structural stability in dairy and frozen desserts. SEM analyses have determined that freezing at cryogenic temperature leads to the formation of more but smaller and more uniform ice crystals, which cause less damage to the structure of frozen fruits and vegetables.

3.3.3.1.2 Transmission Electron Microscopy (TEM)

It gives clear pictures of changes within molecules, such as the aggregation of proteins or the arrangement of bioactive compounds. TEM transmits electron beams through ultrathin samples to visualize internal structures at the molecular level. Its applications involve analysing protein aggregation or denaturation in cryogenically treated foods, studying the crystalline structure of starches and other polysaccharides affected by low temperatures. Observing encapsulated bioactive compounds in functional foods. TEM is used to study the stability of encapsulated probiotics in cryogenically processed nutraceuticals.

3.3.3.1.3 Confocal Laser Scanning Microscopy (CLSM)

It enables low magnification imaging of cross-sections of food samples at micro- and macro-scale, leading to a visualization of microstructures. It is taken to investigate emulsions and gels prepared through cryogenic processing. CLSM uses lasers to generate high-resolution, three-dimensional images of food structures. Its applications involve evaluating fat, water, and protein spatial distribution in cryogenically processed emulsions; monitoring the integrity of cell walls in fruits and vegetables. CLSM is employed to observe the uniform distribution of fat globules in nitrogen-frozen ice creams.

3.3.3.1.4 Cryo-Electron Microscopy (Cryo-EM)

Cryo-EM allows visualization of food samples in their frozen state without thawing, preserving their natural microstructure. The applications are studying protein complexes and molecular arrangements in frozen states, understanding phase transitions in frozen foods. Cryo-EM has been used to observe the vitrification of sugar solutions during cryogenic processing.

Microscopy techniques in cryogenic processing include SEM for analysing ice crystals and cellular damage, TEM for visualizing internal structures, CLSM for three-dimensional (3D) imaging of emulsions and cell integrity, and Cryo-EM for preserving and observing frozen samples. These methods optimize processing by maintaining structural quality and enhancing functionality (Kharbach et al. 2023).

3.3.3.2 Spectroscopic Techniques

3.3.3.2.1 Fourier Transform Infrared Spectroscopy (FTIR)

FTIR quantifies alterations in functional groups, for example, protein unfolding or lipid peroxidation, based on the recognition of a specific vibrational frequency. FTIR measures the absorption of infrared radiation by molecular bonds, providing insights into chemical changes. The applications involve monitoring lipid oxidation in cryogenically treated foods, evaluating protein denaturation or conformational changes in frozen meat or dairy products, assessing the retention of bioactive compounds, such as polyphenols and carotenoids. FTIR analysis of cryogenically frozen fish shows reduced lipid oxidation compared to traditional freezing methods.

3.3.3.2.2 Raman Spectroscopy

Raman spectroscopy can assess protein and polysaccharide hydration to help measure the quality changes during cryo-treatment. Raman spectroscopy measures inelastic scattering of light to identify molecular vibrations. The applications are studying water structure and phase transitions in frozen foods, detecting changes in carbohydrate or protein structures. Raman spectroscopy is used to monitor the structural stability of polysaccharides in cryogenically processed plant-based snacks.

3.3.3.2.3 Nuclear Magnetic Resonance (NMR) Spectroscopy

NMR detects interactions between magnetic fields and atomic nuclei to provide detailed information on molecular structures. The applications are determining water mobility and distribution in cryogenically frozen foods, studying the stability of encapsulated bioactives in functional foods, monitoring the release of volatile compounds in spices and herbs. NMR is applied to evaluate the preservation of volatile aroma compounds in cryogenically ground cinnamon.

3.3.3.2.4 Ultraviolet–Visible (UV–Vis) Spectroscopy

UV–Vis measures light absorption to quantify specific compounds, such as pigments and antioxidants. The applications involve assessing the retention of carotenoids, flavonoids, and other colour compounds in cryogenically processed vegetables, quantifying antioxidant capacity in frozen berries and functional powders. UV–Vis spectroscopy reveals higher anthocyanin retention in cryogenically frozen blueberries than conventional freezing.

Spectroscopic techniques play a vital role in cryogenic processing, analysing molecular changes such as protein denaturation, lipid oxidation, and the preservation of bioactive compounds. FTIR measures infrared absorption to monitor lipid oxidation, protein denaturation, and retention of compounds like polyphenols (Rizzo and Boyarkin 2014). Raman spectroscopy uses light scattering to study water structure, phase transitions, and stability of polysaccharides. NMR detects magnetic interactions to analyse water mobility, bioactive stability, and volatile compounds. UV–Vis spectroscopy measures light absorption to assess the retention of carotenoids, flavonoids, and antioxidant capacity in cryogenically processed foods, optimizing their quality and functionality.

3.3.3.3 Thermal Analysis

1) **Differential Scanning Calorimetry (DSC)**: DSC quantifies phase transition, which is the glass transition and ice-melting points required to determine the thermal stability of cryogenically treated foods.
2) **Thermogravimetric Analysis (TGA)**: TGA measures gains or losses resulting from water loss or thermal degradation because of cryogenic processing or storage.
3) **Rheological Techniques**: Rheometers quantify the rheological characteristics of the lower food material and the shift in the textural as well as flow properties caused by cryogenic treatment.

3.3.3.4 Chromatographic Techniques

High-performance liquid chromatography (HPLC) and gas chromatography (GC) compare the effectiveness of cryogenic processing in the preservation of vitamins, antioxidants, and essential oils. It became clear that the structural alterations that cryogenic treatment affects in foods cause significant impacts on their quality. To single out, due to the maintenance of the cell structure, the protection of perishable nutrients, and the inhibition of deterioration, cryogenic processing provides the final products with improved texturizations, tastes of flavours, and nutritional values. In addition, the fact that the process of ice crystallization and phase change can be easily controlled will authorize the development of innovative food products with improved functionality and extend shelf life. Cryogenic processing is one of the cutting-edge technologies that has remained significant in tailoring the structural and functional properties of foods (Loayza-Salazar et al. 2024). The enhanced qualitative and quantitative description of molecular and microstructural changes is at the heart of its ability to develop superior and sustainable foods that are more responsive to the client's needs. While research is going on to demystify these changes, cryogenics have already set the tone for future food processing.

3.3.4 Sensory Analysis

Sensory analysis is critical for evaluating the consumer perception of cryogenically processed foods, focusing on texture, flavour, appearance, and overall acceptability.

1) **Descriptive Sensory Testing**: Trained panels evaluate specific sensory attributes. The applications are assessing the smoothness of nitrogen-frozen ice creams and evaluating the firmness and juiciness of cryogenically frozen fruits. Descriptive analysis reveals enhanced crunchiness in cryogenically frozen bell peppers.

2) **Consumer Preference Testing**: Consumer panels provide feedback on overall acceptability. The applications are determining consumer liking for cryogenically frozen ready meals, assessing the flavour intensity in cryogenically ground spices. Cryogenically processed herbs score higher in freshness than conventionally dried and frozen samples.
3) **Texture and Mouthfeel Analysis**: It focuses on the tactile sensations experienced during consumption. Their applications involve evaluating the creaminess of nitrogen-frozen dairy desserts, measuring the crunchiness of cryogenically processed snacks. Cryogenically frozen bread retains a softer crumb texture after thawing.

Techniques include descriptive sensory testing for specific attributes, consumer preference testing for overall acceptability, texture and mouthfeel analysis for tactile sensations, and flavour retention studies for aroma and stability (Van Breda et al. 2024). These methods ensure high-quality cryogenically processed foods by preserving crunchiness, creaminess, freshness, and flavour intensity.

3.4 Functional Enhancements and Impacts on Food Quality

Cryogenic processing – a relatively new technique in the food industry – effectively ameliorates the functional attributes of foods through thermal, rheological, flavour, nutritional, and microbiological characteristics. These are enhancements of functional traits that this section seeks to elaborate on in relation to delivering quality food products to consumers.

3.4.1 Texture and Flavour Improvements

Texture and flavour are critical attributes influencing consumer perception and food acceptance. Cryogenic processing plays a vital role in preserving and enhancing these sensory characteristics.

3.4.1.1 Texture Improvements

The use of cryogenic freezing forms small and even ice crystals because of the high freezing rates. This ensures cell structures are retained in foods, giving fruits and vegetables their usual firmness and meat and seafood their normal moisture content. Crater-shaped strawberries frozen by cryogenic processing suffer less from large ice crystals, leading to mushy textures, which is characteristic of processing employing normal freezing techniques.

Rapid freezing minimizes the damage to the plant tissue cell walls, hence the desirable characteristics, that is, crispness in vegetables and low drip loss in thawed products. In cryogenic processing of emulsions like ice creams as well as dairy-based dessert products, this technique of freezing helps to solidify the fat globules so that they do not coagulate, apart from giving a smooth and glossy texture. Cryogenic grinding does not involve the generation of heat that adversely affects the structure of an ingredient such as spices; the technique enhances the structure and application of the food.

3.4.1.2 Flavour Improvements

Compounds responsible for the flavours, such as volatiles in herbs, spices, and fruits, may easily be eliminated by heat as well as oxidation. Cryogenic processing primarily allows these compounds to be sealed off by preventing vaporization. Cryogenically ground spices possess the right characteristics and potency for a longer period than conventionally processed spices. Lipids are protected from oxidation during cryogenic freezing, ensuring flavour stability in high-fat products like nuts, seeds, and fish. Cryogenic freezing preserves raw ingredients' fresh, natural taste, making it ideal for premium frozen products such as ready-to-cook meals and gourmet desserts.

3.4.2 Nutritional Content and Bioavailability

The cryogenic process is much lauded for such reasons as preserving a food's nutrients and improving on such a value. The low temperatures and the short time of blanching and steaming reduce undesirable nutrient content changes and increase bioactive compounds' biological availability.

3.4.2.1 Retention of Essential Nutrients

Cryogenic conditions minimize enzymatic activity and oxidative stress, preserving sensitive vitamins (e.g. vitamin C and B-complex vitamins) and antioxidants such as polyphenols. Frozen berries processed cryogenically retain higher levels of anthocyanins than conventional freezing methods. Minerals remain largely unaffected by cryogenic processing due to the absence of thermal degradation.

3.4.2.2 Preservation of Bioactive Compounds

Cryogenic methods protect phenolic compounds from degradation, enhancing the antioxidant potential of foods like teas, herbs, and spices. The rapid freezing process minimizes oxidative damage, preserving the bioactive compounds responsible for antioxidant activity (Saxena et al. 2016). This ensures that foods retain their health benefits and sensory qualities. In cryogenic encapsulation, probiotic strains are protected in a stable matrix, ensuring their viability during storage and transit through the digestive system. This technique helps to maintain the beneficial properties of probiotics, allowing them to reach the gut, where they can exert their health benefits (Rajam and Subramanian 2022).

3.4.2.3 Improved Bioavailability

Cryogenic grinding reduces particle size to nano or micro levels, increasing the solubility of bioactive compounds and improving their bioavailability. Cryogenically processed turmeric exhibits improved curcumin bioavailability, making it more effective as a nutraceutical ingredient. Cryogenic processing can break down complex matrices, releasing bound nutrients such as carotenoids in plant materials. Cryogenic freezing prevents protein denaturation, preserving amino acid profiles and digestibility in meat, fish, and dairy products. The rapid freezing process minimizes ice crystal formation, which helps to maintain the structural integrity of proteins. This preservation of protein structure ensures that essential amino acids remain intact, enhancing the nutritional value and digestibility of the food products (Liu and Zhou 2015). Additionally, cryogenic freezing inhibits microbial growth and enzymatic reactions, further extending protein-rich foods' shelf life and safety. This method is particularly beneficial for high-quality meat, fish, and dairy products, where maintaining nutritional and sensory properties is crucial.

3.4.3 Shelf Stability and Safety

Cryogenic processing extends the shelf life of food products by mitigating spoilage mechanisms and ensuring microbial safety. This reduces food waste and supports the production of safe, high-quality foods for consumers.

3.4.3.1 Shelf Stability

Cryogenic freezing takes the food to a temperature where necessary micro-activity for spoiling the food stops, making it last longer without any additive chemicals. Seafood frozen cryogenically stays fresh for several months, unlike when it is frozen under the normal freezing methods. High oxygen content is a reason for spoilage in foods with high fat and protein content. But the cryogenic treatment reduces the oxygen content, thus preventing the oxidation. Nuts treated using cryogenic freezing are less likely to get rancid as compared to normally frozen nuts. The minimized phase separation and ice recrystallization during cryogenic freezing result in stable frozen products, reducing texture and quality degradation during storage.

3.4.3.2 Safety

Cryogenic freezing works through the effect of preventing microbial growth by freezing the surface so rapidly that the temperature is lowered to a level at which microorganisms are not capable of surviving. This is especially necessary in ready-to-eat and minimally processed foods. Fresh ground beef processed at cryogenic temperature contains fewer pathogenic bacteria, such as *Escherichia coli*, than ground beef processed using the conventional method of freezing. Cryogenic temperatures can inactivate certain pathogens, contributing to improved food safety. The closed systems used in cryogenic processing reduce the risk of contamination during processing.

3.5 Applications of Cryogenic Processing

Cryogenic is one technology that has been revealed to have many uses, cutting across different food segments. It is now impossible to imagine modern food systems without this technology due to its effectiveness in retaining the quality, shelf life, and functionality of food products. This section discusses its uses in food preservation and other food products such as minced meats and poultry, frozen desserts and baking, and functional foods and nutraceuticals.

3.5.1 Food Preservation and Shelf-Life Extension

Maintaining the quality of food and retaining its shelf life are the major goals of cryogenic processing. The significantly low temperatures bring quick equilibrium in food items, thus reducing spoilage and spoilage mechanisms. Freezing inhibits microbial growth by rapidly freezing it to temperatures that are not liable for its inhabitants. This ensures temperature low enough to preserve perishables such as seafood, dairy, and fruits within the establishment for long time. Evaluating the amount of oxygen to which the food is exposed and lowering lipid oxidation, cryogenic methods retain the colour, flavour, and nutritional qualities of the meat.

Cryogenically frozen avocado retains its green colour and creamy texture, unlike conventionally frozen avocado, which often browns and develops an off-flavour.

Cryogenic processing slows enzymatic activity, preventing undesirable changes like browning in fruits and vegetables or texture degradation in meat. Rapid freezing forms small, uniform ice crystals, preserving the cellular structure of plant and animal tissues. This minimizes texture changes and drip loss upon thawing. Cryogenically frozen strawberries maintain firmness and visual appeal compared to conventionally frozen ones. The extended shelf life and high-quality preservation achieved through cryogenic processing reduce food spoilage, contributing to sustainability and reducing economic losses. Cryogenic systems are widely used for bulk freezing seafood, meat, and fruits for export markets. Products retain their quality even during long transits.

3.5.2 Meat and Poultry Processing

Cryogenic processing is particularly beneficial in the meat and poultry industry, where maintaining quality, safety, and visual appeal are critical. Cryogenic freezing prevents the formation of large ice crystals that can rupture muscle fibres, reducing water loss during thawing and ensuring juiciness. Chicken breasts frozen using cryogenic methods exhibit higher moisture retention than conventional freezing. The rapid freezing process minimizes oxidation of myoglobin, the pigment responsible for the red colour in meat, ensuring a fresh appearance. Cryogenic freezing significantly lowers the microbial load, enhancing the safety of raw meat and poultry products. This is especially crucial for export markets, where stringent safety standards must be met. Cryogenic systems are used to rapidly chill meat during processing, which helps maintain temperature control and prevents bacterial growth. Cryogenic techniques allow for tailored freezing protocols for different cuts of meat, preserving texture and ensuring consistent quality across products.

3.5.3 Frozen Desserts and Baked Goods

The unique properties of cryogenic freezing have led to its widespread adoption in the production of premium frozen desserts and baked goods. Cryogenic freezing produces micro-sized ice crystals, resulting in an exceptionally smooth texture in ice creams and frozen yogurts. Nitrogen-frozen ice creams, popular for their velvety texture and fresh taste, directly result from cryogenic processing. The ultra-low temperatures lock in delicate flavours, ensuring a rich taste profile that does not degrade during storage. Cryogenic systems enable the rapid freezing of customized ice cream flavours or dessert portions, catering to niche markets.

Cryogenic freezing stabilizes doughs, preventing yeast activity during storage and ensuring consistency in baked goods. The freezing process preserves gluten and starch's structural integrity, leading to better volume and texture post-thawing baked goods. Cryogenic cooling is used to temper chocolate, ensuring a glossy finish and a satisfying snap. It also prevents fat bloom during storage. Cryogenically frozen croissants and puff pastries rise better during baking, achieving a consistent and flaky texture. Cryogenic freezing supports the production of high-quality gluten-free and vegan baked goods by stabilizing alternative ingredients.

3.5.4 Functional Foods and Nutraceuticals

Cryogenic processing is a cornerstone in the development of functional foods and nutraceuticals, where the preservation and enhancement of bioactive compounds are paramount. Cryogenic methods protect antioxidants, polyphenols, and flavonoids from thermal degradation, ensuring their bioactivity. Green tea powder processed using cryogenic grinding retains its catechin content, enhancing its antioxidant properties. Cryogenic encapsulation protects omega-3 oils from oxidation, making them suitable for fortified products like beverages and dairy. Cryogenic grinding reduces particles to nano- or micro-scale, increasing solubility and absorption of bioactive compounds. Turmeric processed cryogenically exhibits enhanced bioavailability of curcumin, a potent anti-inflammatory compound. Cryogenic grinding facilitates the extraction of active ingredients from botanicals, improving yield and efficacy.

Cryogenic encapsulation stabilizes probiotic strains, enhancing their viability during storage and transit through the gastrointestinal tract. Cryogenic grinding of legumes, nuts, and seeds preserves their nutritional profile, enabling the production of high-quality protein powders. Cryogenically processed powders, such as matcha and herbal supplements, dissolve easily in beverages, providing functional benefits. Cryogenic freezing stabilizes sensitive ingredients in health bars, ensuring consistent quality and nutritional value. Cryogenic processing does not require chemical preservatives, consistent with consumers' trends towards clean-label and green-label products.

It protects antioxidants; the bioactivity and stability of polyphenols and omega-3 fatty acids are protected from degradation (Ramachandran et al. 2024). In cryogenic grinding, the material is ground under low temperatures, and the size of the particles is reduced significantly. Success in increasing the solubility and bioavailability of compounds like curcumin has been reported. It also facilitates working with cryo-assisted extraction of botanicals, increasing yield of the main product and its efficacy. Probiotics and plant-based proteins benefit from cryogenic encapsulation and grinding, keeping their nutritional value intact and enhancing viability. Thus, cryogenic processing corresponds to consumers' concern about environmentally friendly, clean-label products by reducing the use of chemical preservatives in their products.

3.6 Future Directions and Challenges

Cryogenic food processing improves preservation, quality, and safety for different food products such as plant proteins, non-dairy desserts, and nutraceuticals. This technology enhances texture, flavour, and durability with high fixed costs and legal compliance burdens, though. Environment management techniques like the efficient use of energy, reduced emission of wastes, and environmentally friendly packaging have several costs associated with them, known as setup expenses. Artificial intelligence and nanotechnology enhance efficiency and stabilize the ingredients used, but such technologies demand major capital outlay. Regulatory compliance and worker safety are ensured but incur additional costs. Global collaboration through partnerships and standardized protocols fosters best practices despite coordination challenges. Further details are summarized in Tables 3.1 and 3.2.

Table 3.1 Expansion of applications and sustainability practices.

Section	Details	Impact	Demerits	References
Expansion of applications	Enhances texture, flavour, and shelf life in plant-based proteins, dairy-free desserts, and nutraceuticals. Example: Cryogenic grinding of legumes.	Improves quality and consumer acceptance.	High initial costs and specialized equipment.	Goswami (2017)
Functional foods and nutraceuticals	Cryogenic encapsulation for personalized nutraceuticals. Example: Encapsulation of omega-3 fatty acids.	Tailored health benefits and improved bioavailability.	Regulatory challenges and cost implications.	Rajam and Subramanian (2022)
Convenience foods	Enhances frozen ready-to-eat meals by preserving sensory attributes.	Improves convenience and extends shelf life.	Requires efficient supply chain and storage conditions.	Timmerhaus and Flynn (1989)
Sustainability practices	Energy-efficient systems, minimizing food waste, eco-friendly packaging, water conservation, and carbon neutrality. Example: Circular systems reclaim and reuse liquid nitrogen.	Reduces environmental impact and operational costs.	High initial setup costs.	Goswami (2017)
Reducing energy consumption	Lowers costs and carbon footprint using renewable cryogenic agents.	Sustainable and cost-effective in the long run.	High initial investment.	Timmerhaus and Flynn (1989)
Minimizing food waste	Extends shelf life and repurposes surplus produce. Example: Cryogenically frozen surplus produce.	Enhances food security and reduces waste.	Requires supply chain coordination.	Goswami (2017)
Eco-friendly packaging	Reduces need for chemical preservatives.	Aligns with sustainable consumer preferences.	May increase packaging costs.	Timmerhaus and Flynn (1989)

Table 3.2 Innovations, regulatory concerns, global collaboration, and challenges.

Section	Details	Impact	Demerits	References
Innovations in equipment	Combines cryogenic and mechanical freezing, integrates AI, and uses nanotechnology. Example: Hybrid systems.	Enhances processing efficiency and product quality.	High technology and maintenance costs.	Timmerhaus and Flynn (1989)
Automated processes	Uses AI and IoT for real-time monitoring and optimization.	Improves efficiency and reduces human error.	Requires skilled maintenance.	Goswami (2017)
Nanotechnology	Enhances food additives' functionality and stability. Example: Nanoencapsulated curcumin.	Improves ingredient stability.	Regulatory and consumer acceptance issues.	Shi (2015)
Enhanced safety features	Automated shutdowns and fail-safes. Example: Sensors for nitrogen leakage.	Improves workplace safety.	Increases equipment costs.	Timmerhaus and Flynn (1989)
Regulatory and safety concerns	Complies with international standards and clear guidelines. Example: Probiotic encapsulation.	Ensures product safety.	Regulatory compliance costs.	Rajam and Subramanian (2022)
Worker safety	Strict safety protocols for handling cryogenic agents.	Ensures worker safety.	Training and equipment costs.	Timmerhaus and Flynn (1989)
Global collaboration	Partnerships, knowledge sharing, standardized protocols. Example: FAO and WHO collaboration.	Promotes best practices.	Coordination challenges.	Timmerhaus and Flynn (1989)

3.7 Conclusion

Cryogenic food processing is a revolutionary technology involving many benefits in food preservation, quality improvement, and functionality. For example, ultra-low temperatures of cryogenic processing enhance the texture, flavour, nutritional value, and shelf life of various types of food products, including meats, poultry, frozen desserts, baking products, and functional foods. The application of sophisticated post-treatment methods, including microscopy, spectroscopy, rheological, and sensory analysis, has been instrumental in describing the structural and functional modifications prompted by cryogenic treatment. These provide a way of working, and they are fundamental in streamlining processes as well as achieving standardization, safety, and diversification as may apply to food products.

Future cryogenic food processing industry trends are predicted in novel markets, including plant-based proteins, nutraceuticals, and convenience food. The technology is likely to address issues that concern environmental sustainability through strategies like the use of renewable cryogenic agents and energy-efficient systems. Developments in equipment design and automation, as well as cooperation with facilities globally, may significantly improve the cost and scope of cryogenic processing. However, there are some factors that need to be solved, if we can widely use cryogenic food processing technology. Some of the challenges include high initial costs, issues with consumers' perception, and the fact that there are many regulations that must be met. Meeting these difficulties will require more education, monetary motivation, and a regular framework of official standards. Therefore, vigorous sponsorship of international cooperation to generate and disburse information is imperative in reducing the existing global gap between organizational demands and consumer needs.

Therefore, cryogenic food processing forms a better future for the food industry. It promises better quality, improved performance, and sustainable solutions for catering to the needs of an expanding human population. Such work will continue to be further researched, invented, and pursued so that the food industry can achieve its optimal use and move towards a sustainable food processing.

References

Amini, M. and Benson, J.D. (2023). Technologies for vitrification based cryopreservation. *Bioengineering* 10 (5): 508. https://doi.org/10.3390/bioengineering10050508.

Benson, E.E., Harding, K., Ryan, M. et al. (2018). Alginate encapsulation to enhance biopreservation scope and success: a multidisciplinary review of current ideas and applications in cryopreservation and non-freezing storage. *Cryo Letters* 39 (1): 14–38. PMID: 29734412.

Buzan, V., Boronciuc, G., Balan, I. et al. (2014). Cryogenic changes of lipid during preservation of sperm animal farm. *Scientific Papers. Series D. Animal Science*. Vol. LVII.

Dhanya, R., Panoth, A., and Venkatachalapathy, N. (2023). Comprehensive review on isochoric freezing: a recent technology for preservation of food and non-food items. *Sustainable Food Technology* 2: 9–18. https://doi.org/10.1039/D3FB00146F.

Ghodki, B.M., Chhetri, K.B., and Goswami, T.K. (2020). Numerical modeling of granular flow in star valve type cryogenic precoolerhttps. *Journal of Food Process Engineering* 43 (4): e13376. https://doi.org/10.1111/jfpe.13376.

Giannakourou, M. and Taoukis, P. (2021). Changes during food freezing and frozen storage. *Food* 10 (11): 2525. https://doi.org/10.3390/foods10112525.

Goswami, T.K. (2010). Role of cryogenics in food processing and preservation. *International Journal of Food Engineering* 6 (1): https://doi.org/10.2202/1556-3758.1771.

Goswami, T.K. (2017). Recent trends of application of cryogenics in food processing and preservation. *Journal of Food, Nutrition and Population Health* 1 (3): 27.

Grzybowski, A. (2023). Glass transition and related phenomena. *International Journal of Molecular Sciences* 24 (10): 8685. https://doi.org/10.3390/ijms24108685.

Jafari, S.M. and Capanoglu, E. (2022). *Retention of Bioactives in Food Processing*. Springer https://link.springer.com/book/10.1007/978-3-030-96885-4.

Kharbach, M., Alaoui Mansouri, M., Taabouz, M., and Yu, H. (2023). Current application of advancing spectroscopy techniques in food analysis: data handling with chemometric approaches. *Food* 12 (14): 2753. https://doi.org/10.3390/foods12142753.

Khaydukova, I.V., Ivannikova, V.M., Zhidkov, D.A. et al. (2024). Current state and challenges of tissue and organ cryopreservation in biobanking. *International Journal of Molecular Sciences* 25 (20): 11124. https://doi.org/10.3390/ijms252011124.

Köprüalan Aydın, Ö., Yüksel Sarıoğlu, H., Dirim, S.N., and Kaymak-Ertekin, F. (2023). Recent advances for rapid freezing and thawing methods of foods. *Food Engineering Reviews* 15: 667–690. https://doi.org/10.1007/s12393-023-09356-0.

Liu, B. and Zhou, X. (2015). Freeze-drying of proteins. In: *Cryopreservation and Freeze-Drying Protocols*, Methods in Molecular Biology, vol. 1257 (ed. W. Wolkers and H. Oldenhof). New York, NY: Springer https://doi.org/10.1007/978-1-4939-2193-5_23.

Loayza-Salazar, S., Siche, R., Vegas, C. et al. (2024). Novel technologies in the freezing process and their impact on the quality of fruits and vegetables. *Food Engineering Reviews* 16: 371–395. https://doi.org/10.1007/s12393-024-09371-9.

Lu, X., Ma, X., and Lei, Y. (2023). Deep freezing and thawing modification and its impact on starch properties. In: *Starch: Advances in Modifications, Technologies and Applications* (ed. V.S. Sharanagat, D.C. Saxena, K. Kumar, and Y. Kumar). Cham: Springer https://doi.org/10.1007/978-3-031-35843-2_12.

Pavlyuk, R., Pogarskaya, V., Pogarskiy, A. et al. (2019). Investigation of cryomechanochemistry processes in the development of nanotechnologies of supplements prepared from chlorofill containing vegetables and discovery of hidden chlorophyll forms. *Eastern-European Journal of Enterprise Technologies* 6 (11 (102)): 57–65. https://doi.org/10.15587/1729-4061.2019.185215.

Rajam, R. and Subramanian, P. (2022). Encapsulation of probiotics: past, present and future. *Beni-Suef University Journal of Basic and Applied Sciences* 11: 46. https://doi.org/10.1186/s43088-022-00228-w.

Ramachandran, R.P., Nadimi, M., Cenkowski, S., and Paliwal, J. (2024). Advancement and innovations in drying of biopharmaceuticals, nutraceuticals, and functional foods. *Food Engineering Reviews* 16: 540–566. https://doi.org/10.1007/s12393-024-09381-7.

Rizzo, T.R. and Boyarkin, O.V. (2014). Cryogenic methods for the spectroscopy of large, biomolecular ions. In: *Gas-Phase IR Spectroscopy and Structure of Biological Molecules*, Topics in Current Chemistry, vol. 364 (ed. A. Rijs and J. Oomens). Cham: Springer https://doi.org/10.1007/128_2014_579.

Saxena, S.N., Saxena, P., Rathore, S.S. et al. (2016). Effect of cryogenic grinding on phenolic compounds and antioxidant properties of fenugreek (*Trigonella foenum-graecum* L.) seed extract. *Journal of Spices and Aromatic Crops* 25 (1): 73–78.

Shi, J. (ed.) (2015). *Functional Food Ingredients and Nutraceuticals: Processing Technologies*, 2e. CRC Press https://doi.org/10.1201/b19426.

Timmerhaus, K.D. and Flynn, T.M. (1989). Properties of cryogenic fluids. In: *Cryogenic Process Engineering*, The International Cryogenics Monograph Series. Boston, MA: Springer https://doi.org/10.1007/978-1-4684-8756-5_2.

Van Breda, V.M., van Jaarsveld, F.P., and van Wyk, J. (2024). Pre-fermentative cryogenic treatments: the effect on aroma compounds and sensory properties of Sauvignon Blanc and Chenin Blanc wine—a review. *Applied Sciences* 14 (4): 1483. https://doi.org/10.3390/app14041483.

Wang, X., Wang, E., and Zhao, G. (2023). Advanced cryopreservation engineering strategies: the critical step to utilize stem cell products. *Cell Regeneration* 12: 28. https://doi.org/10.1186/s13619-023-00173-8.

Zheng, O., Zhang, L., Sun, Q., and Liu, S. (2024). Basic theory of ice crystallization based on water molecular structure and ice structure. *Food* 13 (17): 2773. https://doi.org/10.3390/foods13172773.

4

Industrial Design and Engineering Aspects in Cryogenic Applications

Awani Shrivastav and Tridib Kumar Goswami

Department of Agricultural and Food Engineering, IIT, Kharagpur, West Bengal, India

4.1 Introduction

Cryogenics deals with the production and behaviour of fluids and materials at extremely low temperatures below −150 °C (123 K; −238 °F). There are discrepancies in finding out the lowest temperature at which cryogenics begins. According to the Cryogenic Society of America, the lowest cryogenic temperature should be below −153 °C (120 K; −244 °F). However, Gunston (2009) has considered the temperature for cryogenics to be below −180 °C (93.15 K; −292 °F), since the normal boiling point of many cryogenic gases, such as nitrogen, oxygen, hydrogen, helium, and neon, lies below −180 °C (Jawahir et al. 2016). The normal boiling points of the selected cryogens are given in Table 4.1.

Cryogenic technology has a wide range of industrial applications across various fields. It includes food preservation by freezing, spraying liquid nitrogen (LN_2), and metal shrink fitting using LN_2. Liquid oxygen (LOX) can also be used in welding, creating artificial breathing systems in hospitals, cooling body parts before anaesthesia injection, preserving bull semen for artificial insemination to improve the genetic quality or 'creed' of cattle breeds, and in creating cryogenic magnets, superconducting transformers and motors, etc. (Barnwal and Singh 2014). The use of cryogens is also transforming the food industry through rapidly advancing techniques in chilling, freezing, preservation, and transportation. Cryogrinding has tremendous scope among several applications, by which colour, flavour, and nutritional properties of the product can be retained.

Cryogenics' role in controlled and modified atmosphere storage by controlling the temperature and atmosphere is also an important storage aspect nowadays. Cryogenic freezing is now widely preferred over conventional vapour compression refrigeration systems (VCRSs) because of its extremely low temperature, resulting in improved product quality, enhanced shelf life, and better food safety (Goswami 2017). Cryogenic gases have unique physical properties, and therefore, they play a crucial role in automating plants, enhancing versatility and efficiency, and reducing manufacturing cost (Zohuri 2018). Among the

Cryogenic Processing: Advances in the Food Industry, First Edition. Edited by Sumit Sudhir Pathak, Rama Chandra Pradhan, and Sudarshan Ramanathan.

Table 4.1 Key phase transition points of cryogenic fluids.

Cryogenic fluids	Normal boiling point (°C)	Temperature (°C) at triple-point condition	Pressure (MPa) at triple-point condition	Temperature (°C) at critical-point condition	Pressure (MPa) at critical-point condition
Nitrogen	−195.8	−210.01	0.01253	−146.95	3.396
Oxygen	−183.0	−218.82	0.0001503	−118.55	5.053
Hydrogen	−252.9	−259.34	0.00721	−239.85	1.315
Helium-3	−269.9	−273.15	2.94*	−269.85	0.311
Helium-4	−268.9	−270.98	2.53*	−267.95	0.227
Methane	−161.5	−182.48	0.01169	−82.65	4.640
Argon	−185.8	−189.37	0.0689	−122.35	4.870
Neon	−246.0	−248.59	0.0433	−228.75	2.654
Air	−194.2	−213.40	0.005265	−140.55	3.77

Source: Adapted from Kalsi et al. (2010) and Barron and Nellis (2017).
* Minimum melting pressure for liquid helium.

various cryogens available, liquid nitrogen (LN_2), liquid hydrogen (LH_2), liquid helium (L-He), liquid oxygen (LOX), liquid carbon dioxide (LCO_2), and liquid argon (LAr) have a role in a variety of industrial applications. Nitrogen, oxygen, and argon are produced by fractional distillation of air. LN_2 is comparatively cost-effective and can be safely used in most cases like food freezing, cryopreservation, heat treatment of metals, handling explosives, cooling in vacuum systems, etc. LH_2 (and deuterium, an isotope of hydrogen) is used in high-energy physics experiments (Ventura and Risegari 2008).

As cryogenics deals with ultra-low temperatures, the design and engineering aspects play an important role, where material selection, thermal insulation, thermal stress, safety measures and standards, and proper sealing and leak prevention must be carefully considered. Understanding the thermodynamics and heat transfer principles is crucial for the effective design and functioning of cryogenic systems. Zohuri (2018), in his book, has discussed in detail all the thermodynamics and heat transfer principles applied in cryogenics.

4.2 Fundamental Principles of Cryogenics

It is important to carry out the thermodynamics and heat transfer analysis before designing a system based on cryogenics, to ensure safety, proper insulation, and material selection. Cryogenics deals with temperature near absolute zero temperature; therefore, understanding cryogenics is important to understand the flow of heat at these temperatures and to take effective measures to control it. The thermodynamic properties of the material must be known before selecting a suitable material for different cryogenics applications. For the efficient design of heat exchangers, cryogenic systems, and refrigeration cycles, knowledge of heat transfer mechanisms and thermodynamics principles is crucial. Concepts like

Gibbs free energy, exergy, and entropy help us measure how efficiently a system uses energy. However, in real-world applications, we also need to consider the size of the system and find a balance between efficiency and practicality (Radebaugh 2023).

The state (solid, liquid, or gas) of a cryogenic fluid is decided according to the phase rule by looking at its pressure and temperature at which it exists. The triple point refers to the specific condition at which solid, liquid, and gas exist in equilibrium, while the critical point refers to the point at which liquid and vapour phases become indistinguishable. Engineers need to consider these properties of any cryogenic fluid before designing any heat transfer system. Knowing these thermodynamic properties is beneficial for engineers who want to work with cryogenic fluid in a single phase (Jouhara et al. 2023). Table 4.1 shows the key phase transition points of the cryogenic liquids.

Unlike other cryogens, liquid helium-4 behaves differently because it can exist in two distinct liquid phases: helium I and helium II. Helium I is a normal fluid, while helium II acts as a superfluid, exhibiting zero viscosity under certain conditions. The λ-line is the phase transition line between the two phases. The intersecting point between this line and vapour pressure curve is called the λ-point. Above and below the λ-point, the thermal conductivity and heat capacity exhibit non-linear variation in both helium I and helium II (Timmerhaus 2003).

4.2.1 Heat Transfer Mechanism in Cryogenics

Similar mechanism of heat transfer is generally observed at cryogenic temperatures as those at normal temperatures; however, the properties affecting heat transfer change significantly at lower temperatures. Heat transfer occurs through three primary mechanisms: conduction, convection, and radiation. Heat transfer takes place by the transfer of heat inside the solids or fluids by an atomic-scale process in the case of conduction. In the case of convection, macroscopic movement of the fluid results in heat transfer. In radiation, heat is transferred through the effect of electromagnetic radiation (Vandoni 2004). As a result of the change in thermal properties at lower temperatures, Fourier's law of heat conduction can be represented as in Eq. (4.1):

$$Q = k(T)A\frac{dT}{dx} \tag{4.1}$$

where $k(T)$ indicates the change in thermal conductivity with temperature (non-linearity between temperature and thermal conductivity at cryogenics temperatures) (W/m K), Q represents the rate at which heat is transferred over time in watts (W), A denotes the area through which heat is conducted (m^2), and dT/dx indicates the rate at which temperature changes over a distance (K/m). When it comes to convection heat transfer in cryogenics, the general relationship based on Newton's cooling principle remains the same. However, the convective heat transfer coefficient must be carefully evaluated, as fluid thermal properties like specific heat, viscosity, density, and thermal conductivity can vary significantly at very low temperatures.

$$Q = -h\,A\Delta T \tag{4.2}$$

where h is the convective heat transfer coefficient (W/m^2 K), A is the heat exchange surface area (m^2), and ΔT indicates the temperature gradient between the surface and the fluid (K). In radiation heat transfer, a significant change can be observed in the emissivity of the materials at extremely low temperatures. The general equation for the transfer of thermal energy by radiation from one surface to another can be given by the Stefan–Boltzmann law:

$$Q = A\varepsilon\sigma\left(T_1^4 - T_2^4\right) \tag{4.3}$$

where A denotes the surface through which heat is transferred, ε indicates the emissivity of the material (varying between 0 and 1), σ denotes the Stefan–Boltzmann constant (approximately equal to $5.67\times10^{-8}\,\mathrm{W/m^2K^4}$), and T_1 and T_2 represent the absolute temperatures (in Kelvin) of the two surfaces. Understanding and regulating these heat transfer mechanisms are important for the efficient design of cryogenic systems.

4.2.2 Thermodynamic Applications in Cryogenics

For carrying out any industrial design based on cryogenics, it is crucial to know the thermal properties of the cryogenic fluids. The thermal properties of cryogenic fluids can be calculated by considering the aspects of thermodynamics (Van Sciver 2012). The three laws of thermodynamics are particularly interesting in cryogenics, as they can apply to all systems. Applying the first law of thermodynamics, an energy balance is performed to facilitate a better understanding of the process and to design, optimize, and operate any cryogenic system more effectively (Zohuri 2018). The second law of thermodynamics is applied more extensively in cryogenic systems than in most other engineering applications because of the low Carnot efficiencies. This law is of great consideration when carrying out design and performance analysis in the case of magnetohydrodynamic power, LNG, fusion power, superconducting magnet energy storage, and liquid hydrogen storage and transportation involving cryogenic technology (Ahern 1980). The third law of thermodynamics is essential in cryogenic storage and preservation.

4.3 Industrial Role of Cryogenics in the Food Processing

Cryogenic methods have transformed the food processing industry through fast freezing, cold storage, and cryogrinding. Davidge (1981), in his study, has discussed various applications of cryogenics in the food sector. Davidge also studied the ways of distributing and controlling LN_2 such that the cryogenic processes become more practical and affordable for the food industry. Ramakrishnan et al. (2004), in their study, highlighted the capabilities of cryogenic freezing in the futuristic food processing industry. They also highlighted the benefits of using cryogenic freezing technology as low initial capital investment and maintenance costs, higher flexibility, and several other benefits in the food industry. Goswami (2010), in his study, discussed the use of cryogenics for food preservation and processing. The study further discussed the various aspects of utilizing LN_2 for refrigeration and gas applications. Rai and Tassou (2017) consider LN_2 and LCO_2 for cryogenic-based refrigeration systems for applications related to food transportation at controlled temperatures. LN_2 is most commonly used in inert packaging to provide a protective and oxygen-free environment inside the container. According to the recent report on 'Air Liquide', LN_2 is added in a controlled amount in the packaged bottle to keep it sturdy during storage and transportation.

LN_2 can be used for the manufacture of ice cream. The primary goal in producing ice cream cryogenically is to simplify the procedure and cut the expenses by omitting complex equipment. By adapting this method, quick freezing of ice cream can be achieved, in addition to cutting down the usage of electricity. Jones (2005) first patented the idea of manufacturing ice cream cryogenically. Shrivastav et al. (2024) also recently developed a scraped-surface freezer utilizing LN_2 as a refrigerant to produce the ice cream. LN_2 has an extremely low boiling temperature, which makes it a preferred choice in food processing, especially for instantly freezing small food portions in ice cream production.

Grinding, which reduces particle size through friction, results in a significant amount of heat generation, which impacts the volatile compounds or essential oil present in the spices, thereby resulting in the loss of flavour and aroma in the ground spices. Loss of aromatic compounds or essential oils from powdered spices can be efficiently reduced by the cryogrinding technique by lowering the amount of heat build-up. The study conducted on cryogenic grinding of black pepper shows cryogenic grinding retains more volatile oil content in black pepper compared to the traditional grinding methods (Jacob et al. 2000). A cryogenic grinding system specifically for spices like clove and cumin was developed by Singh and Goswami (1999). It consists of a pre-cooler, a LN_2 Dewar, a compressor, a grinder, and a power transmission system. Several studies on the cryogenic grinding of spices reported the superiority of cryogenically ground products compared to other grinding methods (Ghodki and Goswami 2016; Gouveia et al. 2002; Jacob et al. 2000; Murthy and Bhattacharya 2008; Singam et al. 2018; Singh and Goswami 1999; Singh and Goswami 2000). The principle of temperature accumulation by cryogenics is also applied in the meat mixing process, where the high temperature might result in an increased risk of microbial growth (Barham 2021).

Freezing of food involves the use of mechanical refrigeration to lower the temperature of meat, poultry or seafood. It uses cold and dry air blasts for a long period such that it results in dehydration or excessive moisture loss from the surface of the product, affecting the texture and flavour of the end product (James and James 2002). This texture damage and loss of flavour can be prevented with the application of cryogenic freezing technology. LN_2 can be used as a cryogen, which results in the formation of tiny ice crystals that further reduce the texture damage in the end product (Nesvadba 2008; Sanz et al. 1999). A thin layer of ice is formed over the surface with the application of LN_2, which prevents excessive loss of moisture from the surface of the food product. A similar technique of ice layer formation over the product is also applied in the case of ready-to-eat food items like pizzas, pastries, packaged meals, pies, and cakes when it comes to storage and transportation of these products to retain their shape, texture, and mouthfeel. LN_2 is also used to create a glazing ice layer over the frozen surface of the fish fillets to prevent quality loss and dehydration during storage. Truonghuynh and Li (2019), in their research, showed that the quality of aquatic products can be sustained by cryogenic freezing technology. Several studies have been conducted on changes in the physicochemical properties of aquatic products with the application of cryogenic freezing (Benjakul and Bauer 2000; Espinoza Rodezno et al. 2013; Kaale et al. 2014; Oliveira et al. 2017; Shi et al. 2018; Sriket et al. 2007).

Cryogenic technology also helps in improving the appearance of fruits and vegetables, reducing the thawing losses, and maintaining the complete cell structure by quickly reducing the temperature below the threshold value (Razali et al. 2022; Zhao et al. 2020). Quick freezing results in the formation of a layer of ice on the fruits and vegetables, providing mechanical strength to the product, which is a great advantage to be considered for the storage and transportation of fruits and vegetables. The non-agglomeration and smooth movement of the product frozen by cryogenic technologies in the mechanical conveyor belts are some of the other advantages (Forni et al. 1990; Li et al. 2016).

4.4 Industrial Design and Engineering Considerations in Food Processing Applications

Understanding fluid properties is crucial when operating at low temperatures in cryogenic applications. Table 4.2 gives the properties of the selected cryogens at their boiling points and under conditions of normal atmospheric pressure. Thermophysical properties can also be found using graphs, tables, and software like NIST REFPROP, HEPAK, and Cryodata.

In any industrial design related to cryogenics, material selection, insulation, and the maintenance of safety standards are the most important factors to be considered. In the case of applications related to food processing, it is important to consider various parameters like the selection of suitable material, proper insulation system, hygienic design, automation, and control system for precise temperature control, and a data monitoring

Table 4.2 Cryogens' properties at their normal boiling point.

	Hydrogen	Helium	Neon	Nitrogen	Argon	Oxygen	Methane
Normal boiling point (K)	20.4	4.22	27.1	77.3	87.3	90.2	111.6
Liquid density (kg/m^3)	71.0	125	1,205	809	1,393	1,141	423
Ratio of Liquid to vapour density	53	7.4	126	175	241	255	236
Latent heat of vaporization (kJ/kg)	446	20.42	86	199	161	213	512
Vaporization volume resulting from a 1 watt-hour energy input (cm^3)	114	1,410	33	22	16	15	17
Liquid's dynamic viscosity (μNsec/m^2)	13.3	3.3	124	152	260	195	119
Surface tension (mN/m)	1.9	0.10	4.8	8.9	12.5	13.2	13.2
Thermal conductivity (mW/mK)	100	18.7	113	135	128	152	187
Gas volume (at 15 °C) from 1 unit of liquid	830	739	1,412	681	824	842	613

Note: Pressure: 1.01325 bar.
Source: Hands (1997) / with permission of Elsevier.

system. This section discusses various design aspects to be considered while dealing with ultra-low temperature conditions in the case of food processing.

4.4.1 Material Selection for Cryogenics

Material selection for cryogenic applications depends on the required properties of a material for a particular purpose and the material's behaviour at low temperatures. Steel, aluminium alloy, copper alloy, nickel alloy, and titanium alloy are the materials that are suitable for a variety of cryogenic fluids. When materials are exposed to sub-zero temperatures, there is a chance that the material becomes brittle, and therefore can break easily. They also sometimes get stronger and stiffer, which may result in shrinkage. Therefore, it is important to note the reactive nature of the material at sub-zero temperatures. Carbon steels have been used for cryogenics applications since Tobern Bergmann discovered them in 1774. Carbon steels can improve their properties for cryogenic applications by being processed with alloys. The excellent corrosion resistance and mechanical properties of 300-series austenitic stainless steels make them the most suitable choice for cryogenic applications. Inconel alloys (nickel-based superalloys), due to their high specific strength, corrosion resistance, and capability of withstanding low-temperature conditions, make them suitable for cryogenic applications. Aluminium alloys are also used for cryogenic applications due to their ease of manufacture, low cost, and high strength (Bijeta Nayak et al. 2021). Other materials suitable for cryogenic applications include 304 stainless steel, polyimide, oxygen-free copper, Teflon, nylon, Ti-6Al-4V titanium alloy, 6,061-T6 aluminium, and Inconel 718 (Marquardt et al. 2002). In the case of food processing, while selecting the material, it is important to note that the material should resist corrosion, be easy to clean, and be non-reactive. Stainless steel of grades 304 and 316, aluminium alloys, Teflon (PTFE), Ti-6Al-4V titanium alloy, polyethylene (PE), and nylon are widely used in food processing equipment. Among them, stainless steel (Grade 304 and 316), Teflon, polyethylene, titanium alloy (Ti-6Al-4V), and nylon are mostly suitable for even direct food contact and can be used as the construction material of cryogenic equipment used for food processing applications.

The properties of various metals and alloys undergo significant changes under extremely low-temperature conditions and may interrupt the overall performance. Therefore, the thermal, mechanical, electrical, and magnetic properties of materials suitable for cryogenic applications are very important (Timmerhaus 2003). Thermal conductivity, coefficient of thermal expansion, specific heat capacity, and thermal shock resistance are some of the important thermal properties that are important to be considered before selecting a material for cryogenic applications. The performance level of any device strictly depends on the choice of metals, alloys, and substrate materials used for the construction. The thermal conductivity of the material plays a key role in selecting a suitable material for operation at cryogenic temperatures (Jha 2006). Marquardt et al. (2002) reported a general equation to determine the thermal conductivity, 'k', as

$$\log(k) = a + bx + c\,x^2 + d\,x^3 + e\,x^4 + f\,x^5 + g\,x^6 + h\,x^7 + i\,x^8 \quad (4.4)$$

In this equation, the constants a through i represent the fitting parameters, $x = \log T$, while T denotes the temperature.

The material that is to be used for cryogenic applications should remain tough while handling cryogens, as the failure of which can lead to disaster. The mechanical properties that are of interest to cryogenic engineers include toughness, strength, hardness, ductility, fatigue resistance, and elastic modulus. Properties like electrical resistivity and superconductivity, which come under the category of electrical and magnetic properties, are important when selecting the material (Timmerhaus 2003).

4.4.2 Thermal Insulation Technique

Cryogenics is related to low-temperature energy systems, while thermal insulation plays a crucial role in the conservation of energy. To reduce the rate of heat transfer to the environment, thermal insulation plays a crucial role in cryogenics (Augustynowicz et al. 1999). Thermal insulation can be made from a single homogeneous material or a combination of different materials merged to form a composite structure. To deal with extreme temperatures, as in the case of cryogenics, insulation that combines different materials is most suitable. The combination of materials used in cryogenics makes the heat transfer process more complex as it is required to deal with the heat transfer involved between insulation parts having different materials. Therefore, it becomes necessary to take into account the 'apparent thermal conductivity' to evaluate the insulating material's thermal performance (Tien and Cunnington 1973). Considering an insulating slab with an insulating thickness of 'L', 'A' as the cross-sectional area of heat flow, T_1 and T_2 as the boundary temperatures, and 'Q' as the total heat transfer through insulation, the apparent thermal conductivity 'k_a' can be defined as

$$Q = k_a A(T_1 - T_2)/L \tag{4.5}$$

The selection of the proper insulation material is vitally important to avoid heat leakage. The types of insulation used in cryogenics are categorized as vacuum, multilayer insulation (MLI), foam, powder and fibrous, aerogel insulation, and special insulation (Timmerhaus 2003). MLI, consisting of alternating layers of effective shields and spacers within a high-vacuum environment, provides excellent thermal protection for cryogenic insulation. It has applications in guard vessels, cylindrical vessels, vacuum chambers, vent tubes, and filling (Jha 2006). MLI incorporates alternating reflective surfaces and low-conductivity spacer layers within a high-vacuum environment. A lower apparent thermal conductivity value can be achieved when the insulation is applied at optimal density. The extremely low value of apparent thermal conductivity indicates a significant reduction in all modes of heat transfer (Timmerhaus 2003). Conventional insulation methods and materials used for insulation can be categorized into three categories based on their thermal conductivity as about 30 mW/m-K for materials under normal atmospheric conditions, around 1.5 mW/m-K for bulk fill materials under an effective vacuum, and below 0.1 mW/m-K for MLI in high-vacuum environments. The reference boundary temperature condition is between 77 and 300 K for evaluating different materials (Augustynowicz et al. 1999).

The types of insulation used in the area of food processing should be non-toxic in nature, moisture resistant while used in cold pipes, and should not promote the growth of mould. The sealed coating must create a vapour barrier, especially in high-humidity areas

(Baker 2013). There are different methods of insulation, such as spiral wraps, layer-by-layer methods, blankets, and filament-wound techniques. The MLI method might not be appropriate for all shapes; therefore, other materials may also be required to form the most suitable insulation. Materials like fiberglass, foam, glass bubbles, aerogel blankets, or aerogel bulk-fill can also be used (ASTM International 2019).

4.4.3 Safety Measures and Standards

While designing any new cryogenic-based system, safety should be prioritized from start to finish. This involves selecting the right equipment for operation under cryogenic conditions and choosing materials that can withstand extreme temperatures (Edeskuty and Stewart 1996). To properly handle cryogenic fluids, it is important to understand the usage of personal protective equipment (PPE) and the impact of cryogenic temperatures to ensure the safety of the body and eyes. It is important to analyse job hazards and promote a culture of safety to avoid accidents in cryogenic facilities. In job hazards analysis, hazards associated with risk are identified, listed, and steps are taken to mitigate them. The mitigation involves training, technical solutions, PPE, etc. (Peterson and Weisend II 2019). Cryogens are extremely cold and dense, the prolonged exposure of which may result in frostbite (cryogenic 'burns') (Safety British Cryogenics Council 2013). Cryogens can make the material brittle, leading to cracks. They can condense and solidify substances like carbon dioxide and water vapour, and can cause blockages or erosion, which can be dangerous in the case of pressure relief valves and valve seats. Cryogens also have the tendency to expand when evaporated, which may result in the building of high pressure and a reduction in the amount of oxygen present in the air. Some cryogens are also toxic and flammable. The hazards associated with cryogens can be classified as physical, chemical, or physiological hazards (Edeskuty and Stewart 1996). Table 4.3 shows some of the physical, chemical, and physiological hazards and specific risks associated with handling cryogens.

Table 4.3 Types of hazards associated with cryogens.

Cryogen hazards	Sub-hazards	Specific risks
Physical hazards	Freezing tissue	Frostbite (Cryogenic 'burns')
	Material embrittlement	Cracks and breaks in structural materials
	Thermal contraction	Unwanted stresses in materials
	Condensation/solidification of substances	Blockages in safety relief passages Erosion in valve seats
Chemical hazards	Reactivity	Combustion risks
	Toxicity	Harmful effects on health
Physiological hazards	Oxygen displacement	Reduced oxygen content in the local atmosphere
	Pressure build-up	Risk of explosions if expansion is not allowed

While designing any cryogenic equipment to be used in food processing applications, care must be taken to ensure that the selected cryogen is non-toxic and does not possess any potential health hazards. The selected cryogen must be safe to use and not harmful to human health. The properties of cryogens must be carefully considered when selecting them for use in food processing applications. Among the various cryogens available, LN_2 and LCO_2 are popular in the food processing industry, as they are inert and non-toxic, hence appropriate for direct contact food usage.

Everyone working with cryogenics should be informed of the basic rules and standards of cryogenics to ensure safety and health. Regulations and safety information are constantly updated, so staying informed is essential. Due to growing public concern over accidents, many US safety standards were developed between 1894 and 1918. Standards are created by various groups, including:

- American Society for Testing and Materials (ASTM)
- Factory Mutual Engineering and Research (FM)
- American National Standards Institute (ANSI)
- American Society of Mechanical Engineers (ASME)
- Underwriters' Laboratory (UL)
- National Fire Protection Association (NFPA)
- Instrument Society of America (ISA)
- Compressed Gas Association (CGA)

These organizations ensure that safety standards meet the needs of industries, businesses, and the public (Edeskuty and Stewart 1996). The work by Edeskuty and Stewart (1996) provides more details of guidelines, standards, and regulations.

4.5 Role of LN_2 and LCO_2 in Food Processing Applications

LN_2 and LCO_2 are the primary options for various food processing applications, preferred over other cryogens in food processing applications because of their large volume expansion ratio when their phase changes from liquid to vapour form and due to their high enthalpy of vaporization. The high-volume expansion makes them the most suitable option for inert packaging. However, the high-volume expansion coefficient is an important point of consideration when it comes to equipment design and safety. The study conducted by Gas Encyclopedia by Air Liquide (2023) indicates that in the case of LN_2, the volume expansion ratio is 1:694 at a room temperature of 20 °C. This implies that 1 L of LN_2 may result in the volumetric production of 694 L of nitrogen gas. Therefore, it is important to consider the volume expansion ratio while designing any cryogenic storage and handling system. For this purpose, the pressure relief valve and proper ventilation must be carefully installed.

LN_2 and LCO_2 can be effectively used for various food-freezing or chilling operations. The study by Rai and Tassou (2017) shows the least environmental impact of using any cryogen as a refrigerant compared to other refrigerants. In the early twentieth century, chlorofluorocarbons (CFCs), hydrochlorofluorocarbons (HCFCs), and hydrofluorocarbons (HFCs) were used as refrigerants. Later, they were phased out due to their harmful effect on the environment and human beings. Also, they had a high potential for ozone depletion and global warming. Hydrofluoroolefin (HFO) comes under the category of the fourth generation of fluorine-based refrigerants. Refrigerants like HFOs and hydrocarbons (HCs) had excellent refrigeration properties with almost negligible Ozone Depletion Potential (ODP) and minimal Global Warming Potential (GWP). However, they were having issues related to flammability, and there could be a risk of explosion. Some of the common natural refrigerants used nowadays include ammonia, propane, isobutane, and cyclopropane. They have excellent refrigeration properties and have zero ODP and low GWP values. The disadvantage of these modern refrigerants is that they have moderate flammability and, if inhaled for a longer duration, may result in health risks to humans. The best eco-friendly options for carrying out refrigeration include the usage of liquid or dry carbon dioxide (CO_2) and liquid nitrogen (LN_2), as they have no ODP and GWP with no adverse health effects on humans.

4.6 Key Design Elements of Cryogenic Equipment for Food Processing Applications

Different types of cryogenic equipment are utilized in the food industry for chilling, freezing, storage, transportation, or food processing purposes, depending on the available space, throughput capacity, and required quality standards.

4.6.1 Cryogenic Freezers

4.6.1.1 Cabinet Freezers

Cabinet freezers are the batch type of freezers that can be used for a variety of food products. These types of freezers are equipped with multiple nozzles for injecting cryogenic refrigerant and a blower to circulate the gases within the cabinet. It provides consistent temperature spread throughout the cabinet, thereby reducing the loss of moisture and improving the food quality. These types of freezers also occupy less space and are easy to install. The sturdy construction of the cabinet freezer, along with the carefully chosen components, ensures the long-term operation of the freezer with very low maintenance required (Khadatkar et al. 2004).

4.6.1.2 Cryogenic Freezer Tunnel

These tunnel freezers, based on cryogenic applications, are used for continuous operation when chilling and freezing the product. They mainly comprise a pre-cooling portion, where the product gets cooled or partially frozen by counter-current flowing nitrogen, and a freezing section, where a controlled amount of LN_2 is sprayed on the product with the help of uniformly spaced nozzles and the product gets completely frozen; and an equilibrium section where the product is allowed to reach the equilibrium before it discharges (Khadatkar et al. 2004). In the cryogenic freezer tunnel, cooling of the food item occurs through the phase transformation of LN_2 from liquid to vapour, along with the additional heat removal by convection caused by counter-current flowing nitrogen gas (Girardon et al. 2019). The cryogenic freezer tunnel is designed for continuous operation to handle large amounts of products with very little space requirement. The main problem encountered while designing a cryogenic freezer tunnel is figuring out the operating parameters and the right design for handling expected heat loads in a particular space requirement (Shaikh and Prabhu 2007).

4.6.1.3 Cryogenic Fluidized Bed Freezers

Cryogenic fluidized bed freezers consist of nozzles to inject cryogen onto the food product and high-speed convection fans/blowers, which blow the cold air to keep the product moving and getting cooled simultaneously. This enhances heat transfer within the food item and quickly freezes the food product with minimum moisture loss. According to Muthukumarappan et al. (2019), the moisture loss by using fluidized bed freezers is in the order of 2%. The quick-freezing results in a higher throughput rate with minimum space requirement. The fluidized bed freezer is mostly suitable for small and uniformly shaped food products like peas, sliced meat, sliced carrots, and potatoes (Chen and Rosenthal 2009).

4.6.1.4 Cryogenic Spiral Freezers

Cryogenic spiral freezers consist of large circular or square boxes. The design involves the spiral conveyor belts around a central drum, which carries the food product from top to bottom or bottom to top in a cooling apartment. These types of freezers are mostly suitable for handling large quantities of food products. The spray nozzles, along with the cooling fans, are installed on the cooling chamber walls for the even dispersion of cryogenic gases in the cooling apartment. The cryogenic spiral freezer is a more cost-effective solution as compared to the cryogenic tunnel freezer for a similar throughput rate. The cryogen consumption is also less, which reduces its cost of operation (Khadatkar et al. 2004). However, freezing rate is slower in a cryogenic spiral freezer than that of a cryogenic tunnel freezer.

4.6.1.5 Cryogenic Immersion Freezers

Cryogenic immersion freezers are well suited for food items that are challenging to quick-freeze in an individual manner and require rapid freezing, like meat, seafood, and other foods with non-living cells, due to their need for a higher heat transfer coefficient (Khadatkar et al. 2004). The edible item is advanced along the conveyor either by the steady flow of liquid nitrogen (LN_2) or through applied vibratory motion to the trough. The food item moves forward onto the conveyor either by the continuous current of LN_2 or by any external vibratory motion given to the trough. The limitation of using an immersion freezer is its higher consumption of LN_2 (Girardon et al. 2019). Hydro fluidization is also an innovative immersion freezing, where the agitating jet of liquid refrigerant is pumped from the bottom of the freezing chamber or fluidized bed (Fikiin 2003). To overcome the drawbacks of immersion and fluidized bed freezing, hydro fluidization was introduced and patented (Sutariya and Sunkesula 2021).

4.6.2 Cryogenic Storage Tanks

Cryogenic storage tanks are specialized storage facilities for storing extremely cold cryogenic gases like LN_2 and LCO_2. Spherical-shaped cryogenic tanks are mostly preferred over cylindrical-shaped cryogenic tanks because spherical-shaped cryogenic tanks are less susceptible to buckling. Also, the lowest surface area to volume ratio of spherical-shaped cryogenic tanks reduces the heat transfer into the tank and the loss of cryogenic liquid into the environment (Burns and Tien 1979). For laboratory, medical, and small-scale industrial purposes, smaller and portable containers termed 'Dewar' are often used. A proper insulation design is the most important factor to consider to minimize the heat transfer rate for any cryogenic storage system. Porous insulation materials filled with gas, such as perlite filled with nitrogen, are commonly used in insulation design. When the material for insulation, temperature-related boundary conditions, and filler gas are known, heat transfer due to conduction and radiation is fixed. However, the convection depends upon the ability of the porous insulating material to allow fluid flow and the characteristics of the gas filling it. Although the design demands the insulating material of low permeability, the low permeability material adds weight to the insulation (Taghavi et al. 2024). Advanced insulation strategies have now been introduced to minimize heat leakage to the insulation system. Jiang et al. (2021) proposed the combination approach of vapour-cooled shield, foam, and MLI.

4.6.3 Cryogenic Grinding System

Grinding is the most power-consuming operation and generates a lot of heat. Grinding the spices results in the depletion of aromatic compounds or essential oils in the end product. Grinding at low temperatures using LN_2 as a coolant preserves the volatile compounds or essential oils of the spices. Balasubramanian et al. (2012) suggested the design of a cryogenic grinder, which includes the proper selection of materials, a feeding hopper, a pre-cooling system, a grinding mechanism, an LN_2 storage and distribution system, temperature controls, and sensors. Balasubramanian et al. (2012) in their review article also mentioned the major design consideration for any cryogenic system. It includes the determination of refrigeration load by conduction, convection, and radiation; Dewar construction (like the selection of suitable material, the mechanical strength of the Dewar, and the design of other parts used in it); LN_2 distribution system (compressor selection, distribution lines, mechanism for entrance of LN_2 into the system, and valves to regulate the supply of LN_2); and cleaning mechanism for Dewar, pre-cooler, grinder, and compressor.

A cryogenic grinding system mainly consists of a pre-cooling unit and a grinding chamber. A pre-cooler is an important assembly of a cryogenic grinding system to reduce the temperature of the raw material. Different types of pre-cooler or cryogenic freezers, as discussed earlier, can be used in food industries. Lowering the temperature of the product makes it brittle, and therefore, less energy is required for the grinding. Therefore, pre-cooling units in cryogenic grinding systems reduce energy consumption and improve the standard quality of the outcome. Gouveia et al. (2002) homogenized the breakfast cereals by cryogenic grinding. A pre-cooler mainly consist of a screw conveyor system in which the interaction of the raw food item and cryogen liquid results in heat transfer. An important consideration while designing a pre-cooler is the time for which the product stays inside the conveyor. Modified Plank's equation (Barron 1972) calculates the food product's residence time inside the conveyor. The direct immersion of cryogen into the product increases the operational cost, as discussed earlier. Wilczek et al. (2004), in their study, showed the optimization of LN_2 in the cryogenic grinding process.

4.6.4 Cryogenic Encapsulation by Spray Chilling

Spray chilling, also known as spray cooling or spray congealing, is a microencapsulation technique of spraying the liquid feed into the cryogenic liquid's atmosphere to preserve its texture, flavour, aroma, and important nutrients in the food product. A variety of nozzles or atomizers, like vibrating nozzles, pressure nozzles, and spinning disc atomizers, are available to spray the liquid feed (Oxley 2012). Many studies show that encapsulation by spray chilling can preserve vitamin D3, proanthocyanidins, and ascorbic acid throughout storage (Paucar et al. 2016; Tulini et al. 2017; dos Santos Carvalho et al. 2021). The encapsulation by spray chilling can be adapted for small-scale as well as large-scale industrial uses (Consoli et al. 2016). In terms of operation and equipment, a spray chiller closely resembles a spray dryer. On one hand, the spray chilling solidifies the liquid feed by cooling, while the spray drying converts the liquid feed into powder by evaporating the solvent (de Figueiredo et al. 2022).

4.7 Challenges and Future Trends

The extremely low temperatures of cryogens offer several benefits, but if it is too difficult or costly to achieve these temperatures, they won't be widely used. Earlier, the use of liquid cryogens like helium was a common practice, but the use of it required regular refilling and sometimes became inconvenient to use. This is especially a challenge in the case of liquid helium, as the price of helium is increasing rapidly and it is less available. To overcome this, cryocoolers are being used instead of liquid hydrogen. The use of cryocoolers has its own set of challenges, like size, noise, reliability, and cost, but progress is being made to make it more practical and affordable (Radebaugh and Bar-Cohen 2016).

The hydrogen in the gaseous form is liquefied by lowering its temperature below −253 °C, after which it is then stored in large, insulated containers at the liquefaction plant. It takes about 30% of the hydrogen energy content to liquefy hydrogen, and that makes it expensive. Additionally, during the transportation of liquid hydrogen, some amount of energy is lost by evaporation or 'boil off'. This is even more prominent when small storage tanks are used, which have a high ratio of surface to volume, which is the major challenge while transporting liquefied cryogens (Zohuri 2019). In addition to this, storing hydrogen takes up a lot of space. This led to the requirement of special material- or chemical-based storage that can absorb hydrogen. Only a few materials work well for storing hydrogen, but even they have their own set of limitations or problems. Therefore, more studies related to hydrogen behaviour with such materials are required at the microscopic level (Allendorf et al. 2022).

The production of cryogenic fluids has environmental impacts also. Energy-intensive cooling processes rely on fossil fuels, and the production of cryogenic fluids can result in the emission of greenhouse gases (GHGs) when burned (Rai and Tassou 2017). About one-third of the food items transported require refrigeration, and refrigerated trucks are used for this purpose. Most refrigerated vehicles use a system that cools with a vapour compression method powered by an additional diesel engine. These systems use hydrofluorocarbon refrigerants like R134a and R-404A, which have high global warming potentials. According to reports, transporting food with these trucks results in the emission of 12 million tons of CO_2 each year (Schmied et al. 2012). Studies conducted by Bagheri et al. (2017) show that emission of GHGs can be reduced by replacing diesel-operated vapour systems powered by batteries for compression. The cryogenic systems using liquid carbon dioxide and liquid nitrogen can replace reliance on diesel and refrigerants for cooling (Pedolsky and Bau 2010; Tassou et al. 2009). Other challenges in cryogenics include the complexity of cryogenic systems, materials selection and compatibility, energy efficiency, safety and reliability, environmental conditions, etc.

The cryogens can be used for thermal energy storage and can be integrated with power generation systems like solar, wind, and nuclear to store excess energy. This helps in cutting electricity charges and increasing the revenue and lifespan of the power systems. Although the efficiency of cryogenics-based energy storage systems is only 40–60%, still it is profitable (Rehman et al. 2020). The transportation industry is currently oil-based and the oil reserves are limited. There is a demand to find an alternative to non-polluting and combustion-free fuels due to the alarming environmental conditions. Researchers have already started research to find an alternative to fossil fuel. Plummer et al. (1998) had introduced an idea of using a pollution-free vehicle, which uses nitrogen as fuel. Apart from nitrogen, hydrogen has the potential to replace the hydrocarbon fuel as it is present in abundance, pollution-free, and has a heating value about three times that of petroleum (Offer et al. 2010). The special fuel cell vehicles are getting popular in different parts of North America, Europe, and Asia as they use hydrogen as fuel (Müller et al. 2011). However, the price of fuel cells is much more than that of IC engines, which makes the fuel cell vehicles expensive. It is predicted that the expense of fuel cell technology will be similar to that of internal combustion engines by 2030 (Offer et al. 2010).

Cryopreservation has been known in medical science for over 50 years, but we still don't fully understand how to use it to extend life. There is scope in cryonics to develop a technology that can fix and regrow damaged body tissues. Future advancements, like nanomedicine, could be a breakthrough, but it will take years of research. There's still much to explore in this field, especially in developing technologies that can fully regenerate living tissues. Additionally, the evolving nature of genomes results in new viruses being discovered regularly, which remain challenging to treat with modern science (Jain et al. 2021).

There is a tremendous future scope for cryogenics technology in the food and dairy industries as well, as it provides better-quality food products to the consumers. In the case of perishable food products like milk, eggs, meat, fruits, vegetables, etc., cold chain logistics for the supply of food, which maintains the temperature and humidity level around the food product, is crucial. The future trends of cold chain logistics involve minimizing carbon emissions and ensuring the safe and efficient transport and storage of perishable goods while significantly reducing the carbon footprint (Han et al. 2021). LN_2 is mostly used in food and health industries for storage purposes because of its excellent cooling properties. LN_2 can be used to produce ice cream by direct infusion, a method that is becoming increasingly popular in Western countries. Also, LN_2 can be used in heat exchangers for cooling/freezing purposes of various food and dairy products (Shrivastav et al. 2022; Shrivastav et al. 2024). Due to the ongoing demand for the LN_2, there is a need for the optimization of LN_2 systems.

4.8 Conclusion and Summary

This chapter discusses the industrial and design aspects of cryogenics applications. It briefly covers the principles of thermodynamics and heat transfer, which are crucial for the effective design and functioning of cryogenic systems. The chapter highlights the importance of considering thermal, mechanical, electrical, and magnetic properties when selecting materials for cryogenic applications. It is crucial to carefully consider the properties of the material used in cryogenic equipment that is to be used for food processing applications. Proper insulation practices, such as using MLI with alternating layers of shield and spacer in high vacuum, are emphasized as vital for maintaining low temperatures. The chapter also underscores the importance of safety measures and standards, including conducting physical, chemical, and physiological hazard analyses. LN_2 and LCO_2 are the primary options for various food processing applications preferred over other cryogens in the food industry due to their high-volume expansion ratio when their phase changes from liquid to vapour form and due to their high enthalpy of vaporization. LN_2 and LCO_2 are found to have excellent refrigeration properties and have zero ODP and low GWP values. Different types of cryogenic equipment utilized in the food industries and their design considerations are also discussed. While cryogenic technologies offer numerous benefits, they also present challenges related to cost, efficiency, and environmental impact. Although the direct immersion of cryogen in the food material offers several benefits, it results in an increase in operational cost and energy consumption. Continued research and innovation are essential to overcoming these limitations and realizing the full potential of cryogenics in various industries.

References

Ahern, J.E. (1980). Applications of the second law of thermodynamics to cryogenics—a review. *Energy 5* (8–9): 891–897. https://doi.org/10.1016/0360-5442(80)90104-8.

Air Liquide. (2025). Carbonation. Retrieved 19 November 2024, from https://uk.airliquide.com/solutions/carbonation

Allendorf, M.D., Stavila, V., Snider, J.L. et al. (2022). Challenges to developing materials for the transport and storage of hydrogen. *Nature Chemistry 14* (11): 1214–1223. https://doi.org/10.1038/s41557-022-01056-2.

ASTM International (2019). Standard guide for evacuated reflective insulation in cryogenic service (designation: C740/C740M–13). *ASTM International* https://doi.org/10.1520/C0740_C0740M-13.

Augustynowicz, S.D., Fesmire, J.E., and Wikstrom, J.P. (1999). Cryogenic Insulation Systems. In: *Proceedings of the 20th International Congress of Refrigeration*, 1–7. Sydney, Australia: International Institute of Refrigeration (IIR/IIF).

Bagheri, F., Fayazbakhsh, M.A., and Bahrami, M. (2017). Real-time performance evaluation and potential GHG reduction in refrigerated trailers. *International Journal of Refrigeration 73*: 24–38.

Baker, C.G.J. (2013). Hygienic design of food-processing equipment. In: *Handbook of Food Factory Design* (ed. C. Baker), 79–118. Springer https://doi.org/10.1007/978-1-4614-7450-0_4.

Balasubramanian, S., Gupta, M.K., and Singh, K.K. (2012). Cryogenics and its application with reference to spice grinding: a review. *Critical Reviews in Food Science and Nutrition* 52 (9): 781–794. https://doi.org/10.1080/10408398.2010.509552.

Barham, P. (2021). Cryogenics in the Kitchen. In: *Handbook of Molecular Gastronomy: Scientific Foundations, Educational Practices, and Culinary Applications*, 1e (ed. C. Lavelle, H. This, A.L. Kelly, and R. Burkey). CRC Press.

Barnwal, P., and Singh, K.K. (2014). Engineering interventions for global competitiveness of indian dairy industry. *9th Nation Convection of Indian Dairy Engineers Association (IDEA) and NATIONAL SEMINAR* (pp. 48–53), Karnal, Haryana (India).

Barron, R. F. (1972). Cryogenic food processing. Report No. 72-WA/DE-21.

Barron, R.F. and Nellis, G.F. (2017). *Cryogenic Heat Transfer*. United States: CRC Press.

Benjakul, S. and Bauer, F. (2000). Physicochemical and enzymatic changes of cod muscle proteins subjected to different freeze-thaw cycles. *Journal of the Science of Food and Agriculture* 80 (8): 1143–1150. https://doi.org/10.1002/1097-0010(200006)80:8<1143::AID-JSFA610>3.0.CO;2-C.

Bijeta Nayak, B., Jena, H., Dey, D. et al. (2021). Materials selection and design analysis of cryogenic pressure vessel: a review. *Materials Today: Proceedings 47*: 6605–6608. https://doi.org/10.1016/j.matpr.2021.05.095.

Safety British Cryogenics Council (2013). *Cryogenics Safety Manual: A Guide to Good Practice*. United Kingdom: Elsevier Science.

Burns, P.J. and Tien, C.L. (1979). Natural convection in porous media bounded by concentric spheres and horizontal cylinders. *International Journal of Heat and Mass Transfer* 22 (6): 929–939. https://doi.org/10.1016/0017-9310(79)90033-4.

dos Santos Carvalho, J.D., Oriani, V.B., de Oliveira, G.M., and Hubinger, M.D. (2021). Solid lipid microparticles loaded with ascorbic acid: release kinetic profile during thermal stability. *Journal of Food Processing and Preservation* 45 (6): 1–9. https://doi.org/10.1111/jfpp.15557.

Chen, J. and Rosenthal, A. (2009). Food processing. In: *Food Science and Technology* (ed. G. Campbell-Platt), 240–242. West Sussex, UK: WileyBlackwell.

Consoli, L., Grimaldi, R., Sartori, T. et al. (2016). Gallic acid microparticles produced by spray chilling technique: production and characterization. *Lebensmittel-Wissenschaft und -Technologie- Food Science and Technology* 65: 79–87. https://doi.org/10.1016/j.lwt.2015.07.052.

Davidge, H. (1981). Cryogenics in the food industry. *Cryogenics* 21 (5): 287–290.

Edeskuty, F.J. and Stewart, W.F. (1996). *Safety in the Handling of Cryogenic Fluids*. Springer https://doi.org/10.1007/978-1-4899-0307-5.

Espinoza Rodezno, L.A., Sundararajan, S., Solval, K.M. et al. (2013). Cryogenic and air blast freezing techniques and their effect on the quality of catfish fillets. *LWT - Food Science and Technology* 54 (2): 377–382. https://doi.org/10.1016/j.lwt.2013.07.005.

de Figueiredo, J., Silva, C.R., Souza Oliveira, M.F. et al. (2022). Microencapsulation by spray chilling in the food industry: opportunities, challenges, and Innovations. *Trends in Food Science & Technology* 120: 274–287. https://doi.org/10.1016/j.tifs.2021.12.026.

Fikiin, K. (2003). Novelties of food freezing research in Europe and Beyond. *Flair-Flow Europe Synthetic Brochure for SMEs No. 10 (ISBN: 2-7380-1145-4). INRA: Institut National dela Recherche Agronomique*, Paris (France), 55 p.

Flynn, T. (2004). *Cryogenic Engineering*, Revised and Expanded (2nd ed.). CRC Press https://doi.org/10.1201/9780203026991.

Forni, E., Torreggiani, D., Crivelli, G. et al. (1990). Influence of osmosis time on the quality of Dehydrofrozen Kiwi Fruit. *Acta Horticulturae* 282: 425–434. https://doi.org/10.17660/actahortic.1990.282.54.

Gas Encyclopedia by Air Liquide (2023). Nitrogen. https://encyclopedia.airliquide.com/nitrogen#properties. accessed date: December 28, 2025

GeeksforGeeks (2024, April 29). Applications of third law of thermodynamics. https://www.geeksforgeeks.org/applications-of-third-law-of-thermodynamics/. accessed date: December 28, 2025

Ghodki, B.M. and Goswami, T.K. (2016). Effect of grinding temperatures on particle and physicochemical characteristics of black pepper powder. *Powder Technology* 299: 168–177. https://doi.org/10.1016/j.powtec.2016.05.042.

Girardon, P., Moziar, C., and Pathier, D. (2019). Chapter 7.1—Refrigeration. In: *Gases in Agro-Food Processes* (ed. R. Cachon, P. Girardon, and A. Voilley), 223–318. essay, Academic Press https://doi.org/10.1016/B978-0-12-812465-9.00014-1.

Goswami, T.K. (2017). Recent trends of application of cryogenics in food processing and preservation. *Journal of Food, Nutrition and Population Health 1* (3): 1–4.

Goswami, T.K. (2010). Role of cryogenics in food processing and preservation. *International Journal of Food Engineering* 6 (1): 2.

Gouveia, S.T., Lopes, G.S., Fatibello-Filho, O. et al. (2002). Homogenization of breakfast cereals using cryogenic grinding. *Journal of Food Engineering* 51 (1): 59–63. https://doi.org/10.1016/s0260-8774(01)00037-1.

Gouveia, S.T., Lopes, G.S., Orlando, F.F. et al. (2004). Homogenization of breakfast cereals using cryogenic grinding. *Journal of Food Engineering* 51: 36–59.

Gunston, B. (2009). *Cryogenics. Cambridge Aerospace Dictionary*. Cambridge University Press.

Hands, B.A. (ed.) (1997). *Cryogenic engineering*, 1st ed.,e, 494. Academic Press Inc.

Han, J.-W., Zuo, M., Zhu, W.-Y. et al. (2021). A comprehensive review of cold chain logistics for fresh agricultural products: current status, challenges, and future trends. *Trends in Food Science & Technology* 109: 536–551. https://doi.org/10.1016/j.tifs.2021.01.066.

Hurlich, A. (1968). *Summer Study on Superconducting Devices and Accelerators*, 311–325. San Diego, United States: University of California (UC).

Jacob, S., Kasthurirengan, S., and Behra, U. (2000). Development of a cryogenic grinding plant for spices. In: *Cryogenic Engineering Conference and International Cryogenic Materials Conference*, 1731–1738, 45(B), Montr´eal, Qu´ebec, Canada.

Jain, P., Agarwal, P., Mathur, D. et al. (2021). Evolution of cryogenics – a review on applications of cryogenics in medicine. *Materials Today: Proceedings 47*: 3059–3063. https://doi.org/10.1016/j.matpr.2021.05.642.

James, S.J. and James, C. (2002). *Meat Refrigeration*. CRC Press/Woodhead Pub.

Jawahir, I.S., Attia, H., Biermann, D. et al. (2016). Cryogenic manufacturing processes. *CIRP Annals* 65 (2): 713–736. https://doi.org/10.1016/j.cirp.2016.06.007.

Jha, A.R. (2006). Requirements for cryogenic materials and associated accessories needed for various cryogenic coolers. *Cryogenic Technology and Applications* 225–255. https://doi.org/10.1016/b978-075067887-2/50009-5.

Jiang, W., Sun, P., Li, P. et al. (2021). Transient thermal behavior of multi-layer insulation coupled with vapor cooled shield used for liquid hydrogen storage tank. *Energy* 231: 120859. https://doi.org/10.1016/j.energy.2021.120859.

Jones, C. (2005, May 19). Method and apparatus for cryogenically manufacturing ice cream. US Patent 10,950,053.

Jouhara, H., Chauhan, A., Guichet, V. et al. (2023). Low-temperature heat transfer mediums for cryogenic applications. *Journal of the Taiwan Institute of Chemical Engineers 148*: 104709. https://doi.org/10.1016/j.jtice.2023.104709.

Jouhara, H., Chauhan, A., Guichet, V. et al. (2023). Low-temperature heat transfer mediums for cryogenic applications. *Journal of the Taiwan Institute of Chemical Engineers* 148: 104709. https://doi.org/10.1016/j.jtice.2023.104709.

Joy, M. and Vanapalli, S. (2023). Cryogenic nitrogen in the food and beverage industry. In: *Encyclopedia of Refrigeration*. International Institute of Refrigeration https://iifiir.org/en/encyclopedia-of-refrigeration/cryogenic-nitrogen-in-the-food-and-beverage-industry.

Kaale, L.D., Eikevik, T.M., Rustad, T., and Nordtvedt, T.S. (2014). Changes in water holding capacity and drip loss of Atlantic salmon (Salmo salar) muscle during superchilled storage. *LWT - Food Science and Technology* 55 (2): 528–535. https://doi.org/10.1016/j.lwt.2013.10.021.

Kalsi, N.S., Sehgal, R., and Sharma, V.S. (2010). cryogenic treatment of tool materials: a review. *Materials and Manufacturing Processes 25* (10): 1077–1100. https://doi.org/10.1080/10426911003720862.

Khadatkar, R.M., Kumar, S., and Pattanayak, S.C. (2004). Cryofreezing and cryofreezer. *Cryogenics* 44 (9): 661–678. https://doi.org/10.1016/j.cryogenics.2004.03.008.

Li, J., Chotiko, A., Kyereh, E. et al. (2016). Development of a combined osmotic dehydration and cryogenic freezing process for minimizing quality changes during freezing with application to fruits and vegetables. *Journal of Food Processing and Preservation* 41 (1): https://doi.org/10.1111/jfpp.12926.

Marquardt, E.D., Le, J.P., and Radebaugh, R. (2002). Cryogenic Material Properties Database. In: *Cryocoolers 11* (ed. R.G. Ross). Boston, MA: Springer https://doi.org/10.1007/0-306-47112-4_84.

Müller, B., Arlt, W., and Wasserscheid, P. (2011). A new concept for the global distribution of solar energy: energy carrying compounds. *Energy and Environmental Science 4* (10): 4322. https://doi.org/10.1039/c1ee01595h.

Murthy, C.T. and Bhattacharya, S. (2008). Cryogenic grinding of black pepper. *Journal of Food Engineering* 85: 18–28.

Muthukumarappan, K., Marella, C., and Sunkesula, V. (2019). Food freezing technology. In: *Handbook of Farm, Dairy and Food Machinery Engineering*, 389–415. https://doi.org/10.1016/b978-0-12-814803-7.00015-4.

Nesvadba, P. (2008). Thermal properties and ice crystal development in frozen foods. In: *Frozen Food Science and Technology*, 1–25. John Wiley & Sons, Ltd https://doi.org/10.1002/9781444302325.ch1.

Offer, G.J., Howey, D.A., Contestabile, M. et al. (2010). Comparative analysis of battery electric, hydrogen fuel cell and hybrid vehicles in a future sustainable road transport system. *Energy Policy* 38: 24–29. https://doi.org/10.1016/j.enpol.2009.08.040.

Oliveira, F.A., Neto, O.C., Santos, L.M. et al. (2017). Effect of high pressure on fish meat quality – a review. *Trends in Food Science & Technology* 66: 1–19. https://doi.org/10.1016/j.tifs.2017.04.014.

Oxley, J.D. (2012). Spray cooling and spray chilling for food ingredient and nutraceutical encapsulation. In: *Encapsulation Technologies and Delivery Systems for Food Ingredients and Nutraceuticals*, 110–130. https://doi.org/10.1533/9780857095909.2.110.

Paucar, O.C., Tulini, F.L., Thomazini, M. et al. (2016). Production by spray chilling and characterization of solid lipid microparticles loaded with vitamin D 3. *Food and Bioproducts Processing* 100: 344–350. https://doi.org/10.1016/j.fbp.2016.08.006.

Pedolsky, H. and Bau, R.L. (2010). *Reintroduction of Cryogenic Refrigeration for Cold Transport.* International Refrigeration and Air Conditioning Conference, Purdue University.

Peterson, T.J. and Weisend, J.G. II (2019). Cryogenic safety. *International Cryogenics Monograph Series* https://doi.org/10.1007/978-3-030-16508-6.

Plummer, M.C., Koehler, C., Flanders, D.R. et al. (1998). Cryogenic heat engine experiment. *Advances in Cryogenic Engineering* 43: 1245–1252. https://doi.org/10.1007/978-1-4757-9047-4_155.

Radebaugh, R. (2023). Thermodynamics and heat transfer in cryogenics: a personal view. In: *Proceedings of the 28th International Cryogenic Engineering Conference and International Cryogenic Materials Conference 2022. ICEC28-ICMC 2022. Advanced Topics in Science and Technology in China*, vol. 70 (ed. L. Qiu, K. Wang, and Y. Ma). Singapore: Springer https://doi.org/10.1007/978-981-99-6128-3_2.

Radebaugh, R. and Bar-Cohen, Y. (2016). Chapter 14: Low-temperature materials and mechanisms: applications and challenges. In: *Low Temperature Materials and Mechanisms*, 437–476. https://doi.org/10.1201/9781315371962-15.

Rai, A. and Tassou, S.A. (2017). Energy demand and environmental impacts of alternative food transport refrigeration systems. *Energy Procedia* 123: 113–120. https://doi.org/10.1016/j.egypro.2017.07.267.

Rai, A. and Tassou, S.A. (2017). Environmental impacts of vapour compression and cryogenic transport refrigeration technologies for temperature-controlled food distribution. *Energy Conversion and Management 150*: 914–923. https://doi.org/10.1016/j.enconman.2017.05.024.

Ramakrishnan, S., Wysk, R.A., and Prabhu, V.V. (2004). Prediction of process parameters for intelligent control of tunnel freezers using simulation. *Journal of Food Engineering* 65 (1): 23–31.

Razali, N.A., Wan Ibrahim, W.M., Safari, S. et al. (2022). Cryogenic freezing preserves the quality of whole durian fruit for the export market. *Food Research* 6 (3): 360–364. https://doi.org/10.26656/fr.2017.6(3).428.

Rehman, W., Farhan, M., and Rehman, F. (2020). A review of cryogenics applications for power and energy. *ASME 2020 Power Conference*. https://doi.org/10.1115/power2020-16911

Sanz, P.D., de Elvira, C., Martino, M. et al. (1999). Freezing rate simulation as an aid to reducing crystallization damage in foods. *Meat Science* 52 (3): 275–278. https://doi.org/10.1016/S0309-1740(99)00002-9.

Schmied, M., Knörr, W., & Hepburn, L. (2012). Calculating GHG emissions for freight forwarding and logistics services in accordance with EN 16258. Terms, Methods, Examples.

Shaikh, N.I. and Prabhu, V. (2007). Mathematical modeling and simulation of cryogenic tunnel freezers. *Journal of Food Engineering* 80 (2): 701–710. https://doi.org/10.1016/j.jfoodeng.2006.04.065.

Shi, L., Xiong, G., Ding, A. et al. (2018). Effects of freezing temperature and frozen storage on the biochemical and physical properties of procambarus clarkii. *International Journal of Refrigeration* 91: 223–229. https://doi.org/10.1016/j.ijrefrig.2018.04.027.

Shrivastav, A., Goswami, T.K., and Kotra, S.V. (2022). Rheological and adaptive neuro-fuzzy inference system (ANFIS) modeling of ice cream at different gelatin concentrations produced by liquid nitrogen infusion technique. *Journal of Biosystems Engineering* 47 (3): 344–357. https://doi.org/10.1007/s42853-022-00151-z.

Shrivastav, A., Goswami, T.K., Tripathy, P.P. et al. (2024). Performance evaluation and CFD investigation of scraped surface ice cream freezer augmented with LN_2 freezing technique. *Cogent Engineering* 11 (1): https://doi.org/10.1080/23311916.2024.2344732.

Singam, S., Ghodki, B.M., and Goswami, T.K. (2018). Effect of grinding methods on powder quality of king chilli. *Journal of Food Measurement and Characterization* 12 (3): 1686–1694. https://doi.org/10.1007/s11694-018-9784-6.

Singh, K.K. and Goswami, T.K. (1999). Design of a cryogenic grinding system for spices. *Journal of Food Engineering* 39: 359–368.

Singh, K.K. and Goswami, T.K. (1999). Studies on cryogenic grinding of cumin seeds. *Journal of Food Process Engineering* 22: 175–190.

Singh, K.K. and Goswami, T.K. (2000). Cryogenic grinding of cloves. *Journal of Food Processing and Preservation* 24: 57–71.

Sriket, P., Benjakul, S., Visessanguan, W., and Kijroongrojana, K. (2007). Comparative studies on the effect of the freeze-thawing process on the physicochemical properties and microstructures of black tiger shrimp (Penaeus Monodon) and white shrimp (Penaeus Vannamei) muscle. *Food Chemistry* 104 (1): 113–121. https://doi.org/10.1016/j.foodchem.2006.11.004.

Sutariya, S.G. and Sunkesula, V. (2021). Food freezing: emerging techniques for improving quality and process efficiency a comprehensive review. *Innovative Food Processing Technologies* 36–63. https://doi.org/10.1016/b978-0-08-100596-5.23035-7.

Truonghuynh, H.T. and Li, B. (2019). Quality of aquatic products via cryogenic freezing. *Journal of Food Science and Nutrition Research* 02 (04): https://doi.org/10.26502/jfsnr.2642-11000032.

Taghavi, M., Sharma, S., and Balakotaiah, V. (2024). Natural convection effects in insulation layers of spherical cryogenic storage tanks. *International Journal of Heat and Mass Transfer* 220: 124918. https://doi.org/10.1016/j.ijheatmasstransfer.2023.124918.

Tassou, S.A., Hadawey, A., Ge, Y., and Groutte, B.L. (2009). *Carbon Dioxide Cryogenic Transport Refrigeration Systems*. Centre for Energy and Built Environment Research, Brunel University.

Tien, C.L. and Cunnington, G.R. (1973). Cryogenic insulation heat transfer. *Advances in Heat Transfer* 349–417. https://doi.org/10.1016/s0065-2717(08)70065-0.

Timmerhaus, K.D. (2003). Cryogenic process engineering. *Encyclopedia of Physical Science and Technology* 13–36. https://doi.org/10.1016/b0-12-227410-5/00156-3.

Tulini, F.L., Souza, V.B., Thomazini, M. et al. (2017). Evaluation of the release profile, stability and antioxidant activity of a proanthocyanidin-rich cinnamon (cinnamomum zeylanicum) extract co-encapsulated with α-tocopherol by spray chilling. *Food Research International* 95: 117–124. https://doi.org/10.1016/j.foodres.2017.03.010.

Van Sciver, S.W. (2012). Cryogenic principles and applications. In: *Helium Cryogenics. InternationalCryogenicsMonographSeries.*NewYork,NY:Springerhttps://doi.org/10.1007/978-1-4419-9979-5_1.

Vandoni, G. (2004). Heat transfer. In: *CAS - CERN Accelerator School: Superconductivity and Cryogenics for Accelerators and Detectors*, 325–345. https://doi.org/10.5170/CERN-2004-008.325.

Ventura, G. and Risegari, L. (ed.) (2008). Practical, industrial and space applications of cryogenics. *The Art of Cryogenics* 299–306. https://doi.org/10.1016/b978-008044479-6.50015-7.

Zhao, Y., Ji, W., Guo, J. et al. (2020). Numerical and experimental study on the quick freezing process of the bayberry. *Food and Bioproducts Processing* 119: 98–107. https://doi.org/10.1016/j.fbp.2019.10.013.

Zohuri, B. (2018). *Physics of Cryogenics: An Ultralow Temperature Phenomenon*. Elsevier.

Zohuri, B. (2019). Cryogenics and liquid hydrogen storage. In: *Hydrogen Energy*. Cham: Springer https://doi.org/10.1007/978-3-319-93461-7_4.

5

Cryogenic Freezing Process for the Preservation of Fruits and Vegetables

Prateek Gururani[1], *Abhirup Mitra*[2], *Jesni Manjula Sivaprasad*[3], *and Mahipal Singh Tomar*[4]

[1] *Department of Biotechnology, Graphic Era Deemed University, Dehradun, Uttarakhand, India*
[2] *Department of Agriculture, Invertis University, Bareilly, Uttar Pradesh, India*
[3] *School of Food Science and Environment Health, Technological University Dublin, Dublin, Ireland*
[4] *Department of Food Science and Technology, Amity University, Mohali, Punjab, India*

5.1 Introduction

Fruits and vegetables (F&Vs) are vital components of the human diet, which offers various nutritional and health-promoting properties (Jha et al. 2024a,b; Tomar and Pradhan 2024). They are highly perishable and living organisms with a soft and delicate texture that continue to respire even after harvesting. This natural process can be controlled by low-temperature processes such as refrigeration and freezing (Tomar and Pradhan 2023; Loayza-Salazar et al. 2024). Both refrigeration and freezing are effective low-temperature methods that help to maintain the quality and extend the shelf life of F&Vs during post-harvest storage. While refrigeration can prolong the shelf life for few days to months only, freezing is regarded as the most effective technique for the long-term storage of fruits and vegetables (Muthukumarappan et al. 2019; Alabi et al. 2020; Tomar and Pradhan 2023).

Freezing is a low-temperature conventional preservation method subjecting products to reduced temperature below 0 °C or generally to −18 °C. This low temperature slows down the various chemical changes and retards the microbial growth during storage (Espinoza Rodezno et al. 2013). Freezing does not improve the post-harvest quality, but it increases the shelf life by minimizing the nutritional changes during storage. Conventional freezing technology has one disadvantage, which initiated the ice crystals responsible for damage of microstructure. Various freezing technologies are employed in fruits and vegetables freezing industry, including immersion freezing, air-blast freezing, plate freezing, fluidized-bed freezing, and cryogenic freezing. Among them, cryogenic freezing is getting more research attention due to its low initial cost and ultra-fast freezing rate.

The cryogenic freezing process uses low-boiling-point refrigerants, such as carbon dioxide and liquid nitrogen, to achieve the rapid freezing of food products. This method offers

Cryogenic Processing: Advances in the Food Industry, First Edition. Edited by Sumit Sudhir Pathak, Rama Chandra Pradhan, and Sudarshan Ramanathan.

a significantly faster freezing rate compared to conventional mechanical freezing systems, which helps in preserving the microstructure and minimizing cellular damage in fruits and vegetables (James and James 2014). Unlike mechanical freezing, which often leads to the formation of large ice crystals that disrupt cell walls, cryogenic freezing maintains the integrity of the cellular structure due to the formation of smaller, more uniformly distributed ice crystals (Jha et al. 2024a,b). Additionally, cryogenic systems provide a substantially higher heat transfer coefficient than mechanical chillers, enhancing process efficiency and product quality. As a result, cryogenic freezing is particularly well suited for delicate produce and premium-quality applications where quality retention is paramount (Kiani and Sun 2011).

This chapter introduces the basic knowledge, fundamental principles, and practical implementation of a cryogenic freezing system for fruits and vegetables. It explores the different case studies done on cryogenic freezing of fruits and vegetables. It explores a range of case studies and research findings that demonstrate the effectiveness and application of cryogenic freezing across different fruit and vegetable categories. These studies highlight critical parameters such as freezing rate, product morphology, energy efficiency, and quality outcomes. Furthermore, the chapter addresses the economic feasibility and practical limitations of adopting cryogenic freezing technologies in commercial frozen market.

5.2 Overview of Cryogenics in Food Processing

The 'cryogenic' term refers to the field of the production and application of low-temperature refrigerant, commonly known as a cryogenic refrigerant, which changes its phase by absorbing latent heat to rapidly freeze the food materials (Goswami 2010). In food processing, cryogenic has emerged as advanced processing and preservation technique, which plays a pivotal role by enabling ultra-low temperature during processing that significantly enhances product quality, safety, and shelf life. Cryogenic in food processing utilizes solid carbon dioxide (dry ice), liquid carbon dioxide, or liquid nitrogen. Inert, colourless, and odourless properties of carbon dioxide and liquid nitrogen make them suited for food processing operations. Additionally, both cryogenic refrigerants are recognized as generally recognized as safe (GRAS) by regulatory authorities such as the U.S. FDA. Different properties of cryogenic refrigerant are tabulated in Table 5.1.

Table 5.1 Properties of liquid N_2 and CO_2.

Property name	Liquid CO_2	Liquid N_2
Boiling point (°C)	−78.5	−195.8
Density (kg/m^3)	464	784
Specific heat (kJkg^{-1}k^{-1})	2.26	2
Latent heat (kJkg^{-1})	352	358
Total usable refrigeration (kJkg^{-1})	565	690

Application of cryogenic techniques in food processing is gaining more popularity due to extremely low temperatures condition during treatment. It has already been successfully implemented in food processing operations such as grinding and freezing. In cryogenic grinding, low temperature improves grinding efficiency and minimizes the loss of flavour, aroma, colour, nutritional value of heat sensitive materials by preventing heat build-up during grinding of food material. While cryogenic processing has one most critical benefit in freezing food material is uniformly distributed and small ice crystals within the food matrix, which helps preserve the texture and quality of the frozen product.

5.3 Conventional Freezing Methods for Fruits and Vegetables

Freezing is one of the most effective methods for the long-term preservation of fruits and vegetables. It removes the field heat and converts the water present in food products into ice (Alhamdan et al. 2018a; Jha et al. 2024a,b; Li et al. 2022). Freezing generally reduces product temperature to −18 °C or below, which immobilizes the water present in food, lowers the water activity, inhibits the growth of microorganisms, and reduces enzyme activity, resulting in a longer shelf life of fruits and vegetables (Espinoza Rodezno et al. 2013; Goyal 2015; Alhamdan et al. 2018a; Li et al. 2022). Freezing and frozen storage methods are used for longer preservation of some fruits and vegetables. A well-executed freezing process produces superior-quality frozen products at an affordable cost (Muthukumarappan et al. 2019). Frozen fruits and vegetables offer the advantages of preservation of nutrients, aroma, and flavour. Due to these benefits, frozen fruits and vegetables have gain widespread acceptance and established them as a next-generation food products (Jha et al. 2024a,b).

The freezing process has one critical problem: it can never be from ice formation and often results in undesirable and irreversible changes in the colour and textural properties of fruits and vegetables (Poudyal et al. 2019). However, conventional freezing methods such as room freezing, air-blast freezing, and contact freezing have longer freezing times that lead to the formation of large and needle-shape ice crystals (Li et al. 2022). These ice crystals cause structural damage to the cell wall, middle lamella, and protoplast of fruits and vegetables, resulting in texture quality loss (Holzwarth et al. 2012). Despite these limitations, conventional freezing remains a cost-effective and accessible solution, especially in regions with established cold chain infrastructure. However, as consumer expectations for higher-quality frozen products increase, the food industry is progressively exploring advanced alternatives such as cryogenic freezing to overcome the shortcomings of traditional methods.

5.3.1 Air-Blast Freezing

Air-blast freezing employs the mechanical refrigeration systems that use the high velocity of cold air as the cooling medium, which is among the most widely used freezing technologies for fruits and vegetables on an industrial scale. Air-blast freezers operate by circulating cold air at temperatures range between −20 and −40°C. Cover the product at velocities ranging from 0.5 to 10 m/s (Swer et al. 2018). Air-blast freezing can be employed in either batch or continuous modes for freezing of fruits and vegetables. In batch configurations, fresh produces are arranged on trays within thermally insulated rooms or cabinets. While such systems offer notable flexibility in processing different product types, as well as operational advantages such as ease of use and relatively low investment costs, their productivity is generally lower compared to continuous systems. Continuous air-blast systems are commonly configured as either belt or tunnel freezers. Their performance can be adapted to the specific characteristics of various fruits and vegetables by regulating the conveyor belt speed, thereby controlling the residence time within the freezing zone.

Products may be placed on trays mounted on trolleys or directly on conveyor belts that pass through an insulated tunnel or chamber. In belt conveyor systems, plastic or stainless-steel mesh belts are commonly used, and air may be directed perpendicularly through the belt to enhance heat transfer. This approach forms the basis of fluidized-bed freezers, in which cold air is blown vertically upward through a perforated bed at sufficient velocity to suspend and fluidize small-sized products. Due to the combination of mechanical agitation and rapid freezing, fluidized-bed technology is well suited for producing individually quick frozen (IQF) products (Alinovi et al. 2024). Impingement freezing is a related method that employs high-velocity air streams directed perpendicularly onto the product surface through arrays of jet nozzles. In this system, air is applied to both the upper and lower surfaces of the conveyor belt at velocities of approximately 50 m/s, effectively disturbing the insulating boundary layer surrounding the product and thereby markedly enhancing the surface heat transfer coefficient (Chaves and Zaritzky 2018).

5.3.2 Plate-Freezing Systems

Plate freezers are made up of a series of either vertical or horizontal hollow plates. A coolant, usually around −40 °C, is circulated through these plates to freeze products. Fruits and vegetables are placed between the plates and subjected to slight pressure to ensure close contact, which improves the heat transfer efficiency. Due to the loading and unloading process, which can be manual or semi-automated, these freezers typically operate in batch or semi-continuous cycles. While they have a high heat transfer rate, their use is limited by a product's shape and thickness.

5.3.3 Immersion Freezing

In immersion freezing, packed or unpacked food is moved forward through a bath of liquid coolant (propylene glycol, brine solution, glycerol, sugar, or alcohol solutions). Liquid coolant remains in liquid state during freezing; it does not change the state as occurs in case of cryogenic freezing (Fellows 2017; Alinovi et al. 2024).

5.4 Cryogenic Freezing Techniques

Freezing-induced harm can be significantly minimized by employing advanced and higher freezing rates technologies such as air-blast freezing, pressure-assisted freezing, pressure shift freezing, and cryogenic freezing (Xu et al. 2014; Jha et al. 2022). These processes generate numerous small ice crystals in both intercellular and intracellular domains, thereby minimizing the detrimental effects caused by dehydration and destruction resulting from shrinkage. Cryogenic processing is one of the few recent innovative freezing technologies for fruit and vegetable freezing. It involves the spraying or immersion of cryogenic fluids liquid nitrogen (LN_2) and carbon dioxide (CO_2) for the rapid freezing of fruits and vegetables (Jha et al. 2024a,b).

Due to their extremely low boiling points (−196 °C for liquid nitrogen and −79 °C for carbon dioxide), these cryogenic agents facilitate rapid heat removal, resulting in exceptionally fast freezing rates. Such conditions help preserve the texture, colour, flavour, and nutritional quality of fruits and vegetables during freezing (Alinovi et al. 2024). It offers a rapid and efficient alternative to the conventional freezing technique, significantly reducing the freezing times. Cryogenic freezing is typically performed in either batch or continuous mode. Based on the nature of interaction between refrigerants and fresh produce, cryogenic freezing can be broadly categorized into direct-contact and indirect cryogenic freezing system.

5.4.1 Direct-Contact Cryogenic Freezing

In this method of freezing, cryogenic refrigerants come into direct contact with the food materials. This method us is further subclassified into immersion and spray cryofreezers (Khadatkar et al. 2004). Among the two primary cryogenic freezing systems, direct-contact freezing systems are more widely employed in food freezing processing due to superior heat transfer rate, which enables the faster freezing and small-ice formation.

5.4.1.1 Immersion Cryogenic Freezer

Immersion cryogenic freezers are the fastest method of freezing. In this method, uniform-size fruits and vegetables are directly dropped into a bath of cryogenic refrigerant, typically liquid nitrogen (Fellows 2017) (Shown in Figure 5.1). This direct and uniform exposure of fruits and vegetables demonstrated the higher heat transfer rate, resulting in less freezing time. Immersion in boiling liquid nitrogen generates extreme turbulence, which effectively prevents individual food pieces from adhering. This process facilitates the production of IQF products, making it especially advantageous for foods with irregular shape (Fellows 2017). Immersion systems are generally preferred when ultra-fast and complete freezing is critical. This method is particularly suitable for selected food items such as raspberries, shrimp, and diced meat, which can tolerate the associated thermal stress without significant structural compromise.

However, in a broad range of other food matrices, the extremely high freezing rate induces severe thermal gradients, leading to internal mechanical stresses and fruit frost cracking. These stresses often manifest as structural failures, including fissuring or fragmentation of the food tissue, thereby compromising product quality. Residence time must be precisely controlled to avoid over freezing or the development of internal stresses from excessive thermal shock, which can lead to cracking or splitting of the product. It requires more liquid cryogenic refrigerant than spray immersion cryogenic system A primary disadvantage of immersion freezers is their relatively high cryogen consumption, typically requiring 3–4 kg of liquid nitrogen per kilogram of product, compared with approximately 1 kg per kilogram in tunnel freezers (Coulomb 2019).

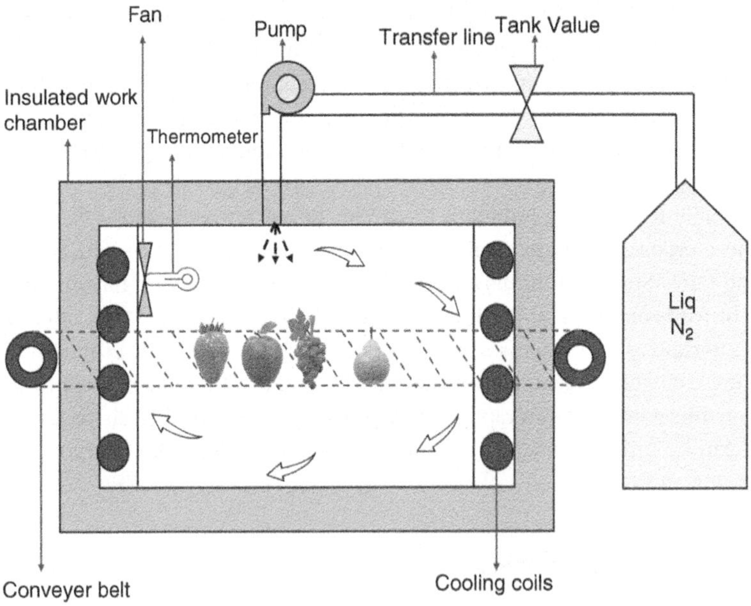

Figure 5.1 Pictorial view of immersion cryogenic freezing system.

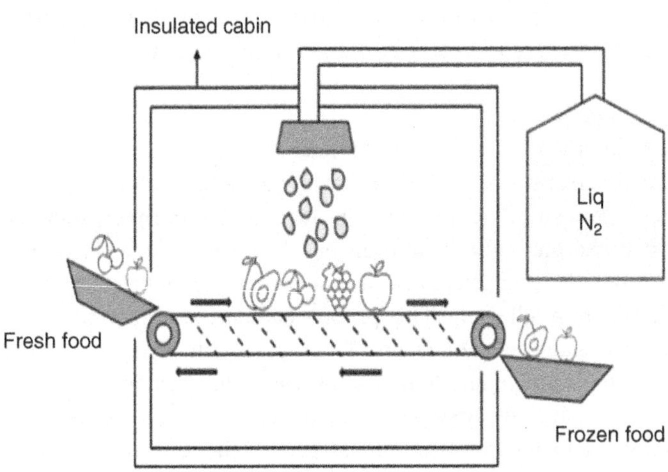

Figure 5.2 Pictorial view of spray cryogenic freezing system.

5.4.1.2 Spray Cryogenic Freezer

Spray cryogenic freezing is another technique of cryogenic freezing, in which fine mist of cryogenic refrigerant is directly sprayed on fruits and vegetables, which are mounted on conveyor belt (In case of continuous freezing system). It consists of a spray system, which atomizes or sprays the liquid nitrogen on fruits and vegetables, as shown in Figure 5.2. The amount of liquid nitrogen consumed during spray cryogenic freezing is influenced by multiple operational and physical parameters, all of which play a significant role in determining the overall cost and efficiency of the freezing system. These factors are mentioned below:

1) Degree of contact between liquid nitrogen droplet and fruit and vegetable surface.
2) Rate of evaporation of liquid nitrogen after leaving the spray nozzle.
3) Droplet diameter at the point of spray.

5.4.2 Indirect Contact Cryogenic Freezing

Indirect contact cryogenic freezing method involves freezing fruits and vegetables by separating them from the extremely cold cryogenic refrigerant (with a barrier, often a metal plate or packaging). This method aims to preserve food quality by minimizing ice crystal formation and cellular damage, which can occur during slower freezing processes.

5.5 Case Studies on Cryogenic Freezing of Fruits and Vegetables

According to Chassagne-Berces et al. (2009), immersion cryogenic freezing resulted in smaller and more uniformly distributed ice crystals in apples compared to conventional freezing at −20 °C. Additionally, immersion freezing showed less degradation of Young's modulus, although it was associated with a relatively higher firmness loss than conventional freezing. This innovation freezing method produced the long and thin tissue crack during immersion in liquid nitrogen. Chassagne-Berces et al. (2010) emphasized that the selection of an appropriate freezing method should be based on the specific type and characteristics of the fruit. In their study, the researchers applied various freezing techniques – including room temperature freezing, nitrogen gas-flush freezing at −80 °C, and liquid nitrogen (LN_2) immersion freezing at −196 °C across different varieties of fresh fruits.

Freezing at temperature −80 °C using nitrogen gas inside the chamber proved to be the most effective option among the tested conditions for preserving small fruit cylinders. LN_2-based cryogenic freezing of blueberries presented higher hardness and similar gumminess and chewiness comparison to freezing done with conventional method of freezing at −20, −40 and −80 °C temperatures. It was observed that supercooling point was not observed in case of freezing at −80 °C and liquid nitrogen method, during different freezing experiments. Extent and controlling of supercooling point are critical in freezing fruits and vegetables to maintain ice crystal size and its distribution. Additionally, the extremely fast freezing rate during LN_2 immersion cryogenic method produced fruit frost cracking after thawing, which resulted in a higher loss and quality deterioration during storage (Cao et al. 2018). Low-field nuclear magnetic resonance (LF-NMR) analysis to analyse the water distribution revealed that relaxation time (T_{23}) was significantly higher in conventional freezing at −80 °C temperature. Results output of this study suggest that freezing of blueberry at −80 °C using conventional treatment could offer better protection to the vacuole membrane (Cao et al. 2018).

Alhamdan et al. (2018b) compared the effect of different freezing techniques such liquid nitrogen cryogenic freezing (LN_2), conventional freezing, and IQF on freezing rate and quality of fresh Barhi. It was observed that fruit's core temperature was reduced from 21 to

−22.1 °C within five minutes (in case of LN_2 freezing), in contrast to reduction from 21 °C to only −2 °C using IQF method. Colorimetric properties of LN_2 frozen, immediately after freezing and during frozen storage, were significantly superior to IQF and conventional freezing methods. Additionally, texture profile analysis properties such as hardness, elasticity, resilience, and chewiness are better in LN_2 than IQF and deep-freezing method. Alhamdan et al. (2018b) concluded that the preservation of textural properties observed with the LN_2 method might be due probably to decrease of the injurious effects of crystallization and recrystallization on the microstructure of Barhi tissues during such quick freezing, subsequent frozen storage, and thawing.

Furthermore, a study reported no overall difference in the quality of frozen raspberry when comparing the LN_2 freezing with commercial conventional freezing methods during storage (Gales et al. 2022). Operating condition of cryogenic freezing was −80 °C in liquid nitrogen tunnel IQF. In addition, Fan et al. (2024) investigated the effect of different liquid nitrogen spray quick-freezing temperatures on various textural, thawing, cell integrity, microstructure, and enzyme activity of bamboo shoot and compared the results with those obtained using conventional refrigeration freezing. The maximum ice-crystal formation time decreased significantly as the liquid nitrogen spray freezing temperature was lowered,

Table 5.2 Comparison of cryogenic and conventional freezing in food processing.

Parameter	Cryogenic freezing	Conventional mechanical freezing
Temperature range	Ultra-low (below −150 °C)	Low (−18 °C to −40 °C)
Freezing rate	Extremely rapid (seconds to minutes)	Slow to moderate (minutes to hours)
Ice crystal size	Very small, uniformly distributed	Large and irregular ice crystal growth
Supercooling	Extend and controlling of super-cooling processing during freezing	There is no control of supercooling curve
Product quality	Excellent texture, flavour, and colour preservation	Moderate; higher risk of cellular damage
Shelf-life extension	Long-term (months to over a year with minimal quality loss)	Long-term but with gradual quality degradation
Energy source	Cryogens (LN_2, CO_2)	Electrical/mechanical refrigeration
Flexibility	Suitable for delicate, irregular-shaped, or high-value products	More suited to bulk, uniform products
Environmental considerations	Requires safe handling; CO_2 emissions possible	High electricity usage; refrigerant leakage risks
Equipment cost	Lower capital cost but higher operational cost	Higher capital cost, lower operational cost
Limitation	High cost of cryogen and higher cryogenic substances consumption. Skin crack can be produced in soft fresh produce like blueberry.	Higher capital cost, higher energy consumption, and texture degradation and drip loss are higher.

showing a 94.66% reduction compared with the conventional freezing method. This resulted in the formation of small, less-damaged ice crystals and more uniform water distribution, thereby preserving the microstructure of bamboo shoot during freezing. LN_2 cryogenic freezing prevents the hardness, thawing loss, and peroxidase activity compared with conventional refrigerator freezing at −20 °C.

Cryogenic freezing, while highly effective due to its rapid freezing rate, often leads to structural damage caused by uneven ice nucleation and growth, particularly in complex biological tissues. Combining cryogenic freezing with additional processing techniques such as ultrasound or high-pressure treatment can help minimize the issue of non-uniform ice formation during the freezing of fruits and vegetables. Xu et al. (2014) demonstrated the potential of high pressure during the liquid carbon dioxide immersion method for carrot slice freezing. Scanning electron microscope images conclude the less tissue damage than liquid nitrogen and ultra-low temperature freezing process.

Based on different case studies, various advantages and comparative study of cryogenic and conventional freezing of fruits and vegetables are summarized and explained in Section 5.6 and Table 5.2.

5.6 Advantages and Limitations of Cryogenic Freezing

5.6.1 Advantages of Cryogenic Freezing

1) Lower capital cost than mechanical refrigeration system.
2) There is no requirement of compressor and evaporator for circulation of refrigerant (as in case of mechanical refrigeration system).
3) Higher freezing rate produces the small ice crystal and prevents texture degradation and drip loss.
4) Lower weight loss and dehydration compared to air-blast freezing.
5) Less cell-wall degradation and lower thawing loss.
6) Minimize the oxidation deterioration due to exclusion of oxygen during freezing (air-blast freezing requires the high velocity of air).

5.6.2 Economic Feasibility, Challenges, and Limitations

Cryogenic freezing technology has shown to have great potential in frozen fruits and vegetables in terms of mitigating the structure damage occurred in conventional freezing process. It offers both opportunities and challenges. On the positive side, the initial capital investment required for cryogenic freezing systems is typically lower than that for mechanical freezing units. However, economic feasibility is significantly influenced by operational costs, particularly the cost of cryogenic gases. Liquid nitrogen and carbon dioxide are consumable inputs that must be replenished continually, making the operating cost much higher than conventional mechanical systems that rely on electricity. According to Alhamdan et al. (2018b), annual production cost for freezing of fresh material using LN_2 is notably higher than IQF. This cost disparity presents a major limitation to the scalability of cryogenic freezing for bulk or low margin produce (James and James 2014). Challenges also arise in terms of supply chain logistics. Continuous access to cryogenic gases is essential for uninterrupted operations, which may not be viable in remote or underdeveloped regions.

From the technical perspective, the implementation of cryogenic-freezing systems presents several challenges. The process necessitates the use of specialized equipment, including insulated storage tanks for cryogenic liquids, controlled vapour-phase distribution systems, and robust safety protocols for the handling and storage of cryogenic fluids. These infrastructural and operational requirements contribute to increased system complexity and elevate both initial capital investment and regulatory compliance costs. This is particularly burdensome in regions where cold chain infrastructure is underdeveloped or absent. Among the most significant limitations of cryogenic-freezing technologies in the commercial frozen fruits and vegetables sector is the relatively high operational cost, primarily due to the continuous consumption of cryogenic agents such as liquid nitrogen or carbon dioxide. In contrast to mechanical freezing systems, which rely on closed-loop refrigeration cycles, cryogenic systems require a constant supply of expendable refrigerants, making them less economically feasible for large-volume, low-margin products such as frozen fruits and vegetables. These economic considerations often hinder widespread adoption, especially in cost-sensitive markets, or in regions where cryogen availability is limited or inconsistent.

5.7 Conclusion

Cryogenic freezing emerges as a highly efficient and innovative technology for preserving quality and extending the shelf life of fruits and vegetables. Its ultra-fast freezing capabilities enable the formation of smaller ice crystals, effectively reducing the structural damage typically observed in conventional freezing methods. Cryogenic freezing process offers notable improvements in textural integrity and nutrient retention, making it suitable for premium or delicate food products. Cryogenic freezing is typically applied through direct contact with fruits and vegetables, using either immersion or spray method. Immersion cryogenic freezing removes the heat fast and presents ultra-high freezing rate than spray cryogenic freezing. However, the immersion method of cryogenic freezing can produce tissue and skin damage to some fresh produce. Spray cryogenic freezing can be used as a controlled freezing method of fruits and vegetables. However, despite its technical advantages, the widespread adoption of cryogenic freezing is constrained by high operational costs, particularly due to the continuous consumption of cryogenic gases and the need for specialized infrastructure and logistics. While not universally feasible for all markets or applications, cryogenic freezing holds considerable promise for high-value, quality-sensitive food segments. Future developments focusing on integration with complementary technologies such as high-pressure processing or ultrasound may further mitigate its limitations, enhancing its viability as a next-generation food preservation method.

References

Alabi, K.P., Zhu, Z., and Sun, D.W. (2020). Transport phenomena and their effect on microstructure of frozen fruits and vegetables. *Trends in Food Science and Technology* 101 (March): 63–72. Elsevier.

Alhamdan, A., Hassan, B., Alkahtani, H. et al. (2018a). Cryogenic freezing of fresh date fruits for quality preservation during frozen storage. *Journal of the Saudi Society of Agricultural Sciences* 17 (1): 9–16. King Saud University & Saudi Society of Agricultural Sciences.

Alhamdan, A., Hassan, B., Alkahtani, H. et al. (2018b). Freezijng of fresh Barhi dates for quality preservation during frozen storage. *Saudi Journal of Biological Sciences* 25 (8): 1552–1561. King Saud University.

Alinovi, M., Paciulli, M., Rinaldi, M. et al. (2024). Freezing of fruits and vegetables. In: *Low-Temperature Processing of Food Products* (ed. S.M. Jafari and H. Rostamabadi), 199–224. Elsevier.

Cao, X., Zhang, F., Zhao, D. et al. (2018). Effects of freezing conditions on quality changes in blueberries. *Journal of the Science of Food and Agriculture* 98 (12): 4673–4679.

Chassagne-Berces, S., Poirier, C., Devaux, M.-F. et al. (2009). Changes in texture, cellular structure and cell wall composition in apple tissue as a result of freezing. *Food Research International* 42 (7): 788–797.

Chassagne-Berces, S., Fonseca, F., Citeau, M. et al. (2010). Freezing protocol effect on quality properties of fruit tissue according to the fruit, the variety and the stage of maturity. *LWT - Food Science and Technology* 43 (9): 1441–1449. Elsevier Ltd.

Chaves, A. and Zaritzky, N. (2018). Cooling and freezing of fruits and fruit products. In: *Fruit Preservation Novel and Conventional Technologies* (ed. A. Rosenthal, R. Deliza, J. Welti-Chanes, and G.V. Barbosa-Cánovas), 127–180. New York, NY: Springer.

Coulomb, D. (2019). Refrigeration process technologies. In: *Gases in Agro-Food Processes* (ed. R. Cachon, P. Girardon, and A. Voilley), 223–318. Elsevier.

Espinoza Rodezno, L.A., Sundararajan, S., Solval, K.M. et al. (2013). Cryogenic and air blast freezing techniques and their effect on the quality of catfish fillets. *LWT - Food Science and Technology* 54 (2): 377–382.

Fan, Z., Shen, H., Hu, T. et al. (2024). Effect of liquid nitrogen spray quick-freezing technology on the quality of bamboo shoots, Dendrocalamus brandisii from Yunnan Province, China. *Journal of Food Engineering* (2023) 368: 111916. Elsevier Ltd.

Fellows, P.J. (2017). Freezing. In: *Food Processing Technology* (ed. P.J. Fellows), 885–928. Elsevier.

Gales, O., Jones, J., and Swarts, N. (2022). An analysis on the impacts of cryogenic freezing on raspberry quality. *Advances in Horticultural Science* 36 (4): 293–301. Firenze University Press.

Goswami, T.K. (2010). Role of cryogenics in food processing and preservation. *International Journal of Food Engineering* 6 (1).

Goyal, S.P. (2015). *Press Information Bureau Government of India Ministry of Power*, 1–3. Government of India, Ministry of Road Transport & Highways.

Holzwarth, M., Korhummel, S., Carle, R. et al. (2012). Evaluation of the effects of different freezing and thawing methods on color, polyphenol and ascorbic acid retention in strawberries (Fragaria×ananassa Duch.). *Food Research International* 48 (1): 241–248.

James, S.J. and James, C. (2014). Food technologies: freezing. *Encyclopedia of Food Safety* 3: 187–195.

Jha, P.K., Le-Bail, A., and Jun, S. (2022). Recent advances in freezing processes: an overview. In: *Food Engineering Innovations Across the Food Supply Chain* (ed. P. Juliano, R. Buckow, M.H. Nguyen, et al.), 187–206. Elsevier.

Jha, P.K., Chapleau, N., Meyers, P.-E. et al. (2024a). Can cryogenic freezing preserve the quality of fruit matrices during long-term storage compared to the mechanical method? *Applied Food Research* 4 (1): 100374. Elsevier B.V.

Jha, P.K., Chapleau, N., Meyers, P.-E. et al. (2024b). Cryogenic vs. mechanical freezing impact on the quality of the sea bass (Dicentrarchus labrax) during long-term storage. *International Journal of Refrigeration* 160 (September 2023): 411–422. Elsevier B.V.

Khadatkar, R.M., Kumar, S., and Pattanayak, S.C. (2004). Cryofreezing and cryofreezer. *Cryogenics* 44 (9): 661–678.

Kiani, H. and Sun, D.-W. (2011). Water crystallization and its importance to freezing of foods: a review. *Trends in Food Science & Technology* 22 (8): 407–426. Elsevier.

Li, X.-Y., Yang, X.-L., Gai, R.-N. et al. (2022). Study of the dynamic characteristics of the food freezing process using a cryogenic immersion freezing tank. *Food Science and Technology* 42: 1–9.

Loayza-Salazar, S., Siche, R., Vegas, C. et al. (2024). Novel technologies in the freezing process and their impact on the quality of fruits and vegetables. *Food Engineering Reviews* 16 (3): 371–395.

Muthukumarappan, K., Marella, C., and Sunkesula, V. (2019). Food freezing technology. In: *Handbook of Farm, Dairy and Food Machinery Engineering* (ed. M. Kutz), 389–415. Elsevier.

Poudyal, R.L., Kobayashi, R., Suzuki, T. et al. (2019). Effect of different freezing and storage condition on the physical properties of protein coagulum (Firm Tofu). *International Journal of Refrigeration* 107: 11–19.

Swer, T.L., Mukhim, C., Sehrawat, R. et al. (2018). *Applications of freezing technology in fruits and vegetables. In Technological interventions in the processing of fruits and vegetables*, 133–162. Apple Academic Press.

Tomar, M.S. and Pradhan, R.C. (2023). Effect of hydro and cold room pre-cooling on cooling kinetics and post-harvest quality of amla. *Journal of Scientific & Industrial Research* 82 (08): 899–905.

Tomar, M.S. and Pradhan, R.C. (2024). Effect of ozone assisted hydro cooling system on cooling characteristics, microbial inactivation, and quality of fresh produce (amla and tomato). *Postharvest Biology and Technology* 218 (July): 113179. Elsevier B.V.

Xu, Z., Guo, Y., Ding, S. et al. (2014). Freezing by immersion in liquid CO_2 at variable pressure. *Innovative Food Science & Emerging Technologies* 22: 167–174. Elsevier Ltd.

6

Cold Revolution: Cryogenic Innovations in Dairy Production and Preservation

Revathi P.[1] *and Sudarshan Ramanathan*[2]

[1] *Department of Food Science and Technology, Arul Anandar College, Madurai, Tamil Nadu, India*
[2] *Division of Food Processing Technology, School of Engineering and Technology, Karunya Institute of Technology and Sciences, Coimbatore, Tamil Nadu, India*

6.1 Introduction

Globally, milk and its derivatives are essential dietary components, yet they spoil quickly. The dairy industry is vital to the world's economy and nutrition since it sustains livelihoods and supplies vital nutrients. Globally, dairy products are consumed, and demand is rising due to dietary changes and population expansion. Conventional preservation techniques, such as refrigeration and traditional freezing, frequently fail to preserve dairy products' ideal sensory and nutritional qualities over time. Energy inefficiency, microbiological contamination, and spoiling are issuing that traditional methods frequently fail to overcome. Supply chain complexity is increased by high operating costs, short shelf life, and transportation issues.

Cryogenic technologies are used in the food industry at low temperatures. In order to improve food quality and produce high-quality products, several industries are using cryogenic technologies these days. These technologies are typically utilized for food processing and preservation. One definition of cryogenics is the area of physics and engineering that focuses on creating freezing 'cold' and studying materials at extremely low temperatures. Cryogenic science is the study of the creation and consequences of extremely low temperatures, according to NIST (National Institute of Standards and Technology, US). The word's origins are in the Greek terms '*Kryos*', which means 'frost', and 'Genic', which means 'to produce'. Depending on the user's definition, it can refer to any temperature below freezing (Kumar et al. 2018).

Cryogenics has several uses in the fields of space research, electronics, automotive production, sports, and musical instruments, biological science, agriculture, and more, according to Balasubramanian et al. (2012). The use of cryogenic freezing in food, specifically in condiments and spices, is crucial. High-quality fine particles are produced as a result of the

Cryogenic Processing: Advances in the Food Industry, First Edition. Edited by Sumit Sudhir Pathak, Rama Chandra Pradhan, and Sudarshan Ramanathan.

technique, according to Kadam (2015). Furthermore, there is a notable increase in process volumes. Cryogenic gases are used to cool heat-sensitive materials during the grinding process to prevent temperature increases brought on by the process's heat. This implies that neither melting nor sticking of the powdered material is possible.

Scientist claims to have researched methods of preservation or processing that extend the shelf life. The engineering field of cryogenics studies the creation of cryogen and the upkeep of low-temperature technologies. It is also utilized for material transportation. Thermal damage was found to be one of the primary drawbacks of the traditional grinding method. Grinding or comminution under regulated temperature conditions is one of the practical ways to reduce thermal damage. Therefore, it is particularly crucial to grind at lower temperatures (Malkin and Guo 2007).

Compared to mechanical freezing, cryogenic freezing requires fewer set-up steps and produces food of higher quality, making it a rapidly emerging food processing technique. The mechanical freezing devices are very different from cryogenic food freezing. Indirect cooling is used in mechanical freezers, sometimes referred to as blast freezers. The products are frozen after the air around them is cooled by a refrigerant. On the other hand, cryogenic freezers use direct cooling. The product is directly sprayed with cold liquidized food-grade gases. These gases include liquid nitrogen, liquid carbon dioxide (CO_2), and solid CO_2 that are used in blast freezers or mechanical freezers that operate within a controlled temperature range of −35 to −40°. Cryogenic freezers are capable of reaching far lower temperatures. You can work in temperatures between −60 and −120° with liquid nitrogen and between −40 and −60° with liquid CO_2. Higher freezing speeds and much smaller ice crystals in the frozen product are the results of the much lower temperatures. Compared to standard mechanical or blast freezers, the product quality is higher due to the smaller ice crystals. The process of employing highly condensed gases to cool items to extremely low temperatures is known as cryogenics. Refrigeration at temperatures lower than 120 K (−153 °C) is required for cryogenics. Typically, a refrigeration cycle is used to reach cryogenic temperatures.

Cryogenic processing is a novel method of producing and preserving dairy products by using temperatures below −150 °C. The science and technology underlying cryogenics in the dairy industry are examined in this chapter, along with the consequences for environmental sustainability, economic efficiency, and quality assurance.

6.2 Cryogenics in Dairy Production and Preservation

Cryogenic freezing systems perform rapid and reliable freezing operations that prevent ice crystal formation, thus protecting milk fats and proteins. Cryogenic freezing technology maintains milk texture and flavour integrity while outperforming standard freezing methods in nutritional preservation. This preservation method works equally well for storing milk in liquid form and for extending the shelf life of dairy products like cheese, yogurt, and ice cream. The precise control of temperature to supervise crystal formation and microbial action at extremely low levels becomes vital for protecting product quality including its structure and taste characteristics. Cryogenic storage ensures that starter cultures in fermented dairy products remain active and potent, thus providing continuous high product quality.

Table 6.1 Cryogenic cooling vs. conventional cooling in milk processing.

Parameter	Cryogenic cooling	Conventional cooling
Cooling time (minutes)	2–5	20–30
Final temperature (°C)	−196	4
Energy consumption (kWh)	0.5–1.5	3–5
Microbial reduction (%)	99.5	85
Shelf-life extension (%)	30–50	10–20

Source: Adapted from Clarke (2020).

Cryogenic freeze-drying has emerged as a popular method to preserve powdered dairy products including newborn formula and milk powder. The process extends the product shelf life by preserving nutritional content and preserving sensory qualities. The combination of cryogenic techniques with advanced freezer and immersion systems has enhanced efficiency and scalability for dairy applications. IoT-enabled systems enable real-time monitoring and control, which enhances operational accuracy while reducing human errors in the process. The implementation of this technology resulted in a 30% reduction of spoilage as well as improved product texture for one major dairy producer, showing how the technology enhances both quality and operational performance. The smooth supply chain in dairy production benefits from cryogenic freezing, which is integrated throughout both the upstream and downstream production processes (Table 6.1).

6.2.1 Cryogenic Technology in Milk Processing and Preservation

Cryogenic technology in milk processing and preservation finds essential applications in the field. Cryogenic cooling extends the expiration date of raw milk and pasteurized milk while maintaining their natural taste. This technology particularly benefits organic and fortified milk by enabling better preservation of their high nutritional value. Cryogenic freezing prevents ice crystal formation, which damages milk proteins and fats through its rapid and uniform cooling mechanism. This technology plays a vital role in the long-term storage and transportation of products. Cryogenic freezing maintains milk's natural properties better than standard freezing methods by preserving its original flavour and texture while minimizing nutrient degradation.

The process of cryogenic freezing involves using liquid nitrogen (LN_2) and liquid carbon dioxide (LCO_2) to achieve rapid temperature reduction. The heat removal process during cryogenic freezing requires either direct or indirect contact between the product and the cryogen. Dairy products benefit from this faster freezing process because it creates smaller ice crystals which protect cellular integrity and maintain their natural quality better than traditional freezing methods.

The storage period for pasteurized milk allows milk to separate and develop slight grittiness when it thaws. It is important to choose small containers for freezing milk because the expansion process needs extra space at the top to prevent damage to the milk. The optimal window for consuming properly stored frozen milk lasts an entire month.

Before you serve the milk, you should shake or stir it properly to restore its original composition. You need to place the milk in the refrigerator for thawing purposes.

Cryogenic technology transformed the milk processing and preservation sectors by implementing advanced methods that rapidly chill and freeze milk while maintaining its quality and safety and nutritional properties. The traditional freezing and refrigeration techniques fail to protect the milk's delicate microbial and biochemical equilibrium despite their widespread usage. The process of rapidly reducing temperatures through cryogenic technology depends on liquid nitrogen or liquid carbon dioxide which operate at ultra-low temperatures. The method decreases heat exposure while preserving the milk's natural characteristics and extending its storage period so it becomes valuable for milk preservation and dairy product manufacturing. Dairy supply chains must adopt cryogenic technology to upgrade their operations while addressing the increasing market need for milk items.

6.2.2 Cryogenic Cooling for Milk Quality Preservation

The nutrient-dense milk composition quickly degrades through time while maintaining high susceptibility to microbial pollution. Milk preservation requires immediate cooling after milking to stop bacterial development and maintain its original quality. Cryogenic cooling provides the fastest method to lower milk temperature since it takes only seconds to reach required levels which exceeds all other cooling methods. The nutritious content and freshness of raw milk remain intact during its storage and delivery through fast chilling processes that prevent harmful bacteria and spoilage organisms from multiplying (Singh et al. 2021). The application of liquid nitrogen or carbon dioxide through cryogenic chilling allows milk proteins and lipids to maintain their original structure. Cryogenic freezing outperforms traditional refrigeration because it prevents lipid oxidation and protein denaturation which would otherwise occur with time. Cryogenic freezing creates extremely low temperatures that halt lipolysis and proteolysis reactions, which would otherwise cause spoilage and develop unpleasant tastes.

6.2.3 Applications in Milk Processing

The field of milk processing receives valuable benefits from cryogenic technology, which enables new dairy product generation. One essential application exists in the production of frozen milk concentrates. Through cryogenic freezing water-content elimination in milk alongside nutrient preservation enables fast milk concentration. High-quality milk concentrates for medical and specialty nutritional products require cryogenic processing for optimal production (O'Connor and Lalor 2022). Milk components benefit from cryogenic technology through their microencapsulation production. The stability and controlled release of beneficial substances, including vitamins and minerals, and probiotics, occurs through their encapsulation inside milk matrices using cryogenic freezing. The functional properties of dairy products and milk-based beverages improve through this method. The technology enables effective preservation of colostrum since this nutrient-rich milk requires meticulous handling to keep its immunoglobulins and bioactive ingredients intact. Cryogenic freezing serves as a main technology for creating unique dairy products including nitrogen-infused beverages. Cryogenic systems provide the essential freezing and infusion capabilities to develop dairy products with innovative textures, which match consumer preferences (Patel et al., 2020).

6.2.4 Energy Efficiency and Environmental Benefits

The energy efficiency of cryogenic technology surpasses that of traditional cooling and freezing systems by a significant amount (Table 6.2). Energy consumption decreases through the direct application of carbon dioxide or liquid nitrogen as cooling agents because mechanical refrigeration systems become unnecessary. The use of cryogenic technologies delivers maximum benefits to dairy farms that operate in remote locations without reliable electric power. The research by Rahman and colleagues published in 2023 demonstrates that cryogenic systems enable fast cooling with minimal energy requirements for optimizing large-scale milk processing in agriculture.

The application of cryogenic technology achieves superior sustainability goals through enhanced food preservation. Sustainable dairy supply chains experience reduced waste because cryogenically processed milk maintains its quality for extended periods, which decreases spoilage. The growing implementation of cryogen recovery and reuse technologies enhances the environmental performance of cryogenic systems compared to traditional refrigeration methods because it minimizes their carbon footprint.

6.2.5 Challenges and Future Perspectives

Milk processing and preservation operations encounter multiple obstacles despite the various advantages of cryogenic technology. Small dairy farms encounter limitations from the expensive start-up expenses of cryogenic systems and the necessity for specialized equipment. The handling and storage of cryogenic substances such as liquid nitrogen needs expert knowledge along with rigorous safety protocols. The dairy industry requires affordable and user-friendly cryogenic solutions to eliminate current restrictions (Mehta and Kulkarni 2021).

Ongoing research and innovation address these limitations through two main approaches: discovering new cryogens and developing scalable cryogenic systems. The next generation of cooling systems uses a combination of cryogenic and conventional refrigeration to achieve better cost management together with operational efficiency. Modern automation technology and digital monitoring systems improve the accuracy and dependability of

Table 6.2 Energy consumption in cryogenic freezing vs. conventional freezing.

Dairy product	Cryogenic freezing energy (kWh/ton)	Conventional freezing energy (kWh/ton)
Milk	25–30	60–80
Cheese	50–60	90–110
Ice cream	40–50	70–100

Source: Adapted from O'Connor and Lalor (2022).

cryogenic processes to achieve stable milk quality standards. Cryogenic technology will gain importance in the dairy industry because customers increasingly seek premium dairy products with minimal processing.

6.3 Cryogenic Processing of Dairy Products

Cryogenics serves to improve the sensory properties of ice cream, yogurt, and cheese products. The precise control over crystallization and microbial activity at ultra-low temperatures maintains the structural integrity and taste profile of these products.

6.3.1 Cryogenic Technology in Cheese Processing

Cryogenic freezing techniques use low temperatures to maintain cheese quality by stopping enzymatic decay and keeping original structure intact. Cryogenic freezing technology revolutionized cheese production and preservation by addressing fundamental quality preservation and extended shelf-life requirements. The cheese rapid freeze process uses cryogenic mediums such as liquid nitrogen at −196 °C and liquid carbon dioxide at −78.5 °C. When cheese undergoes rapid freezing in cryogenic conditions, large ice crystals do not form, which protects the cheese's internal structure and taste. When cheese is stored at extremely low temperatures, it stops microbial growth and enzyme reactions, thus preserving its nutritional content and taste properties during extended duration

The process of using cryogenic cooling on curd both slowed down its acid formation and minimized the loss of whey. The higher cheese yield (4.8% during brining and 3.8% at ripening) occurred only because of better water retention in the product. Cryogenic parameters can enhance cheese production through improved moisture retention as well as better milk component recovery. One of the primary benefits of cryogenic freezing in the cheese-making process is the ability to protect the product's functional characteristics. Cryogenic technology demonstrates an advantage for soft cheeses such as mozzarella and cream cheese because it allows for rapid temperature reduction, which prevents texture problems during standard freezing methods. The preservation method maintains the essential characteristics of cheese including creamy texture and stretchability and meltability, which are vital for industrial and food service applications. The cryogenic process maintains better quality in hard and semi-hard cheeses by preventing drying and cracking, which often occurs during slow-freezing techniques (Goff and Hartel 2013).

Cheese personalization and portioning have seen progress through cryogenic freezing, which has made advanced solutions available. Cryogenic technology allows cheese blocks to undergo partial freezing, thus creating solid structures that enable precise slicing and grating while maintaining product standards. The food industry has found this technology particularly useful because it enables the development of ready-to-use cheese formats for pizzas as well as sandwiches and other prepared meals. Cryogenic freezing provides an optimal solution for handling specialty and artisanal cheeses that possess unique moisture and fat content properties and tend to be more sensitive to cold temperatures.

The development of cryogenic freezing technology focuses on lowering expenses and enlarging production capabilities through energy management systems and automated processing. Hybrid systems that unite cryogenic freezing with mechanical freezing show high effectiveness by maximizing the positive aspects of both methods. Cryogenic freezing benefits from enhanced moisture retention and minimal oxidative harm through novel packaging methods which include vacuum-sealed and modified atmosphere packaging (MAP). The market requires cheese products that are durable while maintaining high quality and providing convenience according to consumer preferences. The cheese production industry will maintain cryogenic freezing as an indispensable method because it continues to evolve in the sector.

6.3.2 Cryogenic Processing in Starter Culture Preservation

The making of fermented dairy products depends on having starter cultures. These microbes' potency and viability are preserved via cryogenic storage, guaranteeing a constant level of product quality. In situations requiring consistent outcomes on a large scale, this approach works very well. The foundation for making cheese, bread, beer, and yogurt is the preservation of starter cultures in the food and fermentation industries. Fermentation depends on starter cultures, including moulds, yeast, or bacteria, to keep final items safe and of good quality. For keeping start cultures alive, working and stable for an extended time, cryogenic processing is now the best technique used.

6.3.2.1 Importance of Starter Culture Preservation

Starter cultures play a part in making food safer, tastier, and easier to eat. We have to keep these cultures healthy and functional by utilizing methods that avoid harming cells, as these cultures are naturally weak to changes in the environment. Even though refrigeration and freeze-drying are often used, they can, over some time, have a negative effect on the microbes' survival or some of their features. By using extremely low temperatures that provide great viability and stability, cryogenic processing overcomes these difficulties.

6.3.2.2 Cryogenic Processing Technique

Cryogenic processing involves the use of cryogens such as liquid nitrogen (LN_2) to rapidly cool microorganisms to temperatures as low as −196 °C. Because temperature is so low, molecular activity within the cells is almost stopped. The project can generally be described as having four key steps.

- Preconditioning: Before freezing, starter cultures are treated with cryoprotectants such as glycerol or trehalose, which prevent the formation of ice crystals and protect cell membranes during freezing.
- Freezing: To achieve rapid cooling, cultures are either immersed directly in LN_2 or exposed to liquid nitrogen vapour, which reduces the formation of large ice crystals that can mechanically damage cell structures. Cultures remain stable for years in LN_2 vapour-phase systems or in freezers that cool with nitrogen gas, preventing genetic changes or any loss of their functions.
- Thawing: Controlled thawing is necessary to prevent thermal shock and ensure maximum recovery of viable cells.

6.4 Cryogenic Drying: A Revolution in Milk Powder Production

A growing number of companies are using cryogenic conditions for freeze-drying to preserve milk powder and infant formula. Using this technique keeps the flavour and vitamins of the product for much longer.

6.4.1 Cryogenic Drying in Milk Powder Processing

During cryogenic drying, the dryer uses cold temperatures and controls the process where moisture turns straight from a liquid to gas to separate it from milk. This method involves freezing the milk into a solid-state using cryogens like liquid nitrogen (−196 °C) and then subjecting it to a vacuum to sublimate the ice directly into vapour without passing through the liquid phase. Quick freezing prevents damage to the parts of the milk that matter for its usefulness and nutrients. This drying method, unlike conventional spray drying, helps keep vitamins, bioactive proteins, and essential amino acids protected from high temperatures. For this reason, milk powder companies are greatly helped during the process of making high-quality milk powders.

There are many ways milk powder production benefits from cryogenic drying. Using this method forms small particles of uniform size, which gives the end product better solubility and ability to rehydrate. Because of the extremely low temperature, the natural flavour, colour, and aroma in milk are preserved, creating superior-tasting cryogenically dried powder. Another plus is that cryogenic drying significantly reduces residual moisture and therefore slows microbial activity and the process of milk. Due to these features, cryogenic drying is the best choice for creating premium infant formula, nutritional supplements and specialty milk powders for those focusing on their health.

Milk powder can be treated with cryogenic drying in many different industries. It's often added to fortified and functional powdered milks to meet the need for health-beneficial foods. The method is also applied to make dairy ingredients for beverages, sweets, and baked goods where solubility and stability are most important. Yet, using frozen storage for drying can be expensive and it requires a lot of energy. Efforts are being made to enhance the overall process and test ways to mix cryogenic drying with traditional approaches for improved cost. Because more people want special and nutritious dairy products, cryogenic drying will likely get more attention in the dairy industry (Table 6.3).

Table 6.3 Cryogenic drying vs. spray drying in milk powder processing.

Parameter	Cryogenic drying	Spray drying
Processing time (minutes)	10–15	30–50
Retained protein (%)	98	85
Energy consumption (kWh/kg)	1.5–2.5	4–6
Moisture content (%)	2–3	4–5
Shelf life (months)	24–36	12–18

Source: Adapted from International Dairy Federation (2019).

6.4.2 Cryogenic Drying Processing in Dairy Products (Milk Powder)

The new methods in cryogenic drying, processing dairy products, mainly milk powder, have been transformed. Although spray drying and freeze-drying are highly effective, they frequently result in less high-quality milk powder. On a different note, drying milk using cryogenics keeps its nutritional, sensory, and useful qualities with help from low temperatures and controlled sublimation. Using this technology, dairy processing gains higher product quality, saves energy, and offers more flexibility in the way operations are conducted. Milk powder is important in the world food industry and commonly appears in infant milk, candy, and baked foods. Ensuring valuable nutrients are preserved in high-quality milk powder depends on carefully drying away the water. Traditional drying uses high heat which may harm vitamins, proteins, and probiotics in milk. Cryogenic drying addresses this issue by employing low temperatures and sublimation to remove moisture, bypassing the liquid phase and preserving the native structure of milk components (Clarke 2018).

To produce better milk powders with easy solubility, dispersibility, and good emulsification and foaming properties, manufacturers often dry liquid or pre-concentrated milk by freezing it, placing it under vacuum at cryogenic temperatures and then processing it further. A dry, powdery product with minimal thermal damage is left behind after the frozen water in the milk sublimates directly into vapour (Nagendra and Radha 2020). Again, the uniformity of particles achieved by cryogenic drying ensures the powder is fit for use in industrial environment.

6.4.3 Preservation of Nutritional and Bioactive Compounds

Maintaining the nutritious value and bioactive parts of milk is a main benefit of cryogenic drying. Under high temperatures, vitamins A, D, and B_{12}, immunoglobulins, lactoferrin, and other similar nutrients usually break down. Using low temperatures in a cryogenic drying process means the molecules are not harmed and help to keep nutrients in better condition. For specialist applications where nutrient retention is crucial, such newborn nourishment and medical meals, cryogenically dried milk powder is very advantageous (O'Leary and Bansal 2016). Adding probiotics and helpful bacteria to milk powders preserves their viability if special low-temperature methods called cryogenic drying are applied. Since probiotics are very easily killed by heat, typical drying methods often reduce their performance. Since cryogenic drying lowers heat and oxidative stress, microorganisms can prosper, so the probiotic activity in the milk powder is maintained. It becomes especially crucial for dairy items aimed at improving your immune system and gut.

6.4.4 Energy Efficiency and Environmental Sustainability

The use of cryogenic drying boosts product quality and also reduces the environmental impact of drying milk for powder. Typically, traditional drying methods such as spray drying need much heat to evaporate water from food. Cryogenic drying, even though it is expensive at first, helps companies save energy later by using less power to dry things quickly. Using low temperatures for milk processing calls for less air conditioning, which helps decrease both energy use and damaging CO_2 emissions (Ahn and Lee 2017). Along with being energy-efficient, cryogenic drying adds to worldwide interest in sustainable ways to process food. By using closed-loop systems, cryogenic drying produces little waste and allows for the recycling of cryogens, raising its eco-friendly status. New ways to power dairy plants with renewable energy and use green cryogens are allowing the industry to become more sustainable and within reach of many processors.

6.4.5 Challenges and Innovations in Cryogenic Drying

Cryogenic drying is valuable, but some issues still keep it from being adopted by many dairy companies. Because the systems needed for cryogenics and the production of liquids like liquid nitrogen are pricey, many small and medium-sized companies find them unaffordable. Besides, finding the right drying settings for various milk types and products takes a lot of work and knowledge. Because cryogenic drying moves liquid milk more slowly than conventional options, this is a challenge for processing large dairy volumes (International Dairy Federation 2019). Yet, new developments are helping to overcome these problems. A blended approach using cryogenic drying and traditional methods for drying such as spray drying is under development to help save money and increase efficiency. Automation and high-tech monitoring now guarantee that drying is done in any needed way, resulting in consistent products. There are currently efforts to develop environmentally friendly new cryogens for the technology. As new innovations develop, cryogenic drying will become an increasingly attractive choice for big milk powder producers.

6.5 Ice Cream and Frozen Desserts

Ice cream has an especially smooth texture because the large ice crystals are avoided. The strategy allows for innovative types of frozen products such as those enriched with nitrogen. Ice-cream and frozen-dessert makers have found that using cryogenic technology helps them achieve much faster freezing than older methods. Applying clothes to liquid nitrogen or carbon dioxide freezes them right away, which helps avoid the formation of large ice crystals. Therefore, cryogenic freezing lets foods keep a smooth and creamy consistency that outperforms what is achieved with regular freezing. By freezing quickly, the air cells kept in the mix are maintained, which results in the proper softness and lightness of the completed product. In addition, the colder temperatures hold onto the compounds responsible for the wine's scent and fruity taste. When freezing desserts with delicate changes such as with fruits, nuts or chocolate, using cryogenic freezing helps them keep their shape.

Benefits of cryogenic technology are available for both small and large manufacturers. Liquid nitrogen is part of the process in creating fresh, individual desserts, bringing a theatrical touch to everyone's culinary experience. Because they offer uniform produce and performance, cryogenic tunnels are great for performing large cooling and freezing operations. However, using this technology comes with issues such as big costs, a lot of needed energy, and safety concerns related to cryogens. Notwithstanding these drawbacks, research into improving cryogenic freezing's sustainability and energy efficiency is motivated by consumer demand for high-end frozen desserts and novel product formats. This type of technology is well-positioned to contribute to more interesting frozen desserts that meet new demands from diners.

6.5.1 Role of Cryogenic Freezing in Texture and Structure Preservation

Customer satisfaction with ice cream and frozen treats depends greatly on how they feel on the tongue. Slow-frozen items often result in big ice crystals which gives the product an uneven texture. Micro-sized ice crystals are guaranteed to form to cryogenic freezing's incredibly quick cooling. This improves the product's overall mouthfeel while maintaining its creamy, smooth texture. The technique permits you to manage aeration exactly which is key to properly mixing in the air for the right texture. The lightness and palatability of frozen desserts are influenced by maintaining a steady and even air dispersion (O'Leary and Bansal 2016). Cryogenic technology is especially useful for making novelty ice-cream products, as looks and texture matter the most. The valuable freezing property of liquid nitrogen allows for the making of many unique shapes and looks in food. What's more, using cryogenic freezing helps stop nuts, fruits, and chocolate chips from migrating or sinking in food items by helping the extras stay in one place.

6.5.2 Enhancing Product Quality and Flavours

Dairy-based sweets and ice creams special are their strong flavour and texture. Cryogenic freezing helps prevent dairy products from developing oxidative deterioration which can cause them to lose their good taste (Table 6.4). By quickly freezing the food, we stop the compounds that give it taste and smell from changing, which means the thawed food will still taste and smell right. It matters a lot for makers of artisanal frozen sweets and luxury ice creams, as their delicate and natural tastes are valued by customers (Nagendra and Radha 2020). Also, using cryogenic freezing stops the rotten ice-lactose crystals from forming, so the dessert tastes and feels better. Because cryogenic handling maintains the quality of milk's proteins and fats, the richness and creaminess are both retained. Using this device, it is possible to add unusual taste notes from fruit purees and alcohol-made infusions without changing their flavour appeal.

Table 6.4 Effect of cryogenic freezing on ice-cream quality.

Parameter	Cryogenic freezing	Conventional freezing
Overrun (%)	50–60	40–50
Ice crystal size (μm)	10–20	50–100
Texture smoothness (%)	90	70
Flavour retention (%)	95	80

Source: Adapted from Sharma et al. (2023).

6.5.3 Energy Efficiency and Operational Flexibility

When it comes to frozen dairy desserts, cryogenic freezing helps save energy and cut down on operations. Unlike previous mechanical methods, cryogenic processes make freezing efficient and very quick. Not needing mechanical refrigeration because of liquid nitrogen or carbon dioxide is better for energy savings and makes everything less complicated. It is especially useful for those who create small quantities and would like better freezing results without being deterred by expensive traditional plants (Ahn and Lee 2017). Such systems make it possible for production processes to be very flexible. Adding modular cryogenic freezing units to current lines means manufacturers can respond to changes in demand and expand their production. They also allow for producing small quantities of frozen desserts to fit the needs of niche diets like those avoiding lactose, animal products, or wanting added probiotics. Because cryogenic technology is flexible, manufacturers can easily expand their product lines to suit any new trends.

6.5.4 Challenges and Innovations in Cryogenic Freezing

Adopting cryogenic technology in desserts and ice cream is not free of difficulties. Because liquid nitrogen and special equipment for cryogenic systems are so expensive, starting small is often impossible. It is also important to closely manage temperature and timing in the rapid freezing because flaws in the product can occur easily. To make sure consumers accept liquid nitrogen, those in the industry need to properly explain its benefits (International Dairy Federation 2019). Innovators keep working to improve cryogenic systems so they are more environmentally friendly and use energy more sensibly. For example, researchers are currently making hybrids of cryogenic and mechanical freezing machines to boost their cost and energy efficiency. To recent advancements, real-time sensor data enable quick adjustments to freezing conditions to assure the same quality of every batch. Work is being done to explore alternative cryogens that are more environmentally friendly and fit the industry's sustainability objectives. As these developments advance, cryogenic technology should grow in popularity and be used by many businesses in the frozen dessert industry.

6.6 Cryogenic Technology in Fermented Milk Processing

Many industries such as food and dairy processing, increasingly depend on cryogenic technology because it uses super-low temperatures (below −150 °C). New technologies in fermented milk have led to a shift in normal production steps for yogurt, kefir, and buttermilk. Due to their sensitivity to change in temperature, meat products should be frozen using advanced methods to ensure they stay fresh, safe and have their probiotic properties. Using LN_2 or LCO_2 from cryogenics lets you quickly cool and freeze products, keeping the flavours, shape, and microorganisms intact. Conventional freezing doesn't keep proteins, fats, and live cultures unchanged, whereas cryogenic freezing does, allowing for long-lasting and quality-rich food (Clarke 2018).

6.6.1 Preserving Product Quality Through Rapid Freezing

Improving the quality of fermented milk by means of rapid freezing using cryogenic technology is a main use of this technology. Larger ice crystals that appear with traditional methods usually harm the microstructure of fermented milk, resulting in changes in texture and the separation of water (whey). Unlike all other methods, cryogenic freezing causes liquids to change extremely fast into a solid state to form microcrystals. It preserves the consistency, smoothness, and the shape and strength of the product. Besides, quickly freezing milk lessens the chance of its proteins denaturing and its fats oxidizing, meaning it keeps its valuable nutrients and pleasant flavour. Frozen yogurt benefits a lot from cryogenic technology, because it avoids making the product grainy when stored (O'Leary and Bansal 2016). Stopping curd instability and phase separation with cryogenic freezing leads to improved texture and appearance in fermented milk products. Maintaining the milk matrix's uniformity guarantees that the finished product will live up to consumer expectations. As a result, this way of making frozen treats lets producers add fermented milk to their products without compromising taste or their beneficial effects. Cryogenic technology allows distinctive goods to stay different from others by preserving their original flavours and textures.

6.6.2 Ensuring Microbial Viability and Probiotic Efficacy

The health benefits of fermented milk depend greatly on the viability of probiotics, including *Lactobacillus acidophilus* and *Bifidobacterium bifidum*. They help maintain the health of your gut, your immune system, and your general health. Still, freezing food for a long time can be very rough for probiotics and may remove most of them. Cryogenic technology uses a specially controlled setting to decrease stress as samples are frozen and stored. Swift cooling during cryogenic freezing saves probiotics from damage caused by ice crystals penetrating their cell walls (Nagendra and Radha 2020).

Cryogenic technologies also promote the establishment of new probiotic-rich products with better shelf life and functionality. For example, freeze-dried probiotic cultures can be processed into fermented milk products and then cryogenically frozen to maintain functionality. This practice provides consumers with probiotic functionality and additional expiration on the product. Furthermore, cryogenic freezing supports live microbial culture preservation during distribution and storage, and allows for global transportation of microorganisms in high-quality fermented milk products.

6.6.3 Energy Efficiency, Sustainability, and Operational Benefits

Cryogenic technologies provide sustainability benefits and energy economy relative to conventional freezing methods. When employed in dairy manufacture, large-scale cooling methods offer significant energy savings from reduced total freezing time, employ less mechanical energy than conventional refrigeration systems, and support the dairy industry sustainability initiatives aimed at reducing greenhouse gas emissions and operating costs (Ahn and Lee 2017). Cryogenic technology's versatility over a broad range of production scales is another important advantage. Cryogenic freezing systems can be customized to satisfy particular needs, whether they are used in industrial-scale operations or artisanal dairy set-ups. For example, automated cryogenic tunnels make it possible to consistently freeze huge quantities of fermented milk products at precise temperatures. Additionally, these systems may be linked with current manufacturing lines to maximize productivity and minimize disruption. Developments like hybrid freezing systems, which blend mechanical and cryogenic refrigeration, increase the technology's viability from an economic standpoint.

6.6.4 Challenges and Future Prospects

Despite its numerous advantages, the adoption of cryogenic technology in fermented milk processing faces several challenges. The high cost of cryogens like liquid nitrogen and the need for specialized equipment can deter small- and medium-scale producers. Additionally, optimizing the freezing parameters for different fermented milk products requires extensive research and technical expertise. Variability in the composition and microbial content of fermented milk adds complexity to the standardization of cryogenic processes. Consumer perception of advanced food processing technologies also necessitates transparent communication about the safety and benefits of cryogenic methods (International Dairy Federation 2019).

To address these challenges, continued research and innovation are focusing on making cryogenic technology more accessible and sustainable. Our research identified a number of emergent trends such as eco-friendly cryogens with less negative environmental impact and smart sensors to capture real-time fluctuations in freezing parameters. Automation and artificial intelligence are also providing more precise control of cryogenic systems irrespective of the product lines. As the demand for functional dairy products increases worldwide, cryogenic technology is likely to be instrumental in satisfying consumers' requests for dairy products and enabling sustainability in the dairy industry.

6.7 Cryogenic Preservation of Cultures and Enzymes

Starter cultures vital for the production of fermented dairy products can also benefit from cryogenic storage to maintain viability and potency to provide quality assurance. The enzymes introduced into cheese and other dairy products can maintain activity when cryogenically stored. The temperature of the environment prevents the biochemical degradation of the cultures and enzymes, so the shelf life can be longer. These advances are necessary for artisanal and premium dairy products, where the consumers' satisfaction is highly influenced by the consistency of microbial and enzymatic properties.

6.7.1 Storage and Transportation of Dairy Products

The processes of cryogenic storage and transportation can revolutionize the dairy industry, since they can provide enhanced storage and transportation processes for perishable products. Cryogenic systems have become an integral part of diverse modern dairy supply chains; current examples in the dairy industry include high-value products such as ice creams, specialty cheeses, and probiotic dairy drinks.

6.7.2 Applications in Cryogenic Storage

For certain dairy products, cryogenic storage provides unparalleled benefits. In contrast to traditional refrigeration, cryogenic systems freeze quickly, avoiding the development of big ice crystals that could alter the texture of the final product. Rapid cryogenic freezing guarantees a smooth, creamy texture for frozen dairy products like ice creams while preserving ingredients like fruits and nuts in perfect condition. Cheeses benefit from cryogenic storage because it reduces biological oxidation processes, especially those aged for particular flavours. When cryogenically preserved, butter and milk powders also maintain their organoleptic and functional qualities, guaranteeing continued high quality (Dey et al. 2022).

Another important use is in the preservation of delicate bioactive products, like supplemented milk and colostrum. Probiotics, enzymes, and immunoglobulins – all of which are abundant in these products – degrade under normal refrigerated conditions and are heat-sensitive. These bioactive ingredients are preserved through cryogenic freezing at extremely low temperatures, allowing for the creation of valuable functional dairy products.

6.7.3 Applications in Cryogenic Transportation

Cryogenic transportation is essential for ensuring that dairy products maintain their integrity during long-distance distribution (Table 6.5). Insulated containers equipped with cryogenic cooling systems ensure that perishable items, such as bulk milk, frozen desserts, and specialty dairy products, are transported under ideal conditions. The ability to maintain consistent ultra-low temperatures throughout transit minimizes microbial growth and enzymatic spoilage, especially for raw milk transported from farms to processing facilities. A particularly important advancement in cryogenic transport is the incorporation of mobile cryogenic containers with real-time monitoring. These containers provide control over retention of both temperature and humidity to maintain the integrity of the product, even with extreme temperature fluctuations from ambient conditions. For frozen dairy exports, especially ice cream and frozen yogurt, this technology is a significant advantage, with product texture and flavour being preserved until it can be sold to consumers in international markets (Yadav et al. 2023).

Table 6.5 Economic impact of cryogenic storage in dairy industry.

Factor	Cryogenic storage	Conventional storage
Initial investment ($/ton)	1,500–2,500	1,000–1,500
Operational cost ($/month)	500–700	800–1,200
Shelf-life extension (months)	24–36	12–18
Reduction in spoilage (%)	40–60	20–30

Source: Adapted from Mehta and Kulkarni (2021).

6.7.4 Benefits of a Cryogenic System

Opened cryogenic storage and transit systems provide the following benefits over conventional refrigeration:

- Goods are quickly brought to ideal temperatures, which also decreases microbial activity, while preservation of quality.
- Steady temperature maintenance preserves the essential texture, flavour, and nutritional value of products at very low temperature.
- Cryogenic systems are considered more sustainable, as they often require less energy than more traditional refrigeration methods.
- Less product is wasted throughout the entire supply chain from longer shelf life due to less spoilage.
- Cryogenic freezing protects material from structural damage, especially for costly items like cheese and frozen desserts (Sharma et al. 2023).

6.7.5 Sustainability and Future Implications

Cryogenic technology upholds the sustainability principles of the dairy industry by both minimizing energy consumption and reducing food wastage. All indications are that consumer demand for high-quality items that are minimally processed will continue to increase, suggesting a broader and broader use of cryogenic systems in larger volumes. Advancements in material sciences will enhance the possibilities for improved insulation in cryogenic containers. In addition, with a growing trend towards renewable resources in cryogen production, the practical efficacy and environmental sustainability of cryogenic technology will only improve.

6.8 Cryogenic Technology in Value-Added Dairy Products

Cryogenic technology can also facilitate the exploration of innovative and value-added dairy products geared towards shifting consumer preferences.

6.8.1 Nitrogen-Infused Milk Beverages

Cryogenic technologies allow for the infusion of nitrogen into brand-new milk-based beverages to achieve creamy texture differentiation and sensory appeal. Nitrogen-infused milk beverages are a novel product combining traditional dairy products and advanced cryogenic technology that provide a unique sensory experience. Infusion of nitrogen gas into milk, gives the beverage a creamy feel and smooth flavour. The infusion process involves adding nitrogen, which produces smaller bubbles than carbon dioxide, creating a less aggressive fizziness and smoother mouthfeel.

The utilization of cryogenic technology in this example is very important. Liquid nitrogen, which is at very low temperature, is added to the milk through a controlled method. As the nitrogen changes from a liquid state to a gas, it creates micro bubbles that get suspended in the liquid matrix. The addition of these bubbles changes the mouth feel and the overall aesthetics of the beverage, giving a swirling visual experience as seen in some nitrogen-infused coffee and teas. The advantages to nitrogen infusion are more than just the sensory effects. As nitrogen is an inert gas, it doesn't readily react with other materials. This allows nitrogen to displace oxygen in the packaging, leading to less oxidation and possibly improved shelf life for the milk beverage. Furthermore, as nitrogen does not dissolve into liquid like carbon dioxide, it won't produce carbonic acids that affect mouth feel and acidity.

That being said, there are several considerations regarding the safety procedures for using nitrogen infusion in dairy products. Liquid nitrogen is extremely cold, and mishandling could result in significant injuries, including frostbite. There are also health considerations for the consumer, as nitrogen vapour could potentially trigger breathing issues, especially for an individual with asthma and possible inflammation of the lungs. So, manufacturers should make sure that the nitrogen is added to the beverage completely and ensure that any residual gas is properly dissipated to avoid problems in this area. The nitrogen-infused milk beverages have plenty of potential in the market, especially with consumers looking for a new and unique dairy product experience. This unique aesthetic coupled with a more desirable texture and mouth feel puts the nitrogen beverages in an enviable market position amongst other value-added dairy products. As consumers continue to be curious and willing to venture into new food/beverage experiences, nitrogen-infused products will more than likely be appealing, especially to millennials who tend to look for products that provide an experience and a bit of surprise.

6.8.2 Nutrient Encapsulation

Achieving cryogenic freezing allows for microencapsulation to occur, which means more bioactive compounds can be used – probiotics, omega-3 fatty acids, vitamins, and so on – which can improve the nutritional content of functional dairy products. Cryogenic processing has developed into a growing practice within the dairy processing sector, mainly pertaining to the encapsulation of nutrients and improving the functional properties of dairy products. Doing this involves extremely low temperatures to encapsulate bioactive compounds, such as vitamins, minerals, and probiotics, to protect them against degradation during processing and storage. Moreover, such encapsulation maintains stability and bioavailability of bioactive nutrients to provide healthier bottled products. The benefits of encapsulating probiotics through milk proteins have improved their viability in fermented dairy products, ensuring a sufficient quantity of live microorganisms to reach the gastrointestinal tract to suggest potential health benefits.

The use of cryogenic encapsulation technologies has many important advantages. As a result of cryogenic conditions leading to quick freezing, smaller ice crystals develop, limiting the damage of the encapsulating material and the encapsulated product (naturally a big concern for preserving structural integrity of bioactive material). In addition, the cryogenic encapsulation process minimizes the sensory impacts of the dairy matrix and encapsulated nutrients, ensuring that no undesired modifications to the flavours, colours, or textures occur. For example, two-stage emulsions, water-in-oil-in-water, can be used with dairy products to encapsulate iron, without the typical oxidation, and off-tastes associated with iron fortification). In addition to the temperature reductions that surrounded by vitamins, nutraceuticals, marine oils associations, etc., some studies have shown it improves the incorporation of the different ingredients into dairy-based matrices. The studies used spray-chilling and spray-cooling processes, which included cryogenic temperatures to prepare lipid-coated active agents for encapsulated flavours and bioactive materials. With these methods, it was look at gastro-intestinal digestive controls of the release and ultimately uptake of the encapsulated nutrients at the target site in the gastrointestinal tracts. Milk proteins have been effective as encapsulating agents because of their biocompatibility and ability to form gels at low temperatures, allowing encapsulated bioactives to experience protection from a gelatinous matrix.

6.8.3 Colostrum Preservation

Colostrum needs to be handled with extreme care since it is full of immunoglobulins. Cryogenic freezing is an effective method of preserving bioactivity as well as the shelf life of colostrum (Mehta and Kulkarni 2021). The importance of cryogenic technology in preserving bovine colostrum is evident in its use to preserve the immunological quality and nutritional value of colostrum. Colostrum (the first milk secreted after calving) is rich in antibodies, growth factors, and nutrients needed by newborn calves. Because colostrum is a perishable product, the use of effective preservation methods is essential for the quality and effectiveness of colostrum.

Freezing is one of the most common cryogenic techniques, as freezing colostrum at less than −18 °C essentially stops the growth of microorganisms as well as enzymes, and this preserves the concentration of immunoglobulins as well as overall nutritional value. Research shows that freezing does not decrease the levels of antibodies, and it is considered a viable, long-term storage option for colostrum; however, a frost-free freezer should be avoided, as it will freeze and thaw colostrum, exposing it to temperatures that may reduce the quality of colostrum. An advanced method of cryogenic preservation is lyophilization, also known as freeze-drying. This method freezes the colostrum and then reduces the normal pressure so that the frozen water can sublimate directly from the solid to the gas state. The bioactive proteins, as well as immunoglobulins, are well preserved by lyophilization. It improves stability and shelf life of the dry colostrum, which does not need refrigeration. Studies have demonstrated that the freeze-dried colostrum is bioactive and safe for human consumption and nutraceuticals (Vasilenko and Zhdanov 2020).

6.8.4 Environmental and Economic Considerations

Cryogenic systems typically consume less energy than conventional refrigeration processes. Consequently, operational costs and carbon footprints are reduced. Quick freezing also minimizes energy waste and an energy-efficient process better supports food systems at scale, which has positive impacts on food security and sustainability. When cryogenic systems also enhance product quality and shelf life, this further minimizes spoilage and waste across the supply chain. Initial capital costs for a cryogenic process can be high, but overall, there are long-term savings at a given facility over the lifespan of the product, as well as productivity gains that further justify diversion to this technology. Less energy consumption leads to reduced energy bills; less spoilage leads to less waste disposal; and better-quality products lead to greater profitability. Carbon-neutral and renewable-energy carbon systems are being researched to eventually address sustainability in respect of liquid gas production and developing energetically neutral systems.

6.9 Technological Evolution of Cryogenic Systems

The evolution of sophisticated cryogenic equipment (i.e. spiral freezers, flash freezers, liquid nitrogen immersion systems) helped to promote the use of cryogenics in dairy production. IoT-enabled cryogenic systems can accurately monitor the food's freezing and temperature parameters accurately. Innovative AI algorithms can provide operational efficiency and real-world time to maintenance prediction. New materials for cryogenic storage, including new insulation and eco-friendly refrigerants, will reduce the total cost of ownership and improve the sustainability and scalability of cryogenic systems. The technological vendors and product producers have created boutique solutions to meet specific brewery/producer needs through collaboration.

6.10 Challenges and Future Prospects

Cryogenic technology has benefits, but is stalled during the adoption stage due to a number of challenges such as the capital cost of equipment, the need for technical expertise, and regulatory frustrations. The capital cost for cryogenic equipment and infrastructure will be prohibitive for many smaller dairy producers without support from a dairy processor. Some act of financing is probably necessary (federal or provincial) to help fruit producers adopt cryogenic storage technology. The technical and operational and maintenance of cryogenic systems will require specialized training. Many companies currently collaborate with technology vendors to help the brewery producer provide specialist training. Environmental constraints and food safety standards may restrict or delay the use of cryogenic technologies in a particular business or facility. If everyone was able to use the same standards in every country, we think that implementation would be much easier; however, let alone the same standards would be easier to adopt a single standard, but the same standards across borders have to be adopted first.

Research to increase the effectiveness of cryogenics, lower its costs, and expand its uses in the dairy business and other fields is ongoing. As new developments appear in cryogenic materials and designs, they will lead to better access for consumers and business users alike. Government funding and partnerships with industry will help encourage the use of cryogenic technology for the dairy sector. The use of partnerships between the public and private sectors will enable funding for pilots and demonstrations to showcase the good effects of using cryogenic technologies. Working to let the consumer know why cryogenically preserved dairy products might be a good option can stimulate further interest in the market. Utilizing effective labelling practices and marketing initiatives that prioritize quality, growth, and sustainability will have a positive effect on consumer demand. Along with partnerships in place with the expansion of dairy in developing markets, there are great opportunities for cryogenic technologies. As a result, companies that utilize cost-effective and scalable ideas in these areas may find opportunities to grow.

6.11 Conclusion

Cryogenics, through its ultra-low temperature applications, delivers clear benefits for product longevity, quality, and sustainability. Rapid freezing, cryo-milling, and freeze-drying offer improved shelf life, protected nutrients, and preservation of unique sensory attributes of milk and milk products, ensuring the consumer receives food of the utmost quality. The addition of cryogenic packages also adds food safety characterized by less spoilage and the chance to take these products globally into export markets, which have strict compliance standards. However more generally, cryogenics, while protecting food quality, is also seeking to tackle ongoing challenges of efficiency and environmental awareness. Lowering energy usage, less waste, and taking the most from our resources are consistent with a larger set of goals on sustainability, which will be especially important for the future of global food systems. Further, if the technology is directly tied to financial issues, we see cost-effective production, increased market competitiveness, and economic development inside the dairy sector from an increased export market. In summary, cryogenic innovation offers a step-change forward in the processing of dairy products. Cryogenics offers unique solutions to old problems, allowing the dairy sector to grow in meeting global demand without giving up food quality and encouraging environmental harmony.

References

Ahn, J. and Lee, J.H. (2017). Cryogenic technologies in food processing: potential and challenges. *Journal of Food Engineering* 6 (5): 34–45. Available at: https://doi.org/10.2202/1556-3758.1771.

Balasubramanian, S.I., Gupta, M.K., and Singh, K.K. (2012). Cryogenics and its application with reference to spice grinding: a review. *Food Reviews International* 52 (9): 781–794. Available at: https://doi.org/10.1080/10408398.2010.509552.

Barnwal, P. and Singh, K. (2014). Application of cryogenic technology in food processing and preservation. *Proceedings of the National Seminar on Engineering Interventions for Global Competitiveness of Indian Dairy Industry*.

Bhatnagar, S. and Singh, R. (2023). IoT-enabled cryogenic systems for dairy supply chains. *Dairy Industry Innovations* 7 (1): 45–58.

Clarke, D. (2018). Cryogenic freezing and drying in dairy processing: advances and applications. *Journal of Dairy Science* 101 (3): 1987–2002.

Clarke, D. (2020). Advances in cryogenic cooling for dairy applications. *Journal of Dairy Science* 103 (4): 2241–2254.

Dey, R., Mandal, S., and Das, A. (2022). Cryogenic applications in dairy product preservation. *Advances in Dairy Science and Technology* 12 (3): 112–126.

Fellows, P. (2017). *Food Processing Technology: Principles and Practice*, 4e. CRC Press.

Goff, H.D. and Hartel, R.W. (2013). Freezing and refrigeration. In: *Ice Cream*, 192–215. Springer. Available at: https://doi.org/10.1007/978-1-4614-6096-1_7.

Goswami, T.K. (2010). Role of cryogenics in food processing and preservation. *International Journal of Food Engineering*. Available at: https://doi.org/10.2202/1556-3758.1771.

International Dairy Federation (2019). Innovations in cryogenic drying for milk powder production. *Dairy Industry* Annual Report.

Kadam, D.M. (2015). Cryogenic grinding technology for spices processing. In: *Spices Handbook*.

Kumar, Y., Tiwari, S., and Kumar, Y. (2018). Cryogenic freezing technology. *International Journal of Pure and Applied Bioscience* 6 (2): 1343–1346. Available at: https://doi.org/10.18782/2320-7051.6458.

Malkin, S. and Guo, C. (2007). Thermal analysis of grinding. *CIRP Annals* 56: 760–782.

Mehta, A. and Kulkarni, S. (2021). Cost-effective approaches to cryogenic preservation in dairy. *Dairy Science Progress* 15 (4): 67–78.

Mohan, T. and Reddy, P. (2022). Sustainable cryogenic storage solutions for dairy products. *Journal of Agricultural and Food Engineering* 5 (4): 67–79.

Mueller, B.W. and Miller, F.K. (2016). Development of a thermodynamic model for a cold cycle ^{3}He-^{4}He dilution refrigerator. *Cryogenics* 79: 85–95. Available at: https://doi.org/10.1016/j.cryogenics.2016.08.004.

Nagendra, P.S. and Radha, C. (2020). Energy-efficient methods in cryogenic drying for the dairy industry. *Sustainability in Food Science and Processing* 9 (5): 289–303.

O'Connor, M. and Lalor, F. (2022). Cryogenic innovations in milk processing: a review. *International Journal of Dairy Technology* 75 (1): 54–68.

O'Leary, M. and Bansal, N. (2016). Low-temperature drying techniques in dairy products: a comparative analysis. *Food Technology and Biotechnology* 54 (2): 123–132.

Patel, S. and Roy, D. (2023). Environmental impact of cryogenic technology in dairy preservation. *Sustainable Food Systems Review* 10 (2): 77–91.

Rahman, S., Akhtar, N., and Chandra, R. (2023). Sustainable cooling methods in the dairy industry: role of cryogenics. *Journal of Sustainable Food Processing* 8 (1): 21–32.

Romain, R., Auber, L., Normand, M. et al. (2009). Influence of cryogenic cooling of cheese curd on yield and quality of semi-hard cheeses. *Dairy Science & Technology* 89 (2): 177–185. Available at: https://doi.org/10.1051/dst/2008038.

Sharma, V., Gupta, S., and Rathore, A. (2023). Enhancing product quality through cryogenic freezing in the dairy industry. *Journal of Food Preservation Science* 18 (1): 23–35.

Singh, P., Kumar, R., and Sharma, A. (2021). Cryogenic preservation of milk: techniques and benefits. *Food Science and Technology Advances* 12 (3): 145–158.

United States Department of Agriculture (USDA), Animal and Plant Health Inspection Service (APHIS) (2019). Bovine colostrum: collection, preservation, and quality control. Available at: https://www.aphis.usda.gov/sites/default/files/bamn01_colostrum.pdf. Accessed date is 18-02-202

Vasilenko, I. and Zhdanov, A. (2020). Lyophilization of colostrum for long-term preservation: techniques and applications. *PMC* 12 (3): 234–241. Available at: https://pmc.ncbi.nlm.nih.gov/articles/PMC8762312/.

Yadav, P., Kumar, N., and Patel, R. (2023). Innovations in cryogenic transportation for dairy products. *International Journal of Dairy Technology Research* 9 (2): 89–102.

7

Advanced Cryogenic Applications in Plant-Based Food Preservation

Mahipal Singh Tomar[1], Rahul Kumar Sharma[2], and Abhirup Mitra[2,3]

[1] *Department of Food Science and Technology, Amity University, Mohali, Punjab, India*
[2] *Department of Agriculture, Invertis University, Bareilly, Uttar Pradesh, India*
[3] *Department of Food Process Engineering, NIT, Rourkela, Odisha, India*

7.1 Introduction

'Veganism' is an ideology that refers to a strictly committed lifestyle that demands to eliminate the utilization of food or clothing derived from animal origin and incorporate foods from non-animal origin. The concept of 'plant-based' or 'veganism' was initiated by Donald Watson in the year of 1944 at Leicester, England. Watson and other members of the Vegetarian Society established a subgroup of vegetarians who do not use or consume any products including meat, fish, eggs, dairy products, and even honey, accompanying eschewing stuffs made out of fur, leather, pearls, coral in addition to other materials derived from animal origin. A plant-based diet is mainly consists of cereals, legumes, vegetables, and fruits, whereas vegetarians prefer few products derived from animal origin, when animals are not killed in obtaining those products, like honey, eggs, milk, and milk products. Even within veganism, there is a group of people who consume fresh vegetables and fruits only, as they do not prefer heat applications (Buckner et al. 2016).

If statistics are considered, plant-based food has continuously grown in popularity in recent years, as shown in Figure 7.1, and is predicted to proliferate in coming years. Market studies from 2016 revealed that the plant-based population of Britain tripled between 2006 and 2016, rising from 150,000 to 542,000. In 2021, the UK-based organization 'Veganuary', an organization based on UK, encouraged consumers to adopt plant-based food lifestyle for the month of January, leading to an unprecedented 560,000 participants. Additionally, in 2019, the United States experienced an 11% increase in plant-based food sales, amounting to approximately $5 billion. Globally, the plant-based food market is estimated to reach $14.3 billion by 2025, up from a sum of $8.9 billion in 2019 (Brouwer et al. 2022).

Cryogenics is a field of engineering that focuses on the production of cryogens and the maintenance of low-temperature technologies. Liquid nitrogen (LN_2), one such cryogen,

Cryogenic Processing: Advances in the Food Industry, First Edition. Edited by Sumit Sudhir Pathak, Rama Chandra Pradhan, and Sudarshan Ramanathan.

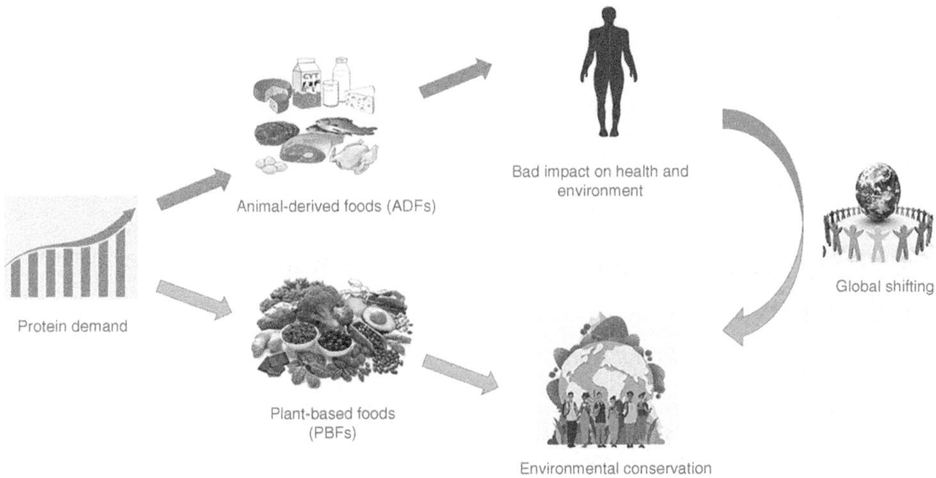

Figure 7.1 Plant-based diet and environment. *Source:* fcafotodigital / Getty Images, cristinn / Adobe Stock Photos.

has significant potential as a total-loss refrigerant. In individual quick freezing (IQF), LN_2 serves as a refrigerant to achieve ultra-rapid freezing (Kumar et al. 2021). This rapid freezing process results in the formation of smaller ice crystals, which helps preserve product quality while minimizing drip loss (Goswami 2010). High-efficiency cryogenic processes often utilize carbon dioxide (CO_2) or nitrogen (N_2) for food preservation. At atmospheric pressure, nitrogen remains in liquid form at −320 °F (−196 °C), while carbon dioxide exists as a solid at −108 °F. These gases are stored as liquids and transported through insulated piping systems to the cryogenic process area within a facility. During freezing, the liquid cryogen is injected into the freezer to extract the necessary heat from the food, ensuring rapid cooling and preservation (Kumar 2013).

Maintaining proper storage conditions is necessary for the long and safe storage of plant-based food products. Plant-based food products can become contaminated and spoiled by microorganisms, like other foods. Ten raw ground meat products' microbiological condition was assessed in a past study in order to determine bacterial makeup and contamination levels. The median result came up at 3.89 $\log_{10}$ CFU/g; the tested bacterial counts ranged from below 1.0 $\log_{10}$ CFU/g to 8.31 $\log_{10}$ CFU/g. The usual 0.5 to 1 $\log_{10}$ differential in lactic-acid bacterial levels below total bacterial counts suggests their predominance among the microorganisms present. Although the detected toxigenic pathogen levels stayed low, several samples had highest counts of about 3.0 $\log_{10}$ CFU/g of presumed *Staphylococcus aureus* and *Bacillus cereus* (Kabisch et al. 2024).

7.2 Emerging Trend for Plant-Based Foods

The plant-based food industry is experiencing a surge in innovation, driven by growing consumer demand for plant-based alternatives that mimic the texture, taste, and nutritional profile of traditional animal products. Emerging trends of plant-based foods include

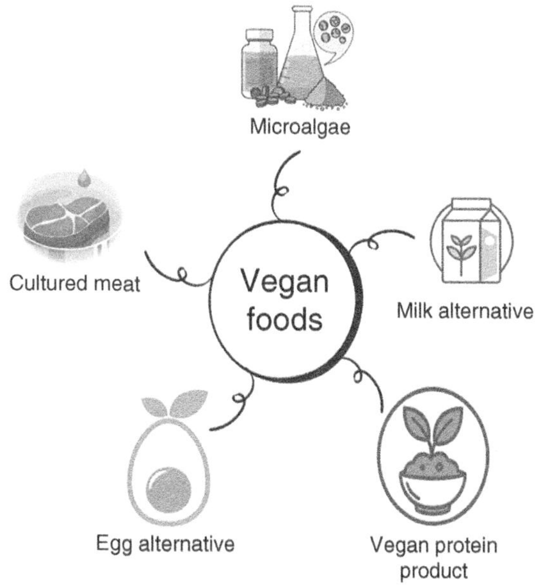

Figure 7.2 Plant-based food products.

plant-based protein, mycoprotein, cultured protein, microalgae, fish, and egg alternatives, as presented in Figure 7.2.

7.2.1 Plant-Based Protein Products

The list of high-protein plant foods includes tempeh and tofu, texturized soy protein, and seitan made from wheat gluten, together with peas, lentils, lupine, and chickpeas, and seeds with rapeseed and canola (Alcorta et al. 2021). The growing significance of plant-based proteins in modern diets is driven by their health benefits and environmental sustainability. Traditionally, legumes and pulses have been key protein sources due to their high nutritional value and affordability. However, consumer acceptance varies based on factors such as taste, texture, and psychological resistance to plant-based alternatives (Moyo 2024).

7.2.2 Fermented Mycoprotein as a Protein Source

Fungal protein derived from the fermentation of *Fusarium venenatum* is a notable alternative protein source (Alcorta et al. 2021). Research indicates that, while some consumers appreciate fungal protein for its sustainability benefits, the palatability of the product does not play a leading role in the acceptance process. The reported sensory experience with fungal protein-based meat substitutes was perceived negatively by participants who participated in the study. Researchers need to conduct extensive sensory evaluations of different product formulations while investigating root causes behind unfavourable perceptions, with food disgust potentially involved (Chezan et al. 2022).

Cultured meat is created by cultivating animal cells in a nutrient-rich environment, removing the requirement for traditional livestock rearing (Alcorta et al. 2021). As commercialization becomes more realistic, research on consumer acceptability of cultured meat has gaining traction (Bryant and Barnett 2018). A recent study used Stoll-Kleemann and Schmidt's model to analyse consumer behaviour, indicating several aspects that influence acceptability of cultured meat, particularly in regard to traditional meat-eating behaviours (Lewisch and Riefler 2023).

7.2.3 Plant-Based Milk: Process and Stability

The extraction of water-soluble plant components through water breakdown methods leads to subsequent homogenization processes. The extracts originate from different plant materials, which include chickpeas and soybeans as legumes together with oats and rice as cereals, along with quinoa and teff and amaranth as pseudo-cereals, in addition to almonds and cashews as nuts and coconut, sesame, and sunflower as seeds (Alcorta et al. 2021).

7.2.4 Plant-Based Cheese Substitutes

The manufacturing of dairy alternatives occurs through complete or partial replacement of milk protein and milk fat with plant-based proteins along with vegetable fats. Soy, nuts, coconut, tapioca, and nutritional yeast are among the ingredients used to develop alternative products. The dairy-fat substitution role falls to partly hydrogenated soybean or palm oil, while peanut or soybean protein functions as common plant-based proteins in these products (Alcorta et al. 2021).

7.2.5 Plant-Based Seafood Mimics

Fish and seafood alternatives are produced through the use of different plant-based ingredients and blends. Several replacement ingredients, such as soy and wheat gluten, along with algae and mushrooms, and vegetables, work to mimic seafood products in view of nutritional content and textural properties (Alcorta et al. 2021).

7.2.6 Functional Plant-Based Egg Replacers

The properties of eggs can be replicated through several plant-based ingredients used to create alternatives to eggs. Hydrocolloids consist of xanthan gum together with Guar gum and Arabic gum, and plant proteins derived from soybeans along with sunflower beans, peas, tomato seed, and wheat, along with white lupin and faba bean, form common egg-replacement alternatives. Multiple natural alternatives for egg replacements include applesauce together with aquafaba and also flax seeds, tofu, ripe bananas, and tapioca starch (Alcorta et al. 2021).

7.2.7 Microalgae

Microscopic algae are nutrient-rich ingredients or products that contain high levels of carbohydrates, protein, lipids, and other bioactive compounds. Notable examples include *Arthrospira* sp., *Chlorella* sp., and *Schizochytrium* sp., which are widely recognized for

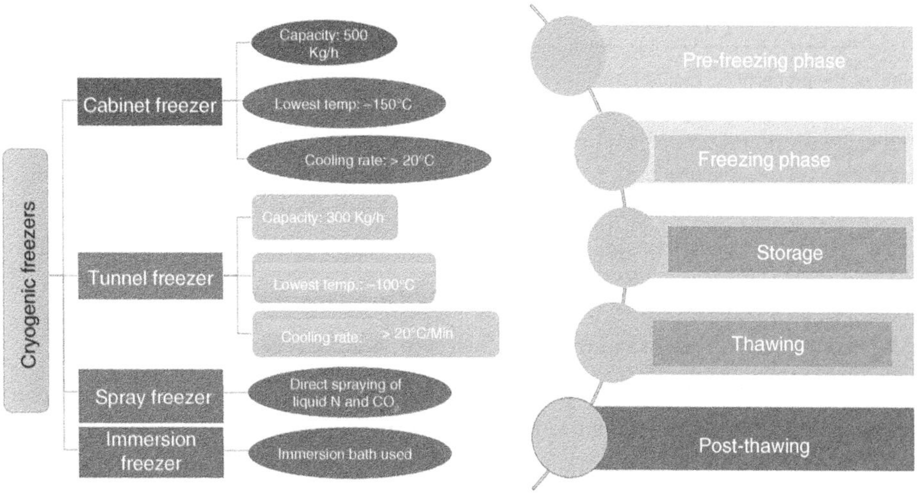

Figure 7.3 Mechanisms and stages of cryogenic freezing.

their potential in sustainable food production and nutritional benefits (Alcorta et al. 2021) (Figure 7.3).

7.3 Basic Principles and Cryogenic Techniques for Plant-Based Product Development

Cryogenic food processing is a preservation method that utilizes extremely low temperatures (below −150 °C) refrigerants to maintain the freshness of food products. This process primarily relies on liquid nitrogen or liquid carbon dioxide (CO_2) to rapidly lower the product's temperature, thereby slowing or halting deterioration and enabling cryopreservation. Unlike conventional refrigeration and freezing methods, cryogenic freezing achieves rapid and uniform cooling, effectively preserving the nutritional and sensory properties of food. This technique employs specialized equipment designed to reach cryogenic temperatures, ensuring minimal ice crystal formation and superior texture retention.

Additionally, various techniques facilitate the transfer of refrigeration energy, enabling efficient cooling and freezing processes. These methods enhance food preservation by maintaining optimal quality throughout storage and distribution and are as follows:

- Liquid nitrogen serves liquid–solid contact methods alongside three additional techniques, which include spraying, immersion, and convection with fan usage.
- The process of using liquid CO_2 for element transformation requires specialized equipment to create carbonated snow through solid–solid contact procedures.

Different physical properties, including density, boiling point, latent heat, and specific heat of liquid N_2 and CO_2, are also considered before selecting based on the product. The properties of the liquid are summarized in Table 7.1.

Table 7.1 Properties of liquid N_2 and CO_2.

Name of the property	Liquid CO_2	Liquid N_2	References
Boiling point	−78.5 °C	−195.8 °C	Kumar (2018) and Nakano (2018)
Density	464 kg/m^3	983.56 kg/m^3	Kumar (2018) and Wang et al. (2015)
Specific heat	2.26 kJ/kg/K	2 kJ/kg/K	Kumar (2018)
Latent heat	500 kJ/kg	200 kJ/kg	Marinhas et al. (2006) and Lucas (2010)

7.3.1 Cryopreservation in Plant-Based Systems

Cryopreservation is the process of preserving live cells and tissues at extremely low temperatures while keeping their structural integrity for an extended period of time (Mandawala et al. 2016). Since unprotected freezing is typically lethal, this process relies on controlled cooling methods to create stable conditions that sustain life. The principal biological consequence of cooling is the freezing of water, which causes an increase in the concentration of solutes in the remaining liquid phase. According to freezing injury theories, damage can occur either directly, mechanically from ice crystals penetrating or splitting cells, or indirectly through secondary effects produced by changes in liquid phase composition. To mitigate freezing damage, cryoprotectants are used to lower ice formation at a given temperature by increasing the overall solute concentration (Jaiswal and Vagga 2022). However, cryomolecules succeed when they fulfil three important criteria, which include biological compatibility with living systems and cellular penetration abilities, together with minimal toxicity levels. The efficient cryoprotectant compounds include glycerol and dimethyl sulfoxide together with ethanediol and propanediol because they minimize the formation of ice and protect cells that freeze. Freezing injury results from ice crystal impact on cells combined with the effects of solution concentration differences depending on cell type and cooling and warming processes. Most experts agree that intracellular ice damages cells severely, but the presence of ice outside of cells typically causes no harm to them. Cell-damage risk exists primarily because of two elements: water-membrane permeability levels and cooling rate. The optimal cooling rate balances the risk of intracellular ice formation with the harmful effects of solute concentration. While extracellular ice is often considered less harmful, densely packed cells may experience mechanical stress, leading to damage. In complex multicellular systems, preserving both cell viability and extracellular structures is crucial. To prevent ice formation altogether, vitrification can be used – this process turns water into a highly viscous, glass-like state without crystallization, allowing for more effective preservation (Uhrig et al. 2022). Vitrification needs high viscosity levels combined with fast cooling/warming rates, together with limited sample volume requirements. The following Eq. (7.1) explains vitrification probability (Pvit):

$$\mathrm{Pvit} = \frac{\mathrm{CR}\left(\mathrm{WR}\right)\eta 1}{\mathrm{V}} \tag{7.1}$$

where Pvit refers to the probability of vitrification, CR and WR are cooling and worming rates, η denoted as viscosity, and V refers to volume (Arav 2022). Similarly, Turnbull created an Eq. (7.2) that computes the propagation speed of ice crystals within vitrification solutions. The following is the equation he developed:

$$u \cong \frac{Ku}{\eta} \int(\Delta\mathrm{Tr}) \tag{7.2}$$

where designed model presents Ku as a constant, while viscosity (η) relates directly to system supercooling ($\int(\Delta\mathrm{Tr})$. The velocity of ice crystal growth in water measures at a speed of $u = 0.158\Delta\mathrm{T}1.69\,\mathrm{cm/s}$ (Arav 2022).

The primary challenge in using vitrification methods is their toxicity. Whether through conventional cryopreservation, which allows freezing, or vitrification, which prevents it, cryoprotectants must effectively penetrate all areas of the system. However, barriers such as membranes can hinder the free diffusion of solutes, leading to temporary or even lasting changes in compartment volumes, which may cause damage. As a result, diffusion and osmosis play crucial roles in several stages of cryopreservation, such as cryoprotectant usage and removal, freezing, and thawing (Murray and Gibson 2022; Amini and Benson 2023). These effects may be experimentally and analytically investigated, allowing the development of effective preservation procedures for a wide spectrum of cells and tissues. Such advances have resulted in major applications in both biological research and medicinal domains (Pegg 2007).

7.3.2 Cryogenic Grinding for Plant-Derived Ingredients

In the food industry, powders serve as either intermediate or final products at various stages of processing. Grinding is a key process that involves different operations using equipment such as crushers, mincers, mills, cutters, shredders, grinders, homogenizers, and disintegrators. This process requires tearing or breaking materials through mechanisms like impact, compression, attrition, or shearing. By applying size-reduction techniques, solid food materials are broken down into fine particles, ultimately producing powders (Jung et al. 2018). During grinding, multiple mechanical forces act on the material, leading to finer particle formation. Achieving a finer end product requires greater force. However, grinding is highly energy-intensive, making efficiency an important consideration in food processing. In conventional grinding processes, temperature rise can be controlled using cryogenic fluids such as liquid nitrogen (LN_2). The extremely low temperature makes the material brittle, facilitating the production of finer particles during grinding. This method also enhances the feed rate and reduces power consumption. Additionally, maintaining lower temperatures during grinding helps prevent undesirable effects like oxidation and discoloration, preserving product quality. Cryogenic grinding, particularly for spices, is an advanced technique with limited commercial adoption in India. To retain volatile oil content in the final product, freezing and chilling methods can be employed. Low temperatures can also be achieved by circulating refrigerants or chilled water around the grinding equipment, or by mixing whole spices with solid CO_2 (dry ice) before grinding (Balasubramanian et al. 2012).

Specific research was done to determine the energy consumption of cryogenic grinding systems, in which, ambient and cryogenic grinding were evaluated in terms of energy

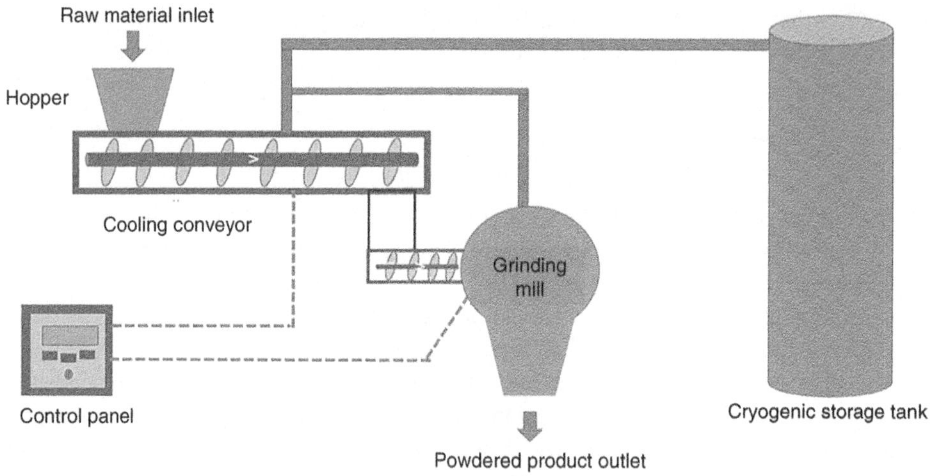

Figure 7.4 Schematic diagram of cryogenic grinding system.

efficiency. In that research, ajwain seed was taken as a function of moisture content – 8, 12, 16, and 20% on dry basis for three distinct types of ajwain, i.e. AA-1, AA-2, and AA-93. The grinding time was taken in rising order, starting from 57.6 ± 0.01 to 180 ± 0.08 s, while the feed rate was taken in decreasing order, starting from 6.25 ± 0.07 to 2.00 ± 0.12 kg/h. This investigation demonstrated increased power consumption, starting from 0.012 ± 0.08 to 0.037 ± 0.007 kW/h in all three kinds for the previously specified moisture range in ambient and cryogenic grinding process. However, the specific energy consumption was proven to be greater in ambient grinding as compared with cryogenic grinding. Simultaneously, different energy models of size reduction like Kick's law, Rittinger's law, and Bond's law work index were showcased to be in increasing order ($p < 0.05$) with increasing moisture of the samples – starting from 1.02 to 13.22 kJ/kg, 0.55 to 21.92 kJ/kg, and 15.31 to 19 t grinding 5.31 kJ/kg, individually. In this research, it was found that the cryogenic grinding was superior in terms of energy efficiency as compared with ambient grinding (Singh et al. 2022) (Figure 7.4).

7.4 Engineering Challenges in Cryogenic Plant-Based Food Processing

Cryogenic technology has plethora of challenges, as demonstrated in Figure 7.5, as it needs proper infrastructure to maintain ultra-low temperature, including well-equipped storage facilities. It requires facilities or special containers to store the dry ice or liquid nitrogen, which should be capable enough to maintain the ultra-low temperature of below −78 °C for dry ice, as well as below −196 °C for liquid nitrogen during processing and storage of plant-based foods. Apart from this, it should have advanced monitoring and control

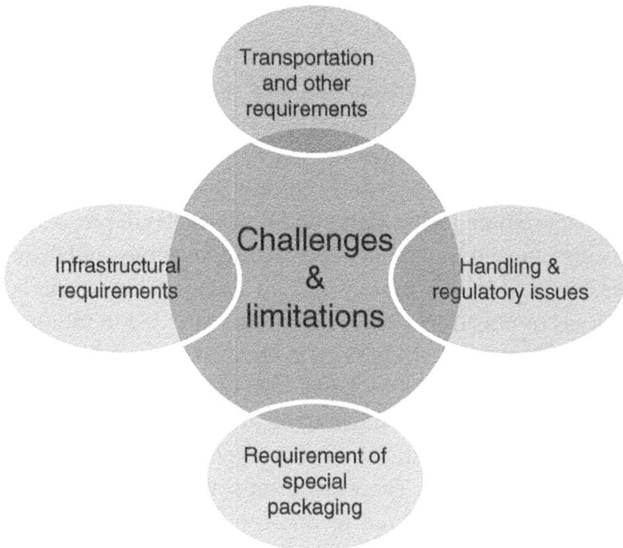

Figure 7.5 Challenges and limitations.

system to maintain cryogenic temperatures. Food products processed by cryogenic technologies attract high numbers of cryogenic microorganisms due to nutrient abundance, leading to fast multiplication inside the system. The quality and safety of food production are adversely affected by cryogenic microorganism colonies, which become very difficult to eliminate or remove from processing equipment (Xu et al. 2024).

It was also demonstrated that, grinding operation in low temperature, as well as ultra-low temperature, increased the grinding zone temperature from 40.30 to 47.45 °C, 21.2 to 39.95 °C, and −3.33 to 23.13 °C. Such increasing grinding zone temperature was also influenced by the grinding time, which increased from 155 to 245 sec, and specific energy also increased from 24.79 to 46.62 kJ/kg °C (Aradwad et al. 2021). Apart from these, liquid nitrogen may also release oxygen-depleting gasses during evaporation, and it needs to be considered for safety of the workers. Apart from that, the process needs to be adhered with all the associated regulatory guidelines and standards implemented by regulatory bodies like FDA (Food and Drug Administration) and IATA (International Air Transport Association).

Few limitations make it challenging to adopt this technology due to its high initial expenditures related to installation and associated cost, running expenses, material-to-liquid nitrogen ratio, and availability of liquid nitrogen (Aradwad et al. 2021). Beyond all the above-mentioned challenges, the cost of final products, including processing, packaging, storage, and shipping, is usually higher as compare with the other products. Due to the higher cost, the plant-based food products processed with cryogenic technology have selected and targeted consumers and are not suitable for regular consumers.

7.5 Conclusion and Future Perspective

Compared to non-vegetarian food products, the processing and storage of plant-based products present unique challenges in terms of nutrition, texture, and overall quality. As a result, special attention is required at every stage, from processing to packaging and storage.

Cryogenic technology is futuristic technology as it has multiple applications, starting from processing to storage and transportation. It gives better-quality products in processing and storage, as it has a faster freezing rate that leads to better quality after storage and thawing, as compared with the other low-temperature processing and storage. Thus, it can be stated that the future of cryogenics is promising, with advancements expected in multiple areas including food processing and storage. It will be a great weapon to minimize food waste throughout the globe. Though cryogenic technology may currently be cost-prohibitive for widespread use in plant-based food production, ongoing engineering-based research focused on system development, particularly the utilization of cryogenic vapour generated during processing, could offer promising solutions. These advancements may pave the way for more commercially viable adoption of cryogenic technologies soon. Support and encouragement from governments and different bodies will also help to adopt this technology on a commercial scale.

References

Alcorta, A., Porta, A., Tárrega, A. et al. (2021). Foods for plant-based diets: challenges and innovations. *Food* 10 (2): 293. https://doi.org/10.3390/foods10020293.

Amini, M. and Benson, J.D. (2023). Technologies for vitrification based cryopreservation. *Bioengineering* 10 (5): https://doi.org/10.3390/bioengineering10050508.

Aradwad, P.P., TV, A.K., Sahoo, P.K., and Mani, I. (2021). Key issues and challenges in spice grinding. *Cleaner Engineering and Technology* 5: 100347. https://doi.org/10.1016/j.clet.2021.100347.

Arav, A. (2022). Cryopreservation by directional freezing and vitrification focusing on large tissues and organs. *Cells* 11 (7): https://doi.org/10.3390/cells11071072.

Balasubramanian, S., Gupta, M.K., and Singh, K.K. (2012). Cryogenics and its application with reference to spice grinding: a review. *Critical Reviews in Food Science and Nutrition* 52 (9): 781–794. https://doi.org/10.1080/10408398.2010.509552.

Brouwer, A.R., D'Souza, C., Singaraju, S., and Arango-Soler, L.A. (2022). Value attitude behaviour and social stigma in the adoption of plant-basedism: an integrated model. *Food Quality and Preference* 97 (November 2021) https://doi.org/10.1016/j.foodqual.2021.104479: 104479.

Bryant, C. and Barnett, J. (2018). Consumer acceptance of cultured meat: a systematic review. *Meat Science* 143: 8–17. https://doi.org/10.1016/J.MEATSCI.2018.04.008.

Buckner, C.A., Lafrenie, R.M., Dénommée, J.A. et al. (2016). We are IntechOpen, the world' s leading publisher of open access books built by scientists, for scientists top 1%. *INTECH* 11 (tourism): 13. https://www.intechopen.com/books/advanced-biometric-technologies/liveness-detection-in-biometrics.

Chezan, D., Flannery, O., and Patel, A. (2022). Factors affecting consumer attitudes to fungi-based protein: a pilot study. *Appetite* 175 (April): 106043. https://doi.org/10.1016/j.appet.2022.106043.

Goswami, T.K. (2010). Role of cryogenics in food processing and preservation. *International Journal of Food Engineering* 6 (1): https://doi.org/10.2202/1556-3758.1771.

Jaiswal, A.N. and Vagga, A. (2022). Cryopreservation: a review article. *Cureus* 14 (11): 1–6. https://doi.org/10.7759/cureus.31564.

Jung, H., Lee, Y.J., and Yoon, W.B. (2018). Effect of moisture content on the grinding process and powder properties in food: a review. *PRO* 6 (6): 6–10. https://doi.org/10.3390/pr6060069.

Kabisch, J., Joswig, G., Böhnlein, C. et al. (2024). Microbiological status of plant-based ground meat products from German retail. *Journal für Verbraucherschutz und Lebensmittelsicherheit* 19 (1): 33–40. https://doi.org/10.1007/s00003-023-01461-w.

Kumar, S. (2013). Effect of cryogenic freezing on food: a review. *Journal of Postharvest Technology* 01 (01): 91–096. www.jpht.info.

Kumar, Y. (2018). Cryogenic Freezing Technology. *International Journal of Pure & Applied Bioscience* 6 (2): 1343–1346. https://doi.org/10.18782/2320-7051.6458.

Kumar, A.K., Hari Chandra, B.P., and Kulkarni, G.V. (2021). Application of cryogenics in grinding of spices for value addition: a review. *Volatiles & Essent. Oils* 8 (5): 10580–10593.

Lewisch, L. and Riefler, P. (2023). Cultured meat acceptance for global food security: a systematic literature review and future research directions. *Agricultural and Food Economics* 11 (1): https://doi.org/10.1186/s40100-023-00287-2.

Lucas, S.A. (2010). Latent heat of vaporisation of liquid nitrogen. March, 1–5. Department of Physics and Astronomy, University College London. Available from: www.ucl.ac.uk

Mandawala, A.A., Harvey, S.C., Roy, T.K., and Fowler, K.E. (2016). Cryopreservation of animal oocytes and embryos: current progress and future prospects. *Theriogenology* 86 (7): 1637–1644. https://doi.org/10.1016/j.theriogenology.2016.07.018.

Marinhas, S., Delahaye, A., Fournaison, L. et al. (2006). Modelling of the available latent heat of a CO_2 hydrate slurry in an experimental loop applied to secondary refrigeration. *Chemical Engineering and Processing: Process Intensification* 45 (3): 184–192. https://doi.org/10.1016/J.CEP.2005.08.002.

Moyo, H.N. (2024). Innovations in plant-based proteins: nutritional value and consumer acceptance. *International Research Journal of Engineering and Technology (IREJOURNALS)* 8 (2): 412–434.

Murray, K.A. and Gibson, M.I. (2022). Chemical approaches to cryopreservation. *Nature Reviews Chemistry* 6 (8): 579–593. https://doi.org/10.1038/s41570-022-00407-4.

Nakano, T. (2018). Uncertainty estimation for extrapolation of the ITS-90 down to the boiling point of nitrogen from the triple point of argon. *Journal of Physics: Conference Series* 1065 (12) https://doi.org/10.1088/1742-6596/1065/12/122025: 122025.

Pegg, D.E. (2007). Principles of cryopreservation. *Methods in Molecular Biology (Clifton, N.J.)* 368: 39–57. https://doi.org/10.1007/978-1-59745-362-2_3.

Singh, H., Meghwal, M., Prabhakar, P.K., and Kumar, N. (2022). Grinding characteristics and energy consumption in cryogenic and ambient grinding of ajwain seeds at varied moisture contents. *Powder Technology* 405: 117531. https://doi.org/10.1016/J.POWTEC.2022.117531.

Uhrig, M., Ezquer, F., and Ezquer, M. (2022). Improving cell recovery: freezing and thawing optimization of induced pluripotent stem cells. *Cells* 11 (5): https://doi.org/10.3390/cells11050799.

Wang, S.X., Xue, X.D., Zhang, X.L. et al. (2015). The application of cryogens in liquid fluid energy storage systems. *Physics Procedia* 67: 728–732. https://doi.org/10.1016/j.phpro.2015.06.123.

Xu, Z., Li, Y., Liu, Z. et al. (2024). Current knowledge on cryogenic microorganisms and food safety in refrigerators. *Trends in Food Science & Technology* 146: 104382. https://doi.org/10.1016/j.tifs.2024.104382.

8

Frosty Enhancements: Cryogenic Approaches in Meat, Poultry, and Marine Product Processing

Sourav Misra[1], Sanchita Biswas Murmu[1], Sivashankari Manickam[2], Rahul Kumar Rout[3], Ankit Kumar[4], Sitesh Kumar[5], and Ritu Bharat Kukde[6]

[1] *ICAR – National Institute of Natural Fibres Engineering and Technology, Kolkata, West Bengal, India*
[2] *ICAR – Central Rice Research Institute, Cuttack, Odisha, India*
[3] *ICAR – National Research Centre for Makhana, Darbhanga, Bihar, India*
[4] *ICAR – National Research Centre on Litchi, Muzaffarpur, Bihar, India*
[5] *Department of Agricultural Engineering, Rabindra Nath Tagore Agriculture College, Birsa Agricultural University, Deoghar, Jharkhand, India*
[6] *ICAR – Central Institute of Post-Harvest Engineering and Technology, Ludhiana, Punjab, India*

8.1 Introduction

Meat is a good source of protein for the human body and is widely consumed in various processed forms. However, it has high water activity and is highly perishable in nature. Preserving the meat in its raw and processed form requires methods to reduce its water activity. In this regard, drying and freezing are two commonly followed methods of meat preservation (Cheng et al. 2017). Some common methods of freezing involve air-blast, cryogenic, plate (both single and double plate), and immersion freezing, to name a few. The quality of frozen products is greatly determined by the freezing rate. Slow freezing results in large, unevenly distributed, and irregular ice crystals, causing severe drip loss in meats. On the other hand, quick freezing results in small, uniformly distributed ice crystals, which avoids the occurrence of drip loss (Kaale et al. 2013). Technologies following quick freezing include electrically assisted, high-pressure, magnetically assisted, and ultrasound-assisted freezing processes.

8.2 Cryobiology: Principles and Methods

Research on low-temperature tissue storage began in the late nineteenth century. Since then, there have been many developments in the field of cryopreservation, and research on protocol optimization is still ongoing. Numerous cells and tissues can be banked through cryopreservation for use in medical and scientific research, such as *in vitro* fertilization,

Cryogenic Processing: Advances in the Food Industry, First Edition. Edited by Sumit Sudhir Pathak, Rama Chandra Pradhan, and Sudarshan Ramanathan.

blood transfusion, bone marrow transplantation, artificial insemination, hepatocyte and pancreatic islet transplantation, and more (Whaley et al. 2021).

The area of biology known as cryobiology examines how low temperatures affect biological systems and living things. Cells, tissues, and organs are preserved at below-freezing temperatures with the goal of preserving their structural and functional integrity when they defrost (Misra et al. 2022; Priyadarsini et al. 2022). There are important uses for this field in conservation, agriculture, and medicine. Preserving biological materials without sacrificing their viability is the main goal of cryobiology (Misra et al. 2021, 2023). Along with the physicochemical mechanisms involved in the freezing and thawing processes. Controlling ice production, which can harm cells, is a basic problem. While extracellular ice can result in mechanical stress and osmotic imbalances, intracellular ice development is more harmful and can induce cell death. Cryoprotective agents (CPAs), which shield cells from freezing and thawing, are used to lessen these effects. CPAs work by stabilizing cellular structures, decreasing ice production, and lowering the freezing point of water. Glycerol and dimethyl sulfoxide (DMSO) are common CPAs (Boafo et al. 2022).

8.2.1 Principles of Cryopreservation

The method of cryopreservation and cell recovery involves four main phases. The cells are first protected from the freezing process using CPAs. After that, the cells are carefully cooled until their temperature falls below −130 °C, at which point all of their metabolic activities cease. The reverse process is used for cell recovery: the cells are first thawed using a defrost approach, and then the CPAs are extracted. Each of these actions can affect the cells' vitality. Numerous experimental results show that the freezing procedure has harmed or even killed cryopreserved living cells; this damage is known as cryoinjury (Misra et al. 2024). Cells must pass through an intermediate zone of temperatures between −15 and −60 °C twice, once during cooling and once during thawing, which damages them even though the cells are not harmed by their incapacity to withstand low-cryogenic storage temperatures (Mazur 1963). The phase shift of water in the intracellular and extracellular solution is what causes damage at high temperatures since the primary component of all living cells is water. Cellular metabolism halts when all the water has turned to ice. To freeze the cell's outside without creating intracellular ice, the refrigeration mechanism must deliver gradual chilling. Water is discarded from the cell because of an osmotic imbalance between the intracellular and extracellular media caused by ice formation in the cell's extracellular environment. This leads to a rise in solute concentrations both inside and outside the cell, which frequently results in cell death (Nannou et al. 2016).

There are several penetrating agents (PAs) on the market today, including ethylene glycol (EG), propanediol (propylene glycol), DMSO, and glycerol (the first agent found). As the name implies, they have a protective effect outside of the cell because they do not penetrate intracellularly. Usually larger, they are covalently bound as trimers, dimers, or polymers. Among the frequently utilized substances in this class are trehalose, raffinose, polyvinylpyrrolidone (PVP), sucrose, and polyethylene glycol (PEG) (Bartolac et al. 2018).

8.2.2 Methods of Cryopreservation

Cryopreservation methods are designed to preserve materials of bio-origin, *viz.* organs, tissues, and cells, at extremely low temperatures to maintain their viability for future use. The main methods include (Jang et al. 2017):

- Slow freezing
- Vitrification
- Subzero non-freezing storage
- Preservation in the dry state

8.2.2.1 Slow Freezing

By dehydrating cells, slow freezing inhibits intracellular ice formation (IIF) and has a low cooling rate. Because slow freezing is effective regardless of low concentrations of toxic CPAs and the amateur ability of operator, it is frequently utilized for scattered single cells (Ishizaki et al. 2023). The main characteristic is the gradual, regulated drop in temperature, often between 1 and 2 °C/min, which enables appropriate dehydration of cell and inhibits the production of ice crystals within the cells. Before freezing, cells are treated to a cryoprotectant solution (such as glycerol or DMSO) to help shield cell membranes and reduce ice crystal damage. Sperm, embryos, ovarian tissue, and certain cell lines are among the biological materials that are frequently cryopreserved by slow freezing (Marques et al. 2019).

8.2.2.2 Vitrification

It is a cryopreservation method which causes rapid cooling of cells or tissues to extremely low temperatures, effectively turning them into a glass-like state. This method mitigates cell injury by employing a substantial amount of cryoprotectant compounds and extremely rapid cooling rates. Because of its significantly higher survival rates after thawing, it is thought to be superior to traditional slow freezing techniques (Fahy and Wowk 2015). By decreasing the freezing point of the solution and enabling more water to remain in a liquid state, special chemicals such as DMSO, ethylene glycol, or propylene glycol are added to the sample prior to freezing to avoid the formation of ice crystals. Compared to gradual freezing, vitrification necessitates a higher concentration of CPAs to produce a glass-like condition. Vitrification dramatically lowers the danger of cellular harm linked to ice crystal formation by quickly chilling the sample without permitting ice crystals to form (Yavin and Arav 2007).

8.2.2.3 Subzero Non-Freezing Storage

It keeps biological materials at temperatures below 0 °C without allowing ice crystals to form. This keeps the material in a supercooled liquid state, minimizing damage from freezing processes (De Vries et al. 2020). Cryoprotectant solutions and careful temperature control are frequently used to prevent ice nucleation. This approach aims to maintain the substance in a liquid condition beneath its freezing point, in contrast to conventional cryopreservation techniques, which cause ice crystals to form during freezing. It is based on the idea of 'supercooling', which states that if there isn't a nucleation site for the creation of ice crystals, a liquid can be chilled below its typical freezing point (Matsumoto et al. 2002).

8.2.2.4 Preservation in the Dry State

To provide prolonged storage at exceedingly low degrees without the crystallization of ice, which can harm cells, dry state cryopreservation – also referred to as desiccation or freeze-drying – involves removing most of the water from a biological sample prior to freezing, usually by subjecting it to a controlled drying process under vacuum. Cryoprotectants are frequently used in this process to further stabilize the sample during the drying and storage phases. To prevent damage, the first step is to thoroughly remove water from the sample using a controlled drying process, typically under vacuum. Cryoprotectants, viz. trehalose or sucrose, are frequently included to cells prior to drying to stabilize their membranes and shield them from harm during dehydration and rehydration (Wolkers and Oldenhof 2021).

8.3 Cryogenic Freezers for Meat, Poultry, and Marine Products

These methods fundamentally depend on the sublimation of N_2 or liquid CO_2 directly exposed to the sample for achieving rapid freezing rates. The system consists of a sealed conveyor equipped with sprayers containing refrigerant regulated with a fixed contact duration. The system is continuous in operation; nevertheless, the high cost of refrigerants necessitates their usage for valuable products like shrimp and prawns. The quality is improved through the elimination of oxygen and reduction of product dehydration; but the extremely low temperatures induce thermal shock to the product, which causes fractures and alterations in texture. The vaporization of liquid nitrogen takes place at −196 °C, while the vaporization of liquid CO_2 occurs at −78 °C.

Liquid nitrogen (LN_2): In this system, heat leaks through tanks and pipes due to the significant temperature difference between the liquid nitrogen and its environment while storage and transportation. These losses can be reduced with the help of a compact system with superior insulation. Standard liquid nitrogen (LN_2) storage capacities range from 900 to 12,000 gal, maintained at pressures between 20 and 250 psi, with a consumption rate of 1.2–2.0 pounds of LN_2 per pound of fish frozen.

Liquid carbon dioxide (CO_2): It is stored in tanks with a capacity ranging from 3,500 to 12,000 gal at 300 psi pressure and −18 °C temperature. After storing in this combination of temperature and pressure, heat leakage is minimized to almost zero. The application of liquid CO_2 parallels that of liquid N_2, with 1.2–2.0 lb/lb for freezing of fish.

8.3.1 Tunnel Freezers

It operates as a counter-current heat exchanger employing cryogen on one end, while the gas flows counter-currently to the product with the opposite moving belt. Because of the movement, the cryogen temperature might reach at 20 °C lower than that of temperature at which the product enters. Consequently, the product could be precooled to experience the complete impact of the refrigeration effect. In this way, the thermal load of the product can be reduced up to 50% before cryogen spray. Spiral or straight tunnels are employed. In general, tunnels for carbon dioxide have a more extended capacity than tunnels for liquid nitrogen.

The operation of tunnel freezer is done by the conveyance of the food products within a restricted gap at the discharge of cryogen. The optimization of application of cryogen is carried out by isolating as well as insulating the tunnel with the regulation of inner cold gas. The passage of the product in the tunnel occurs at three different zones, such as pre-cool, freezing, and post-cool zones (Khadatkar et al. 2004).

The effective design of the tunnel freezers is for efficient usage of the cryogenic liquid to freeze or cool objects quickly. They are also applied to process substantial volumes of items within a compact footprint, enhancing product yield relative to other methods. Cryogenic tunnel freezers minimize product dehydration while preserving the texture of the product. Despite these notable benefits over traditional (mechanical) freezing methods, the primary concern for numerous food processors continues to be the expense of freezing the product. Consequently, it is essential to build a cryogenic tunnel freezer that operates with maximum efficiency while integrating the configuration of the processors and necessary production capacity. A cryogenic tunnel freezer fundamentally comprises four essential components:

- Thermally insulated structure containing a conveyor belt for transporting products
- System for the injection of cryogenic liquid
- Circulatory fans used to enhance thermal transfer within the items
- System for expelling surplus gases

Maximizing the use of the energy accessible from the supplied cryogen is crucial for developing the most effective cryogenic freezing system. The energy accessible in the cryogen is contingent upon several elements; still, the primary determinants are the storage pressure of the cryogen and the temperature of the exhaust vapour from the freezer. Approximately 85% of the refrigeration capacity for carbon dioxide (CO_2) is derived from the latent heat supplied by solid CO_2 sublimation, while 15% is gained from the sensible heat supplied by cold vapours. About half of the refrigeration capacity in a nitrogen system comes from the cold vapours (sensible heat) and the other half from the liquid's evaporation (latent heat).

An effective CO_2 distribution within a freezer plays a crucial role in optimizing the refrigeration capacity of the refrigerant. In the N_2 distribution, it is essential for positioning of point of injection to facilitate the spraying of the liquid while simultaneously achieving optimal recirculation of nitrogen gases. The exhaust gas temperature primarily indicates the efficiency of a nitrogen system. Since the temperature of exhaust gas is insufficiently low, there is a chance of a decrease in the sensible heat of the nitrogen gases available for product freezing. This leads to an increase in the consumption of nitrogen and a reduction in the efficiency of the freezing process.

8.3.2 Rotary Freezers

This is generally employed for individual quick freezing of products, *viz.* meatballs/cubed poultry meat, and shrimp. The product surface is frozen by a fine spray of cryogen, which prevents individual parts from adhering to one another before they enter the rotary cylinder. This separation is maintained throughout freezing (Khadatkar et al. 2004).

8.3.3 Immersion Freezers

These are cost-effective and simple methods of utilizing cryogens to preserve food items. Immersion freezers are suitable for large quantities of products, where crust freezing is necessary or where the materials are resilient enough to endure the thermal shock from a significant differential in temperature. In this type of freezer, the heat transfer is rapid but inefficient since it utilizes only the heat generated during phase transformation and loss of cold vapour.

8.3.4 Cabinet Freezers

This freezer has the application of freezing precooked and prepared foods, especially for airline meals. The foods stored on racks in the cabinet are cooled by injecting liquid cryogen at intervals; circulatory fans are typically employed to ensure effective and uniform freezing (up to one hour).

8.3.5 Pellet Freezers

In this type of freezer, semi-solids and liquid foods are frequently frozen into pellets. This process is executed through contact freezing, which has belts, or via cryogenic freezing, which employs liquid N_2. Belt freezers utilize a comparable contact freezing mechanism alike plate freezer. Freezers having single and double bands have been developed for freezing in thin layers, typically semi-liquid or liquid substances, e.g. egg yolk, fruit pulps, vegetable purees, soups, and sauces. The product is immobilized in freezer having double bands within two continuous belts, with the lower belt corrugated and the upper belt flat (Khadatkar et al. 2004).

8.3.6 Combined Freezers

The combined freezer serves as an affordable alternative to exclusive cryogenic freezing. The process entails cryogenic freezing for the initial surface freezing, succeeded by mechanical freezing for lowering the temperature of the foods (Khadatkar et al. 2004).

8.4 Effect of Cryogenic Processing on Quality Changes of Meats

The physical characteristics of muscular tissues are notably complex. Certain characteristics, such as colour and softness, might intuitively indicate meat quality, while from the storage and processing point of view, moisture content is an essential parameter. Consequently, these characteristics affect the market value of meat-based food items. Simultaneously, these characteristics typically indicate the amalgamation of physical and chemical alterations (Leygonie et al. 2012). This leads to the change in the structures of frozen samples at the micro level due to ice crystal formation, which is the primary driver of changes in the quality during the freezing process.

Cryogenic technology shows better results than regular freezer programs by preserving poultry products. The freezing technology for poultry reaches subzero levels, made possible by cryogenic agents such as liquid nitrogen and carbon dioxide. Cryogenic processing helps maintain poultry quality features better than regular freezing techniques. Over the past 10 years, scientists have studied poultry product reactions to freezing at extremely cold temperatures. Processing at extreme cold temperatures affects the physical features of the product.

8.4.1 Physical Properties

8.4.1.1 Microstructure

The quality of poultry meat depends significantly on its microstructure, which influences its texture, juiciness, and appearance after thawing (Mir et al. 2017). Cryogenic freezing rapidly lowers the temperature to extreme levels, reaching −196 °C with liquid nitrogen and −78 °C with liquid carbon dioxide, by directly exposing the meat to cryogenic agents (James and James 2023). This rapid freezing process prevents water from remaining in a liquid state long enough to form large ice crystals, instead creating small and evenly distributed ice crystals throughout the sample (Jia et al. 2022). Unlike conventional freezing methods, cryogenic freezing prevents the slower temperature decreases that can compromise the muscle fibre structure and disrupt the reticulum network supporting it. Both Xie et al. (2023) and Lee et al. (2024) demonstrated that cryogenic freezing preserves muscle fibre alignment and maintains proper cellular spacing after thawing. Advanced microscopy techniques reveal that cryogenically frozen samples exhibit a superior muscle structure, free from the tears and fissures commonly observed in samples frozen using traditional methods.

8.4.1.2 Loss of Moisture

As moisture is the primary constituent of the food items, researchers have consistently focused on the macroscopic impacts of internal and exterior quality parameters (Schäfer et al. 2002). In muscle, the predominant moisture content is located either in the interstitial space between the myofibrils or in the interspaces of muscle bundles and cells. During the freezing process, the reduction in the moisture has negative impacts on the attributes of frozen meats arising from thawing loss and surface desiccation resulting from the damage of the structures. Consequently, reducing weight loss and maintaining meat integrity are essential factors for consumer appeal (Sun 2005). Numerous variables can influence the loss of moisture in tissues, including the initial state of meats, ionic strength, adenosine triphosphate depletion, pH reduction, post-mortem proteolysis, and the steric effects resulting from myofibril shrinkage due to rigor mortis and conditioning (Huff-Lonergan and Lonergan 2005).

Currently, various metrics are used to assess meat's potential to keep the water content, such as water holding capacity, drip loss, cooking loss, and overall moisture content (Leygonie et al. 2012). Various procedures involving pressing, centrifugation, and heating are employed for measuring these indicators (Petracci and Baéza 2011). During the freezing process, the moisture reduction in foods is being subjected to thermal as well as moisture migration to the surroundings (Campañone et al. 2002).

Simultaneously, in relation to meat fibres, the freezing process would generate substantial surface fractures, thereby exacerbating water loss. Anese et al. (2012) discovered that the cryogenically frozen pork samples exhibited negligible drip loss compared with the air-frozen sample. The application of radio frequency (RF) pulses during cryogenic freezing significantly reduced drip loss. This suggests that RF may mitigate fracturing because of the substantial differential temperature between the cryogenic fluid and food products.

8.4.1.3 Tenderness

Muscle tenderness, defined as toughness or resistance to cutting, is crucial for customer acceptance and significantly influences satisfaction of consumers. Conversely, the predominant objective method for assessing the sensory hardness (tenderness) of muscle tissue, is the Warner–Bratzler shear force (WBSF) (Lyon and Lyon 2000; De Huidobro et al. 2005). Consumer preference for tenderness largely depends on muscle fibres remaining intact within their connective tissue framework. Cryogenic freezing performs exceptionally well in this regard, resulting in lower shear force readings because of the ice crystals formation around the myofibrils. The firmness of RF cryo-frozen pork meat samples was nearly equivalent to control meats, and significantly less firmer compared to the cryo-frozen and air-frozen samples (Anese et al. 2012).

8.4.1.4 Colour

In addition to softness, the colour attributes of meats significantly influence consumer acceptance (Faustman and Cassens 1990) and serve as a primary factor in marketing (Jeyamkondan et al. 2000). The colour of muscle tissue in commonly consumed meat and certain myoglobin-rich fish is intricately linked to iron ions in myoglobin, existing in three states: metmyoglobin, oxymyoglobin, and deoxymyoglobin, and molecular species having six coordination bonds. Their ratio dictates the colour of the muscle tissue (McKenna et al. 2005). Additionally, their inner structure also influences its optical properties (Swatland 2004). During the freezing, storage in frozen conditions, and thawing processes, myoglobin denaturation occurs, resulting in heightened sensitivity of muscle colour stability.

8.4.2 Biochemical Properties

8.4.2.1 Protein Denaturation

It is the most important parameter to be considered during the cryogenic processing of meat, as freezing affects the drip loss, tenderness, and colour of the product (Joo et al. 1999). The myofibrillar proteins constitute around 70% of total proteins, which play a major structural role in interacting with other protein constituents (Sun et al. 2013). In frozen storage and during freezing, protein denaturation may take place due to the concentration of buffer salts and an increase in intracellular ionic bonds, which results in forming and propagation of ice crystals in the extracellular spaces, causing drip loss (Zhan et al. 2018).

Enzyme inactivation during freezing and correlative functionality losses are the major reasons for protein denaturation during freezing. To study protein exudate patterns, methods like differential scanning calorimetry (DSC), sodium dodecyl sulfate–polyacrylamide gel electrophoresis (SDS–PAGE), and capillary gel electrophoresis (CGE) are used. Proteomics is also used to assess drip composition in meat proteins (Zhan et al. 2018). A study by Fernández-Martín et al. (2000) on the DSC of traditional meats showed three typical peaks at 58–58.5, 64–70, and 81–81.5 °C for myosin, collagen, and actin, respectively. The samples were treated with a high pressure of 200 MPa, and DSC thermographs showed the disappearance of actin endotherm, while that of myosin and sarcoplasmic proteins was greatly reduced. It was inferred that actin was pressure-sensitive and that high-pressure treatment results in protein denaturation. Collagen, on the other hand, was a stable protein under high-pressure treatments due to hydrogen-bonded triple α-helix.

8.4.2.2 Lipid and Protein Oxidation

During the storage of frozen meat, biochemical reactions continue to occur until the sample has free unfrozen moisture. Here, the lipid and protein oxidation lead to qualitative deterioration. Lipid oxidation occurs due to the generation of lipid free radicals, which induces the production of toxic compounds like malonaldehyde (Morrissey et al. 1998). Drip loss during freezing results in the release of cell contents like haem iron, which is a pro-oxidant primarily responsible for oxidation. The extent of lipid oxidation is measured by the presence of thiobarbituric acid reactive substances (TBARS) and free fatty acids (FFA) (Garcia-Lomillo et al. 2016). The changes in side-chain groups attached with amino acids are the major results of protein oxidation.

8.4.3 Microbial Properties

Microbiological contamination significantly impacts poultry product quality and industry profitability (Alum et al. 2016). Since, meat is a high-moisture commodity, they are vulnerable to microbial contamination. However, when it is frozen, it reduces the risk of microbial contamination due to a reduction in the water activity and low-temperature environment (Coombs et al. 2017). A study by Vieira et al. (2009) on frozen beef storage showed that after ageing, the samples were free from microbial contamination till 90 days of frozen storage. Research by Tomar et al. (2024) demonstrated that cryogenic freezing reduces the total viable bacteria count in poultry by 2 log CFU/g (Tomar et al. 2024). The quick freezing damages the cellular membranes of spoilage microbes like *Pseudomonas* spp. (Sebranek 2019). It is noteworthy that frozen storage cannot eliminate the psychotropic and lactic acid bacteria (LAB), but a combination of high-pressure processing followed by frozen storage below −35 °C can significantly reduce the bacterial count (Fernández et al. 2007). Therefore, it can be said that a combination of high-pressure processing with frozen storage is helpful in the shelf-life extension of meat products.

8.5 Effect of Cryogenic Processing on Quality Changes of Marine/Aquatic Products

The physical, chemical, and physicochemical behaviour of frozen marine products depends on the species, freezing rate, temperature, techniques, and frozen storage duration. Various dynamic mechanisms contribute to the deterioration of the product, including autolysis

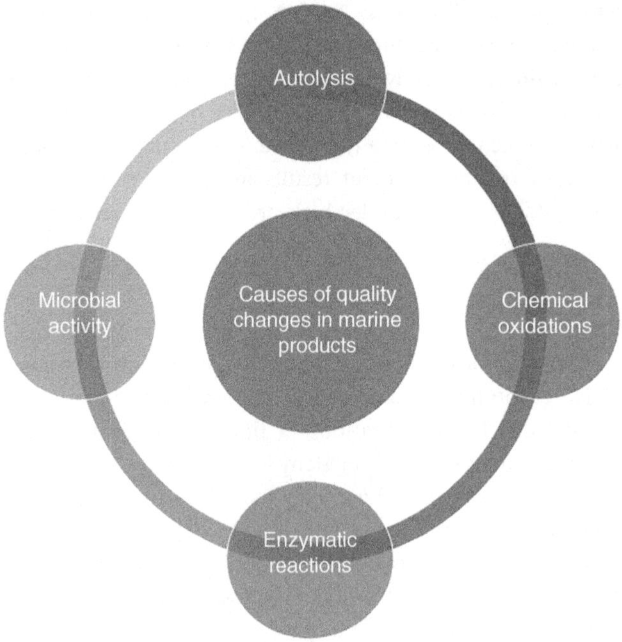

Figure 8.1 Factors contributing to quality deterioration of marine products during long-term frozen storage.

due to degradation of tissue to accommodate ice crystals, chemical oxidation of fat, proteins, enzymatic reactions, and microbial activity (Figure 8.1). Cryogenic freezing helps minimize loss of moisture, prevent dehydration and drip loss, and helps to retain food product texture because of its quick-freezing rate. In the following section, we shall discuss the consequences of cryogenic freezing on the quality attributes of marine products.

8.5.1 Freezing Loss

One measure that takes into consideration the product's weight loss is freezing loss during the freezing process and contributes to the product's potential financial losses during processing. Cryogenic freezing causes lower freezing loss compared to other freezing techniques like air-blast freezing or freezing in refrigerator. At the start of freezing, faster rate of freezing results in the creation of an ice crust on the product's surface, which serves as a coating to lessen the moisture transfer from marine product to the surroundings and contributes to lessen moisture loss during freezing. For catfish fillet, it was observed that cryogenic freezing reduced freezing loss (0.37%) compared to blast freezing (freezing loss 0.45%). This suggested that fillets that are cryogenically frozen retained greater moisture compared to air-blast frozen fillets.

8.5.2 Weight Loss During Storage

Water sublimation causes a porous and dehydrated layer to form on the food product's surface, changing the frozen food's physical and organoleptic properties and causing it to reduce weight while being stored in a frozen state. Hence, it is another crucial quality metric for frozen food products, such as weight loss, which also can be used to calculate the possible monitory losses for the product while it is frozen. The reduced storage loss of cryogenically frozen marine products like catfish and shrimp resulted from their quicker rate of energy removal compared to sluggish freezing rates; faster freezing rates led to less weight loss.

8.5.3 Drip Loss

During freezing cell membranes rupture. Massive formation of ice crystals is caused by delayed freezing, whereas there is production of tiny ice crystals that are evenly spaced beneath the cryogenic conditions, as seen in Figure 8.2. When slow-frozen flesh is thawed the cells that got burst during freezing start leaking water (known as drip) and other dissolved chemicals that give the fish its distinct flavour. When thawing, fish that travel through the zone of greatest ice crystal formation in less than an hour will often have the least amount of drip loss due to less cellular structural damage, which lowers extracellular fluid outflow and some intracellular fluid outflow. The higher freezing rate provided by the cryogenic freezing approach (Figure 8.2) results in the creation of small crystals that are equally dispersed and have a larger specific surface area, which makes it easier for water to be reabsorbed after thawing. However sometimes numerous biological alterations brought on by freezing and thawing, including protein denaturation, fibre shrinkage, and cell injury, might reduce matrices' ability to bind water, hence promoting exudates leakage.

8.5.4 Water Content

Water-retaining capacity is a crucial characteristic that influences how muscles look before cooking, behave when cooking, and maintain freshness after eating. It is an important

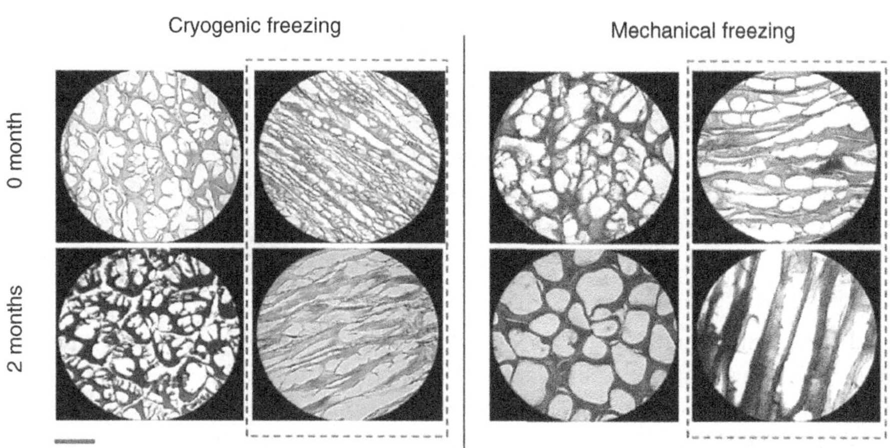

Figure 8.2 The size of ice crystals in frozen marine product as observed under SEM. *Source:* Jha et al., (2024) / with permission of Elsevier.

metric for both customers and the industry. Fish fillets sometimes exhibit inability to bind water as a result of agglomeration, reduced protein dissolution, denaturation of proteins, fibre loss of volume, and cell damage that occur both before and after freezing. After six months of frozen storage, marine product fillets show minimal relative moisture loss and a minor drop in moisture content, demonstrating the role of cryogenic freezing techniques in the ability of frozen marine fillets to retain water.

8.5.5 Protein Denaturation

Denaturation of fish muscle protein is prevented by cryogenic freezing. The primary indicator of protein denaturation is a reduction in the myofibrillar fraction's solubility or extractability, especially those of salt-soluble proteins. The reduction of the total sulfhydryl group level in fish muscles is another sign of protein oxidation. Myosin molecules denature more quickly as a result of the sulfhydryl groups that are reactive on being exposed to oxidation, which cause them to disappear and are the foundation for the rise of the disulphide bond concentration. Furthermore, reduction in the Ca^{2+}-ATPase activity is also a result of protein rearrangement through protein–protein interactions. In aquatic immersion frozen meat, such as bighead cryogenically frozen carp (*Aristichthys nobilis*), differences between initial and final total sulfhydryl group concentration and Ca^{2+}-ATPase activity assess protein oxidation.

At the termination of the freezing period, the myosin and sarcoplasmic components of the protein fractions show the most alterations. Prooxidants may be released as a result of membrane breakdown brought on by an increase in crystal size, and cryo-concentrated fluid may reorganize around the meat fibre. Protein oxidation and denaturation are probably encouraged by these processes. The amount of salt-soluble protein in the northern snakehead (*Channa argus*) did not significantly change from the fresh one after it was frozen using liquid nitrogen and kept for 14 days at −20 °C. However, further freezing northern snakehead at −20 °C in a freezer capable of maintaining an ultra-low temperature (−80 °C) decreased the amount of salt-soluble protein after 14 days of storage. In another case, during 180 days of frozen storage at −18 °C, the freezing in immersed cryogenic medium decreased the amounts of extractable proteins from the frozen bighead carp (*A. nobilis*) compared to air-blast freezing. The freezing techniques employed affected how much the enthalpy value of fresh fish changed. After two days of freezing, pressure shift freezing showed a decrease in enthalpy, but air-blast freezing produced no change.

8.5.6 Lipid Oxidation

Fish and other marine products are particularly vulnerable to oxidation because they are naturally high in haemoproteins like haemoglobin, polyunsaturated fatty acids, and pro-oxidants such as enzymes and transition metals. Variations in pH, heat, packing conditions, slaughtering stress, and pressure are some examples of external influences that affect the oxidation of lipids in seafood. During storage, the increase in lipid oxidation in frozen marine products is measured by the thiobarbituric acid (TBA) content. Lower TBA value is observed in the cryogenically frozen samples compared to mechanical freezing or refrigerator freezing. This suggests that lipid oxidation is effectively reduced by a rapid pace of energy removal and brief freezing durations, as applicable for cryogenic freezing.

It's possible that the ice layer that formed on the catfish fillets' surface during the quick freezing serves as an obstruction to diffusion of oxygen between the fillet and its surroundings. Nonetheless, the impact of freezing methods on fillet oxidation is most likely caused by varying degrees of damage to certain cellular components, particularly membrane lipids, which is manifested as a reduction in phospholipids that permit direct interaction between enzymes and substrates. It is clear that prolonged frozen storage and a low energy removal rate seem to work against lipid oxidation prevention. The TBA value of the catfish fillets frozen cryogenically after six months was 0.94 mg MDA/kg fillet, much lower than that of the fillets frozen by air-blast freezing (1.25 mg MDA/kg fillet). Higher TBA values in conventionally frozen marine products indicates increased lipid oxidation caused by recrystallization, which exacerbates cellular structure damage and releases pro-oxidants and enzymes.

8.5.7 pH and Colour

Cryogenic-frozen fish products first show a minor pH reduction throughout the frozen storage period, followed by a progressive pH increase (0.2–0.6 units) from 60 days and beyond. The freezing by liquid N_2 delays the start of rigour mortis since the pH value of the cryo-frozen marine product did not alter considerably from that of the fresh meat during the first 14 days of frozen storage.

A number of mechanisms, including pigment degradation, hemo-protein denaturation, and ice crystal acceleration, may contribute to the colour variations of cryogenic-frozen fish samples. Fish muscles would temporarily lose their colour, nevertheless, due to the temperature of freezing and pace rather than the concentration of pigments. However, it cannot be ruled out that the extended frozen storage period would cause the β-carotene, canthaxanthin, and astaxanthin in fish muscles to degrade. Additionally, cryogenic freezing, which produces smaller ice crystals and less aggregation, as well as less lipid oxidation, may cause a little colour change during storage at frozen conditions. The rise in greenness may be due to biochemical processes like the way muscle proteins coagulate, the formation of peroxides from lipid oxidation, the oxidation of ferrohaemochrome to ferric state, and the generation of green sulfmyoglobin. A positive b^* value indicates that fish flesh is yellow, which is associated with the presence of carotenoid pigments. The deterioration of carotenoid pigments following freezing may be the cause of the decrease in yellowness.

Because carotenoid pigments, particularly astaxanthin and canthaxanthin, predominate in darker fish like salmon, their flesh is coloured. A drop in a^* value results from the degradation of these pigments during frozen storage. However, because xanthophylls are the main pigment, catfish have lighter flesh. Muscle contraction and texture are also altered by freezing; therefore, it stands to reason that the muscle's light-reflectance characteristics are likewise altered.

8.5.8 Morphological Characteristics

The inherent biological characteristics of fish species and muscle fibre density determine the texture of their muscles. The morphology of frozen food products can be partially reflected in the size and distribution of the tissue's crystals. When fish muscle freezes and is stored at or below its freezing point, the nucleation of ice expands in the external gaps, causes the enlargement of external ice crystals. The temperature during freezing influences the site of growth, shape, size of ice crystals, and their numbers during storage. A refrigerator's slow freezing rate causes an uneven distortion of tissue ultrastructural integrity, whereas liquid nitrogen's rapid freezing rate results in a significant number of tiny ice crystals in the goldfish's skeletal muscle.

8.5.9 Sensory Attributes

The sensory qualities and consumer perception of frozen foods are greatly influenced by their morphology. The unfrozen water remains accessible for lipid oxidation and chemical reactions during the frozen storage period. Several compounds that produce unpleasant tastes and smells can covalently link with muscle proteins, changing the texture. Sensory attributes are reflected by tenderness, flavour, juiciness, and overall attractiveness of the cooked frozen meet after thawing to the judges of the sensory panels.

8.6 Cryogenic Freezing of Shrimp

8.6.1 Harvesting

Aquaculture is swiftly supplanting natural harvest as a reliable source of highly qualitative produce. The shrimps are supplied by the culturist to the processing facility in the live form but packed with ice. Generally, the evaluation of shrimp quality, as well as the checking of non-living populations during transportation, is carried out. The higher number of dead populations during the supply shows the stress during the harvest and deterioration of desirable attributes, such as softening of texture at the post-freezing stage. The shrimp feed influences the colour, flavour, and storability of the product, which is meticulously regulated in the effective culture operations. Due to the possible contamination risk from pathogenic microorganisms, environmental pollutants, routine testing of incoming cargoes is done. Shrimp obtained through wild harvesting are stored on ice or in refrigerated seawater (RSW). The highly qualitative product can be kept alive in the storage tanks and transported alive to the shops rather than being frozen.

8.6.2 Holding

Shrimp undergo processing both before and after rigor mortis. Temperature regulation during the holding phase before the processing is essential. In numerous tropical regions where shrimp are cultivated, refrigeration is limited, and there is scarcity of ice. The quality of the product is also markedly inconsistent. Certain wild-harvested shrimp are treated with sulphating agents in the post-harvesting operations to mitigate 'black spot', a condition caused in the tissues of shrimps by the enzymatic action. The quantities of a limit of 10 ppm in the product are allowed to be added with clearly labelling.

8.6.3 Peeling

The processing of shrimp is generally carried out in either a fresh or cooked state. Proper sanitation and good manufacturing practices are the essential components. Occasionally, freezing of the entire head-on shrimp is practised. The head is removed either manually or automatically for frozen shrimp. Visceral tissue adjacent to the cranium is excised. The products are served with the shell or exoskeleton, in attached or detached form. The entire shell is eliminated, excluding the small tail fan, followed by the elimination of the digestive tract. It is also customary to either retain the vein or excise it by incising the exoskeleton, resulting in deveined shell-on shrimp. Ready-to-eat products coated with batter can be prepared from the cooked shrimps.

8.6.4 Freezing

Generally, the individually quick-frozen shrimps are frozen using spiral blast freezers in Asian regions, whereas these are frozen by carbon dioxide snow in the plants of Southern American parts. The rapid freezing rate results in a better-quality end product.

8.6.5 Glazing

The glazing is carried out with the application of water sprays after the freezing operations. The shrimp are dipped in the medium containing phosphate during or before the glazing in order to increase the retention of water. Afterwards, the shrimp are commonly packed in airtight plastic bags and subsequently placed in a cardboard master case.

8.6.6 Packaging

The individually quick-frozen shrimp using the cryogenic freezing process are generally encased in plastic bags. The commonly used product from the frozen shrimp block is 2–5 kg. It minimizes the product surface area, hence reducing dehydration from the surface and preventing the physical injuries of fragile shrimp pieces.

8.6.7 Frozen Storage

A properly regulated cold storage facility preserves the quality of value-added products. During storage, the variable temperature of the freezing process will lead to a reduction in dehydration and glaze. Furthermore, the digestive enzymes may transfer from the vein, resulting in disintegration. The development of ice crystals is influenced by variations in temperature, as detailed below: When the freezer temperature rises, small ice crystals melt; when the temperature drops, the sizes expand due to water refreezing. A reduced quantity of larger ice crystals then occupy their position. These substantial crystals induce tissue injury. Storage freezer temperatures should ideally be maintained at −20 °C, although achieving this temperature consistently may not always be feasible.

8.7 Major Considerations for Design and Safety

Following are the major considerations for design and safety, which discussed in brief below:

1) The primary cost component that affects the total cost of ground spices in the cryogenic grinding is the amount of liquid nitrogen used. It increases the entire cost of cryogenic grinding.
2) The amount of liquid nitrogen required in cryogenic grinding is the key cost component that influences the total cost of powdered spices; thus, cryogen consumption should be optimized.
3) Preventing needless cryogen use requires precise monitoring of the glass transition temperature.
4) Another important factor to consider is human safety when handling cryogens, and precautions must be taken before handling such low-temperature liquids.

8.8 Safety Measures in Cryogenics

Cryogens must be handled and operated with caution due to their very cold temperatures. Cryogenic liquid can cause burns and loss of vision if it comes into contact with exposed skin or eyes. Proper personal protective equipment, such as face shields, gloves, and goggles, as well as suitable safety mechanisms, including warning sensors, should be used when working with cryogenics. The often-mentioned cryogen risks are listed in Table 8.1.

8.9 Conclusion

This chapter examines the impact of cryogenic freezing on the qualitative characteristics of foods, including meat, fish, and related items. This freezing technique seeks to enhance the quality of frozen products and increase their market worth. It has a beneficial impact on meat samples, particularly in regulating ice crystals. From a micro viewpoint, this freezing technique can produce smaller, more uniform, and consistent ice crystals than conventional freezing processes, such as air-blast freezing. The enhanced dimensions and dispersion of ice crystals can significantly maintain superior microstructure, hence augmenting the associated physicochemical characteristics. Proteins are the fundamental constituents of muscular tissues, and their conditions directly influence the quality of these tissues. The influence of cryogenic freezing on frozen muscle tissues is intricate due to its capacity to regulate the crystallization process. This cryogenic preservation method enhances the size and dispersion of ice crystals, hence preserving tissue integrity in frozen muscle tissues. Furthermore, the available research reveals a lack of diversity in the varieties of frozen meats, predominantly concentrating on pork and some aquatic species. Consequently, substantial efforts must be undertaken, and further research should be conducted to enhance the understanding of their mechanisms and facilitate their applicability in the food sector.

Table 8.1 Precautionary and safety measures in handling of cryogenics.

Risk	Factor	Symptoms	Precautions	Rescue procedures
Enrichment of O_2	Air condensation (containing 40% O_2)	A higher danger of fire	Using the right materials for cryogen handling, as well as handling, transporting, storing, and exhausting cryogens	Pipelines and equipment should be properly insulated, and leaks should be checked often
Extreme cold	Prolonged exposure to a cold environment, direct contact with a cryogen or cold surface, frostbite, a cloud of vapour, and brittle fraction	Skin adhering to cold surfaces, frozen tissues, localized cell structure damage, slower mental and physical reactions, trouble speaking and seeing	Wear a full face mask, safety shoes, sleeve cuffs, non-absorbent gloves, a non-absorbent apron, and warm clothing. Cryogens should be handled, transported, stored, used, and exhausted properly. Appropriate materials for handling cryogens	Sufficient first aid facilities, if at all possible, isolation of the N_2 supply, tepid water flushing of the cold burn region, releasing tight clothing, and abstinence from all alcohol and tobacco use
O_2 deficit	Dilution and displacement of air can lead to reduced oxygen levels and inadequate ventilation	Symptoms may include fast breathing, shortness of breath, exhaustion, nausea, vomiting, difficulty moving, and strange behaviour	To prevent asphyxiation, monitor O_2 levels regularly, avoid working in confined spaces, ensure adequate ventilation, and label the area with a 'asphyxiant' sign	Provide adequate first aid, isolate N_2 supply, if possible, employ a lifeline or 'buddy' system, and have artificial respiration available
Integrating pressure	Transition from a liquid to a gas		Pipelines bursting	Utilizing pressure relief valves that are rated and sized
Cryogenic burns	Skin exposure to low temperatures and contact with cold, uninsulated surfaces	Burn-like skin injury and even flesh tearing when adhered to cold surfaces	Evaluate the risks and wear loose, non-absorbent gloves (leather or PVC) in accordance with EN 511 guidelines; do not wear damp gloves. If splashing is expected, wear goggles or face shields, and make sure your pants are over your boots and free of any lace-up or open footwear. When it is required, aprons should be worn	Move the victim to a warm location (around 22 °C) and wash the afflicted area with tepid water without using direct heat. Keep the patient warm and at rest, remove any constricting clothing, and cover frozen regions with sterile, dry bandages that don't impede blood flow. Avoid giving oxygen if it is accessible, giving painkillers, and telling medical staff about the condition and treatments
Hot burns	Leakage of flammable liquids or gases (ethane, hydrogen, natural gas, LPG, and ethylene)	Burns caused by thermal radiation or direct flame impingement	Regularly maintain cryogenic plants, promptly address any leaks, and stay alert for nearly invisible hydrogen flames	Treat affected areas with cold water for at least 20 minutes, apply a sterile gauze dressing, and seek immediate medical attention for serious cases

References

Alum, E.A., Urom, S.M.O.C., and Ben, C.M.A. (2016). Microbiological contamination of food: the mechanisms, impacts and prevention. *International Journal of Scientific and Technology Research* 5 (3): 65–78.

Anese, M., Manzocco, L., Panozzo, A. et al. (2012). Effect of radiofrequency assisted freezing on meat microstructure and quality. *Food Research International* 46 (1): 50–54.

Bartolac, L.K., Lowe, J.L., Koustas, G. et al. (2018). Effect of different penetrating and non-penetrating cryoprotectants and media temperature on the cryosurvival of vitrified in vitro produced porcine blastocysts. *Animal Science Journal* 89 (9): 1230–1239.

Boafo, G.F., Magar, K.T., Ekpo, M.D. et al. (2022). The role of cryoprotective agents in liposome stabilization and preservation. *International Journal of Molecular Sciences* 23 (20): 12487.

Campañone, L.A., Roche, L.A., Salvadori, V.O., and Mascheroni, R.H. (2002). Monitoring of weight losses in meat products during freezing and frozen storage. *Food Science and Technology International* 8 (4): 229–238.

Cheng, L., Sun, D.W., Zhu, Z., and Zhang, Z. (2017). Emerging techniques for assisting and accelerating food freezing processes: a review of recent research progresses. *Critical Reviews in Food Science and Nutrition* 57 (4) https://doi.org/10.1080/10408398.2015.1004569: 769–781.

Coombs, C.E.O., Holman, B.W.B., Friend, M.A., and Hopkins, D.L. (2017). Long-term red meat preservation using chilled and frozen storage combinations: a review. *Meat Science* 125. https://doi.org/10.1016/j.meatsci.2016.11.025: 84–94.

De Huidobro, F.R., Miguel, E., Blázquez, B., and Onega, E. (2005). A comparison between two methods (Warner–Bratzler and texture profile analysis) for testing either raw meat or cooked meat. *Meat Science* 69 (3): 527–536.

Fahy, G.M. and Wowk, B. (2015). Principles of cryopreservation by vitrification. In: *Cryopreservation and Freeze-drying Protocols*, 21–82. New York, NY: Springer New York.

Faustman, C. and Cassens, R.G. (1990). The biochemical basis for discoloration in fresh meat: a review. *Journal of Muscle Foods* 1 (3): 217–243.

Fernández, P.P., Sanz, P.D., Molina-García, A.D. et al. (2007). Conventional freezing plus high pressure-low temperature treatment: physical properties, microbial quality and storage stability of beef meat. *Meat Science* 77 (4) https://doi.org/10.1016/j.meatsci.2007.05.014: 616–625.

Fernández-Martín, F., Otero, L., Solas, M.T., and Sanz, P.D. (2000). Protein denaturation and structural damage during high-pressure-shift freezing of porcine and bovine muscle. *Journal of Food Science* 65 (6) https://doi.org/10.1111/j.1365-2621.2000.tb09407.x: 1002–1008.

Garcia-Lomillo, J., González-SanJosé, M.L., Skibsted, L.H., and Jongberg, S. (2016). Effect of skin wine pomace and sulfite on protein oxidation in beef patties during high oxygen atmosphere storage. *Food and Bioprocess Technology* 9 (3) https://doi.org/10.1007/s11947-015-1649-y: 532–542.

Huff-Lonergan, E. and Lonergan, S.M. (2005). Mechanisms of water-holding capacity of meat: the role of postmortem biochemical and structural changes. *Meat Science* 71 (1): 194–204.

Ishizaki, T., Takeuchi, Y., Ishibashi, K. et al. (2023). Cryopreservation of tissues by slow-freezing using an emerging zwitterionic cryoprotectant. *Scientific Reports* 13 (1): 37.

James, S.J. and James, C. (2023). Chilling and freezing. In: *Food Safety Management*, 453–474. Elsevier.

Jang, T.H., Park, S.C., Yang, J.H. et al. (2017). Cryopreservation and its clinical applications. *Integrative Medicine Research* 6 (1): 12–18.

Jeyamkondan, S., Jayas, D.S., and Holley, R.A. (2000). Review of centralized packaging systems for distribution of retail-ready meat. *Journal of Food Protection* 63 (6): 796–804.

Jha, P.K., Chapleau, N., Meyers, P.E. et al. (2024). Cryogenic vs. Mechanical freezing impact on the quality of the sea bass (*Dicentrarchus labrax*) during long-term storage. *International Journal of Refrigeration* 160: 411–422.

Jia, G., Chen, Y., Sun, A., and Orlien, V. (2022). Control of ice crystal nucleation and growth during the food freezing process. *Comprehensive Reviews in Food Science and Food Safety* 21 (3): 2433–2454.

Joo, S.T., Kauffman, R.G., Kim, B.C., and Park, G.B. (1999). The relationship of sarcoplasmic and myofibrillar protein solubility to colour and water-holding capacity in porcine longissimus muscle. *Meat Science* 52 (3) https://doi.org/10.1016/S0309-1740(99)00005-4: 291–297.

Kaale, L.D., Eikevik, T.M., Bardal, T. et al. (2013). The effect of cooling rates on the ice crystal growth in air-packed salmon fillets during superchilling and superchilled storage. *International Journal of Refrigeration* 36 (1) https://doi.org/10.1016/j.ijrefrig.2012.09.006: 110–119.

Khadatkar, R.M., Kumar, S., and Pattanayak, S.C. (2004). Cryofreezing and cryofreezer. *Cryogenics* 44 (9): 661–678.

Lee, S., Jo, K., Jeong, H.G. et al. (2024). Freezing-induced denaturation of myofibrillar proteins in frozen meat. *Critical Reviews in Food Science and Nutrition* 64 (5): 1385–1402.

Leygonie, C., Britz, T.J., and Hoffman, L.C. (2012). Impact of freezing and thawing on the quality of meat. *Meat Science* 91 (2): 93–98.

Lyon, B.G. and Lyon, C.E. (2000). Meat quality: sensory and instrumental evaluations. In: *Poultry Meat Processing*, 107–130. CRC Press.

Marques, L.S., Fossati, A.A., Rodrigues, R.B. et al. (2019). Slow freezing versus vitrification for the cryopreservation of zebrafish (*Danio rerio*) ovarian tissue. *Scientific Reports* 9 (1): 15353.

Matsumoto, N., Yoshizawa, H., Kagamu, H. et al. (2002). Successful liquid storage of peripheral blood stem cells at subzero non-freezing temperature. *Bone Marrow Transplantation* 30 (11): 777–784.

Mazur, P. (1963). Kinetics of water loss from cells at subzero temperatures and the likelihood of intracellular freezing. *The Journal of General Physiology* 47 (2): 347–369.

McKenna, D.R., Mies, P.D., Baird, B.E. et al. (2005). Biochemical and physical factors affecting discoloration characteristics of 19 bovine muscles. *Meat Science* 70 (4): 665–682.

Mir, N.A., Rafiq, A., Kumar, F. et al. (2017). Determinants of broiler chicken meat quality and factors affecting them: a review. *Journal of Food Science and Technology* 54: 2997–3009.

Misra, S., Pandey, P., and Mishra, H.N. (2021). Novel approaches for co-encapsulation of probiotic bacteria with bioactive compounds, their health benefits and functional food product development: a review. *Trends in Food Science & Technology* 109: 340–351.

Misra, S., Pandey, P., Dalbhagat, C.G., and Mishra, H.N. (2022). Emerging technologies and coating materials for improved probiotication in food products: a review. *Food and Bioprocess Technology* 15 (5): 998–1039.

Misra, S., Pandey, P., Panigrahi, C., and Mishra, H.N. (2023). A comparative approach on the spray and freeze drying of probiotic and Gamma-aminobutyric acid as a single entity: characterization and evaluation of stability in simulated gastrointestinal conditions. *Food Chemistry Advances* 3: 100385.

Misra, S., Mandliya, S., Pandey, P. et al. (2024). Effect of spray-and freeze-dried microcapsules containing probiotics and γ-Aminobutyric acid on nutritional, physicochemical, textural, pasting, rheological, and microstructural characteristics of composite dough. *Food and Bioprocess Technology* 17 (2): 464–478.

Morrissey, P.A., Sheehy, P.J.A., Galvin, K. et al. (1998). Lipid stability in meat and meat products. *Meat Science* 49 (SUPPL. 1) https://doi.org/10.1016/S0309-1740(98)90039-0: S73–S86.

Nannou, T.K., Jouhara, H., Trembley, J., and Herrmann, J. (2016). *Cryopreservation: Methods, Equipment and Critical Concerns*. International Institute of Refrigeration.

Petracci, M. and Baéza, E. (2011). Harmonization of methodologies for the assessment of poultry meat quality features. *World's Poultry Science Journal* 67 (1): 137–153.

Priyadarsini, D., Rayaguru, K., Misra, S., and Dash, S.K. (2022). Effect of drying techniques on physicochemical properties of oyster mushroom (*Pleurotus sajor-caju*). *Journal of Food Processing and Preservation* 46 (7): e16598.

Schäfer, A., Rosenvold, K., Purslow, P.P. et al. (2002). Physiological and structural events post mortem of importance for drip loss in pork. *Meat Science* 61 (4): 355–366.

Sebranek, J.G. (2019). Poultry and poultry products. In: *Freezing Effects on Food Quality*, 85–108. CRC Press.

Soyer, A., Özalp, B., Dalmiş, Ü., and Bilgin, V. (2010). Effects of freezing temperature and duration of frozen storage on lipid and protein oxidation in chicken meat. *Food Chemistry* 120 (4) https://doi.org/10.1016/j.foodchem.2009.11.042: 1025–1030.

Sun, D.W. (2005). *Handbook of Frozen Food Processing and Packaging*. CRC Press.

Sun, W., Zhou, F., Sun, D.W., and Zhao, M. (2013). Effect of oxidation on the emulsifying properties of myofibrillar proteins. *Food and Bioprocess Technology* 6 (7) https://doi.org/10.1007/s11947-012-0823-8: 1703–1712.

Swatland, H.J. (2004). Progress in understanding the paleness of meat with a low pH: keynote address. *South African Journal of Animal Science* 34 (6): 1–7.

Tomar, G.S., Seri, M., Gundogan, R. et al. (2024). Freezing of meat, poultry, and seafoods. In: *Low-Temperature Processing of Food Products*, 225–258. Woodhead Publishing.

Vieira, C., Diaz, M.T., Martínez, B., and García-Cachán, M.D. (2009). Effect of frozen storage conditions (temperature and length of storage) on microbiological and sensory quality of rustic crossbred beef at different states of ageing. *Meat Science* 83 (3) https://doi.org/10.1016/j.meatsci.2009.06.013: 398–404.

de Vries, R.J., Tessier, S.N., Banik, P.D. et al. (2020). Subzero non-frozen preservation of human livers in the supercooled state. *Nature Protocols* 15 (6): 2024–2040.

Whaley, D., Damyar, K., Witek, R.P. et al. (2021). Cryopreservation: an overview of principles and cell-specific considerations. *Cell Transplantation* 30: 0963689721999617.

Wolkers, W.F. and Oldenhof, H. (2021). Principles underlying cryopreservation and freeze-drying of cells and tissues. In: *Cryopreservation and Freeze-drying Protocols*, 3–25. New York, NY: Springer US.

Xie, Y., Zhou, K., Tan, L. et al. (2023). Coexisting with ice crystals: cryogenic preservation of muscle food—mechanisms, challenges, and cutting-edge strategies. *Journal of Agricultural and Food Chemistry* 71 (49): 19221–19239.

Yavin, S. and Arav, A. (2007). Measurement of essential physical properties of vitrification solutions. *Theriogenology* 67 (1): 81–89.

Zhan, X., Sun, D.W., Zhu, Z., and Wang, Q.J. (2018). Improving the quality and safety of frozen muscle foods by emerging freezing technologies: a review. *Critical Reviews in Food Science and Nutrition* 58 (17) https://doi.org/10.1080/10408398.2017.1345854: 2925–2938.

9

Spice Chills: Cryogenic Technologies in Spice Processing

Gintu Sara George[1], *Akshay D. Sonawane*[2], *and Shivanand S. Shirkole*[3]

[1] *Plant Lipids Private Limited, Ernakulam, Kerala, India*
[2] *Leibniz Institute for Agricultural Engineering and Bioeconomy (ATB), Potsdam, Brandenburg, Germany*
[3] *D. Y. Patil Agriculture and Technical University, DYP Educational City, Kolhapur, Maharashtra, India*

9.1 Introduction

India is globally recognized as the largest exporter, consumer, and producer of spices in the world (Spice Board of India 2019). Indian spices are recognized for their unique taste, aroma, and medicinal properties. Approximately 80–85% of the spice production of India is consumed domestically (Singh and Solanki 2015). The internal and extrinsic qualities of any spices can be used to assess its overall quality. Besides, the distinctive quality is mostly due to particular volatile oils or chemical compounds that they possess, and these compounds can be readily rendered available after grinding and are found in the cells of the plant matrix (Zachariah et al. 2012). This is why it is preferable to ingest main spice crops in powder form, such as turmeric, coriander, and red chilli. There is a lot of potential for both local and overseas markets for the pulverized spice products, which leads to pulverization as a key unit operation in spice processing.

The ancient grinding process reduces particle size, producing powders that can be used as intermediate or final goods. The goal of grinding is to decrease the particle's size mechanically through cutting, compression, impact, and shear (Sahay and Singh 1996). Ball mills, plate mills, hammer mills, pin mills, and roller mills are among the most commonly used tools for grinding spices (Balasubramanian et al. 2016). In the energy-intensive grinding process, only 1% of the total supplied energy is used for size reduction of particles, whereas the rest of the energy is dissipated as heat (Jung et al. 2018). Subsequently, it causes a significant rise in the temperature of the pulverized product, the mill, and the surrounding air (Balasubramanian et al. 2012). Moreover, a rise in temperature during the grinding operation significantly degrades the quality of the ground product, as the flavour, aromatic, and medicinal components in spices are naturally heat-sensitive (Liu et al. 2013). Because of different agroclimatic conditions, India provides a wide range of spices,

Cryogenic Processing: Advances in the Food Industry, First Edition. Edited by Sumit Sudhir Pathak, Rama Chandra Pradhan, and Sudarshan Ramanathan.

including annuals like coriander, garlic, turmeric, fenugreek, chili, cumin, nigella, fennel, dill, ajwain, as well as perennials like black pepper, clove, cinnamon, cardamom (both large and small), and nutmeg. These spices are used in a variety of amounts in most Asian and Bharatiya dishes, either whole or ground. Cumin, coriander, as well as black pepper are necessary ingredients for dishes that have a spicy touch because of their strong aroma. Though Bharat is the largest producer and exporter of spices in the world, most of its exports are in raw form.

Ground powders of different spices, either alone or combined, demonstrate enormous potential for commercial exploitation among value-added products. One of the main concerns for value addition is the product's intrinsic and extrinsic features. When it comes to spices, external qualities include things like size, appearance, colour, flavour, as well as odour. Although these traits of spices differ based on the agroclimatic conditions, effective grading, sorting, and packing during the harvest and post-harvest processes are crucial in enhancing these traits. The inherent quality of the spices is determined by their moisture content, volatile content, and fixed oil content. All of these factors are addressed by value addition in order to maximize advantages. Adding value to a processed spice product relies mostly on the spices' inherent flavour and therapeutic qualities. Depending on the intended purpose, the particle size of ground spice powder might range from 50 to 850 μm. For this reason, hammer mills, attrition mills, or pin mills are used in spice grinding (Singh and Goswami 1999a; Goswami and Singh 2003; Murthy and Bhattacharya 2008).

Low-temperature grinding methods for spices undoubtedly enhance the flavour and therapeutic qualities of the resulting spice powders. This chapter emphasizes the importance of cryogenic grinding in the size reduction of spices and the efforts to develop a domestic cryogenic grinder specifically designed for these spices.

9.2 Overview of Conventional Spice Processing Techniques

Spices are being produced from various plant parts, such as fruits, leaves, seeds, barks, and rhizomes. Though a lot of technological innovations are reported, conventional post-harvest spice processing practices are still being used at many places across the world. Processing techniques vary to some degree because of the raw material's notable variety. Challenges during conventional spice processing practices are discussed shortly hereunder. Nonetheless, a few fundamental unit procedures (Figure 9.1) are frequently used with most spices.

9.2.1 Conventional Post-harvest Practices Responsible for Contamination of Spices

Various post-harvest practices, such as drying, microbial inactivation, storage, and transportation, play a vital role in the production of quality spices and their by-products. The estimated post-harvest spoilage at different post-harvest stages is given in Table 9.1.

Open Sun-Drying: It is widely adopted to bring the moisture content of spices up to 8–10%, which helps to extend the shelf life and ensure safe storage of spices. In Bharat, spices are dried at a large scale on open ground, which takes one to three weeks (depend upon climatic conditions) to reduce moisture to a safe level (Oberoi et al. 2005). Although open sun-drying is widely used, it has drawbacks such as being a time-consuming process; the product becomes contaminated with soil, dust, sand particles, and insects, leading to a dried product with a high microbial load (Öztekin et al. 1999). The colour of spices, which is the most important quality parameter, is degraded significantly by open sun-drying. The open sunlight accelerates the browning reaction in spices (Condori et al. 2001; Ramesh et al. 2001).

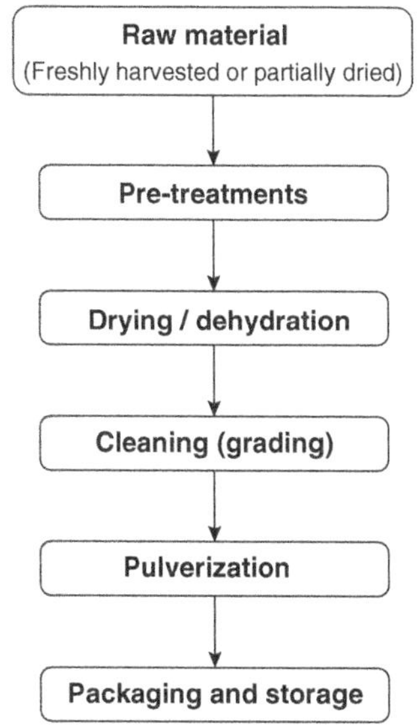

Figure 9.1 Conventional spice processing methods.

Transportation and Distribution: Unhygienic transportation and distribution of spices lead to high microbial load and physical contamination. In underdeveloped and developing countries, transportation and distribution practices using gunny bags with improper fumigation result in insect and microbial contamination. Therefore, hygienic transportation, storage, and proper distribution practices should be adopted to meet the quality requirements for the export of the spices.

Storage of Spices: Storage of dried spices at an appropriate moisture content is required to maintain quality and fetch higher prices in the market. Aflatoxins can contaminate spices because of delayed drying and high moisture levels beyond critical values for mould growth during storage (Cho et al. 2008). For microbial growth, a certain amount of moisture content or water activity is required (Sahar et al. 2015), and spices assist in making a favourable environment for the growth of fungi due to their high nutritional profile (Tripathi and Mishra 2009). Product storage conditions such as moisture content (8–12%), temperature (5–25 °C), and relative humidity (60–70%) are required to maintain properly for safe storage (Scharnow 1986).

Concern of Microbial Contamination in Powdered Spices: Spices become contaminated with animal waste, dust, filth, insects, and pathogenic microorganisms during different stages of post-harvest practices. Powdered spices are easily susceptible to microbial contamination as increased surface area due to pulverization results in increased

Table 9.1 Estimation of spice losses at various post-harvest stages.

Causes	Losses (% of total production)
High moisture content	15–25
Spoilage at the field level	01–10
From the farm to the processing facility	05–10
From the processing facility to the distribution	02–05

Source: Adapted from Marketing of chillies in India, 2002, DMI, Nagpur.

hygroscopicity of the spices. Therefore, increased moisture content in powdered spices during storage or transportation is considered to be a major concern among spice processors and the spice industry. Many investigators have reported their studies on microbial contamination of spices. The microbial status of red pepper powder collected from retail shops was examined by Banerjee and Sarkar (2003). Their results showed that the samples had counts of total aerobic mesophilic bacteria (TAMB) and mesophilic bacterial spores (MBS) that were 66% in 6.1–9.0 and 67% in 5.1–7.0 range of log cfu g^{-1}, respectively. Also, they found the highest mould contamination of 33% in the 4.1–6.0 range of log cfu g^{-1}.

Foodborne bacterial pathogens detected include *Bacillus cereus, Staphylococcus aureus, Clostridium perfringens,* and *Enterobacteriaceae*. They found that microbial contamination was high compared to contamination in dried whole red peppers. Prevalence and characteristics of *Salmonella* spp., *Escherichia coli, Bacillus* spp., *and S. aureus* in condiments or spices, including red pepper, were investigated by Mankee et al. (2005). They inspected 196 samples of condiments or spices, including hot pepper, and detected 50% of the total samples contaminated with *Staphylococci, E. coli*, and *Bacillus* spp. Hara-Kudo et al. (2006) tested *Salmonella* prevalence, spore populations, and total microbial count in 40 types of spices total of 259 samples, and found 6.93 log CFU/g as an average aerobic bacteria count in the two *Salmonella*-isolated samples. They recommended microbial inactivation for food safety, as spices have been found to serve as a source of contamination in the products.

Sterilization Practices and Current Issues: There are different sterilization practices used worldwide for microbial inactivation purposes, and it's obligatory to diminish the microbial load to ensure safety towards consumption. Various advantages and disadvantages associated with different sterilization methods are discussed hereunder in abridged form:

Ethylene oxide (EO) fumigation was initially considered a successful method for microbial destruction in spices, but it leaves some mutagenic and carcinogenic residues, which led to its ban in the European Union (Tateo and Bononi 2006). Also, additional aeration is required to remove EO gas, which is a time-consuming practice.

Irradiation, comprising the application of X-rays, gamma rays, and electron beams, has shown the capability of the destruction of pathogenic microorganisms in enormous numbers without any side effects and is also an officially accepted method. However, it is practiced barely in developing countries because of the high equipment cost. Also it showed minor alterations in antioxidative and sensorial characteristics (Farkas and Mohácsi-Farkas 2011).

Since the application of EO has been banned and irradiation is very expensive, steam-assisted microbial inactivation is widely practiced in the European and Indian spice industries. The steam heating involves further exertion of drying to eliminate condensed moisture to restrict unwanted mould growth. Steam treatment is problematic, particularly for powdered spices, especially red pepper, because it yields to discoloration and quality degradation, and subsequently fetches a low price in the export market (Almela et al. 2002). Recent studies on the different methods for microbial destruction have focused on a quick and efficient heating method for spices. The combination of different methods can be used to eliminate drawbacks associated with individual ones so that the new method would have features such as quick heating, rapid microbial destruction, short processing time, and minimum quality degradation. The new technology should afford challenges in shelf-life extension with better microbial stability, superior quality retention, and energy efficiency.

9.2.2 Treatment Based on Harvest and the Post-harvesting

Cultivation methods and the raw material's ripening stage, as well as other elements like the weather, are crucial for producing high-quality spices. When the active component contents are at their peak, harvesting can be done mechanically or manually (Krajayklang et al. 2000; Pruthi 2003). To improve colour features, reduce storage and transport volume, and speed up the drying process, a number of post-harvest treatments are used before drying, such as punching and curing capsicum fruits (Bosland et al. 2012; Pruthi 2003). Additionally, to enhance colour qualities and manage microbial and insect infestations in cardamom and ginger, fumigation with ethylene oxide, along with other chemical treatments such as sodium hypo chloride, sulphur dioxide, and hydrogen peroxide, is occasionally used (Subbulakshmi and Mridula Naik 2002). But in the European Union, many treatments are out-of-date, restricted, or even prohibited by legislation.

9.2.3 Process of Drying the Raw Material

In order to create spice products, the raw material is typically dried to a hygienically safe moisture content (about 10%, wet basis), either whole or in a cut form. The plant material is often stretched out on open surfaces in tropical and subtropical areas, occasionally on elevated platforms or racks, as well as sun-dried for a few days. Even though it might be challenging to maintain consistent temperatures, stoves that employ heated iron pipes or an open fire are another crucial drying technique (Subbulakshmi and Mridula Naik 2002). To minimize discoloration and the loss of volatile oils, various mechanical dryers are being used to develop hygienic conditions, speed up the drying process, and maintain consistent drying temperatures between 45 and 60 °C (Díaz-Maroto et al. 2003, Ibrahim et al. 1997). Solar drying is a cutting-edge, promising technology that is being widely used for the drying of spices in industrialized nations, but it is not suitable for small farms in developing countries. Nonetheless, natural sun-drying continues to be the most popular technique, with very few people using artificial or solar dryers (Das and Sarma 2001). Moreover, the selection of proper drying techniques with respect to the types of spices, with special emphasis on its heat sensitivity, is very important as the type of drying with involvement of air along with temperature and humidity during the drying process directly affects volatile and phenolic characteristics of spices (Guclu et al. 2021).

9.2.4 Reduction in Particle Size

Following drying, air classifiers, gravity separators, magnets, destoners, as well as colour and sieve graders, are used to separate the contaminated spices and the raw materials that have been damaged or discoloured (Subbulakshmi and Mridula Naik 2002). The majority of spices are subjected to a reduction in particle size, as it may be used directly in food products. A variety of grinding tools are being used in the spice industry, such as plate mills or hammer mills (Pruthi 1980; Klieber and Bagnato 1999). Cooling techniques such as air blowing through the grinders, pre-chilled feed material, nitrogen liquid, jacketed water-cooled units are frequently used to reduce losses and degradation of valuable compounds that frequently occur during the grinding operation (Duxbury 1991; Bera et al. 2001). The flavour ingredients are better preserved because of the low temperature. Moreover, reduced particle size of whole spices results in increased hygroscopicity and thereby elevated moisture content during storage and transportation, which eventually causes higher microbial contamination in powdered spices as compared to whole spices. Therefore, in the dried spices sector, proper attention must be given to maintaining good storage and hygienic practices, along with process controls and Food Safety Management Systems (FSMSs) (Karam et al. 2021).

9.2.5 Storage and Packaging

Following the process, the spices are often placed in gunny bags and kept in storage for a period of between one and three months before being supplied to the processing sector and traders. Storage is another critical control point, besides the drying process, where temperature and humidity must be carefully managed for safe storage of spices (Peter et al. 2003; Pruthi 2003). Microbial contamination in spices is directly influenced by the types of packaging material used and storage conditions maintained during the complete supply chain from processor to consumers, as a lot of temperature fluctuations and along with alterations in storage conditions result in a rise in microbial growth beyond unacceptable limits for consumption (Choi et al. 2021; Esmaeili et al. 2021; Rahman et al. 2021).

9.2.6 Limitations with Conventional Spice Grinding

Grinding generates a considerable amount of heat, which results in the loss of product quality. The damage varies depending on the type of volatile oil as well as the moisture level of the seeds during grinding, but the final product loses a significant amount of volatile oil, up to 30–40%, as well as its original colour. During the standard grinding process, the decreased volatile oil in several spices, such as nutmeg (37%), mace (14%), cinnamon (17%), oregano (17%), and caraway seed (32%), has been reported (Pruthi and Misra 1963; Andres 1976; Pesek and Wilson 1986). Such produce will not be accepted by the importing countries since it does not meet international quality requirements. Most spices include a considerable quantity of fatty oils in addition to volatile oils, which melt when heated and prevent the grinder from operating continuously because the melted fat causes the ground powder to adhere to the grinder's surface. Heat-induced fat melting is another issue that prevents the grinder from operating continuously. Slow-revolving roller mills, on the other hand, can lessen volatile loss, but they diminish the grinder's energy efficiency and are not appropriate for materials with high fat content and heat sensitivity. Moreover, conventional spice grinding methods (Burr plates, granite plates, grinding stones, and grinders) and frequencies (single grinds, double grinds, and triple grinds) are greatly impacted by the grinding process, as conventional grinding equipment results in contamination of iron content in spices (Adejumobi et al. 2025).

9.3 Cryogenic Grinding Technology

The International Institute of Refrigeration defined cryogenics as 'the scientific and technological fields that work with temperatures below −153 °C'. A very cold fluid (preferably in gas form) is typically used in cryogenic grinding and other related fields, which can cause

Table 9.2 Cryogen temperature.

Cryogen	Triple point	Critical point	Normal boiling point
Argon	−189.40	−122.30	−185.90
Helium	−271.00	−268.00	−29.00
Hydrogen	−259.40	−240.00	−252.80
Oxygen	−218.80	−118.60	−183.00
Nitrogen	−210.00	−147.00	−19.90

Source: Balasubramanian et al. (2012), Qiu et al. (2022), Ashok et al. (2021), and Sossa et al. (2021).

the sample to become brittle when ground (Balasubramanian et al. 2012). Cryogens such as liquid nitrogen, argon, xenon, oxygen, liquefied natural gas (LNG)/methane, neon, hydrogen, krypton, helium, and carbon dioxide help reduce the heat generated during the grinding process. These fluids serve various purposes: Liquid hydrogen and liquid oxygen are used as cryogenic liquid propellants for rocket engines; liquid nitrogen and liquid oxygen are employed in metallurgical and chemical processes; and liquid hydrogen functions as a clean energy carrier in transportation. In biological applications, cryogenics is used for artificial insemination (Ostashko 1963), freezing biological systems, embryo preservation (Steponkus et al. 1990), pulverization of soft tissues (Chan et al. 1996), pulverization of teeth for the extraction of DNA (Sweet and Hildebrand 1998), along with electron cryotomography (Iancu et al. 2006). These are just a few of the biological applications of cryogenics. Kamogawa et al. (2001) used cryo-milling to grind a hair sample to identify heavy metals. Current cryopreservation methods, cryobiological variables, cryoinjury mechanisms, and cryoprotectant action were all reported in 2009 (Sieme et al. 2008). The temperature profile of different cryogenic fluids being used is given in Table 9.2.

The recent developments in cryogenic technology for high-quality spices have been reported by Qiu et al. (2022), Ashok et al. (2021), and Sossa et al. (2021).

9.3.1 Benefits of Cryogenic Grinding for Spices

Cryogenic applications revolutionized the food sector by enabling the processing of foods at low temperatures and rapid freezing for the preservation of various fresh food commodities. Additionally, industrial gases (not limited to cryogens) are used for the promotion of plant and floral growth, support seed germination, and enhance greenhouse environments. In aquaculture, oxygen is introduced to the water to increase output. Spice quality is determined by both extrinsic characteristics, such as appearance, texture, form, impurities, and colour, and intrinsic characteristics, such as the retention of oleoresins, alkaloids, and volatile oils. Furthermore, several health needs, such as heavy metals, aflatoxin, pesticide residue, sulphur dioxide, solvent residues, as well as microbiological quality, were envisioned by specific export quality standards.

The physicochemical characteristics are greatly influenced by the agroclimatic conditions of the producing area, as well as by harvest and post-harvest processes. However, as these characteristics define its grade in the market, the physicochemical quality of spices continues to be the most important factor for export requirements. Therefore, higher technology is required for production, processing, packing, and shipping due to the current strict laws and quality preferences. As a result, cryogenic grinding of spices exhibits encouraging qualities for export or domestic market promotion (Balasubramanian et al. 2012). Moreover, cryogenic grinding results in minimum loss of volatile oils, moisture, and colour degradation, along with retention of flavour compounds. It is also reported that cryogenically ground spices are of superior quality compared to conventional grinding technologies being used for spices (Ashok et al. 2021; Sossa et al. 2021).

9.3.2 Benefits of Cryogenic Grinding over Conventional Grinding

The issue with spices is typically their fat content, but other factors to take into account include the particle size of the product, yield, homogeneity, freedom from contamination, economy, and operating dust. The resulting cryogenically ground product has double the flavour power per unit mass as the traditionally ground spice product. According to Pruthi and Misra (1963) and Ashok et al. (2021), during routine grinding, the product's temperature rose from 42 to 95 °C, while a considerable amount of its volatile oils or flavouring components were lost. In conventional grinding mills, the amount of heat production is somewhat decreased by the use of cooling water, circulating air, etc.

According to Bera et al. (2001), when cumin seeds were ground conventionally, the product temperature increased to 95 °C, and a significant amount of volatile oil and its contents were lost. At room temperature, producing the finest powders from viscoelastic and plastic materials is extremely expensive, energy-intensive, and sometimes impossible. Cryogenic grinding, also called freezer grinding, freezer milling, or cryo-milling, is the process of grinding material near its glass transition temperature. This technique enables cost-effective grinding of heat-sensitive, thermoplastic, and elastic materials into extremely fine particle sizes. Table 9.3 lists the possible distinctions between traditional and cryogenic spice grinding. The use of cryogens, such as liquid nitrogen or liquid carbon dioxide, has recently been encouraged in the milling industry to preserve food quality.

Besides, it has also been found that cryogenic cooling fluids are among sustainable methods that involve minimal use of cutting fluid compared to conventional cutting fluids, which pose significant environmental and biological risks (Manikanta et al. 2024).

Table 9.3 Comparison between cryogenic grinding and conventional grinding.

Characteristics	Cryogenic	Conventional
Consumption of energy	Low	High
Throughput	High	Low
Clogging of the mill	Not found	Frequent
Losses of volatile	Minimum	Higher
Capacity of motor	Low	High

Source: Adapted from Balasubramanian et al. (2012).

9.4 Design and Engineering of Cryogenic Grinding Equipment

Analysis of physico-mechanical properties of the feeding material helps in designing the feeder type, determining residence time, selecting grinder capacity, screw diameter and length, choosing the conveyor, setting the feed rate and rotor speed, optimizing cooling load and energy consumption, and assessing the deformation pattern and fineness modulus of the cryogenic grinder's final product (Balasubramanian et al. 2012). Spices were ground using varying feed screw speeds. The LN_2 Dewar was removed during grinding to perform standard grinding. Compared to ambient grinding, cryogenic grinding produced lower grinding characteristics and values for energy constants, specific energy consumption, as well as particle size of spices such as turmeric, fenugreek, coriander, cinnamon, as well as black pepper (ICAR 2014). The thermal conductivity as well as diffusivity of cryo-ground black pepper powder were discovered to have quadratic correlations with moisture content (Barnwal et al. 2013). Temperature and moisture content have a major impact on the specific heat. Additionally, Barnwal et al. (2015) studied certain physical and mechanical properties of fenugreek, cinnamon, and turmeric, and it was found that cryogenically ground turmeric with variable moisture contents exhibited improved colour and higher thermal conductivity.

Meghwal and Goswami (2014) compared the grinding methods for black pepper and fenugreek seeds using various types of mills. While a ball mill was found to be suitable for small-scale grinding, rotor, pin mills, and hammer mills were observed to be suited for continuous pulverization on a large scale. During cryogenic grinding, they reported a much lower specific power usage. Seed oil and powder become lodged in the grinder's cover and clog the sieves when grinding at room temperature. For the same sieve opening, it was discovered that the clogging of the sieve reduced as the grinding temperature dropped, and vice versa (ICAR 2014).

Since liquid nitrogen evaporates quickly at room temperature, insulating material was prioritized in the design of the cryogenic pre-cooling unit. Feed and building materials cause heat loss during cooling. In order to prevent cooling heat losses, the outside casing was appropriately insulated. Singh and Goswami (1999b) calculated that 20% of the cooling heat loss occurred during the cryogenic pre-cooling process, with the cooling load in the liquid and gaseous zones being approximately 70% and 30%, respectively. In addition to this retention period, the cryogenic grinder's design considered the amount of time needed to reach the proper temperature of the spices. A modified version of Plank's formula was used to determine the retention or freezing duration, and a standard relation was used to determine the pre-cooling screw conveyor's capability (Singh and Goswami 1999a).

The cryogenic spice grinding system is a small, innovative, and one-of-a-kind system that was scientifically designed to accommodate both hammer and pin mills. It has a control panel that allows precise control of all operational parameters, and additionally, it contains a feature that allows used LN_2 vapour to be recycled to the pre-cooling system's inlet, saving LN_2 consumption and making the system cost-effective. Satisfactory safety precautions have been included to shield the grinder from accidents.

9.4.1 Construction Methods

Materials behave differently at lower temperatures than they do at room temperature. According to Marquardt et al. (2002), some of the most popular materials for cryogenic systems include G-10CR fiberglass epoxy, oxygen-free copper, 718 Inconel, 6061-T6 aluminium, Kevlar 49, beryllium copper polyamide (nylon), 304 stainless steel, niobium titanium (NbTi), polyimide, Ti-6Al-4V titanium alloy, and Teflon. When choosing the right building material for cryogenics, material qualities like specific heat, thermal conductivity, and thermal expansion are primarily taken into account. Cryogenic treatment can extend the service life of drill bits, milling tools, and piston rings by 100–1,200%, while carbide tools and copper electrodes last 400% longer. Treated dies may have a lifespan 200–800% greater than that of untreated ones.

9.4.2 Pre-coolers

Various types of pre-coolers and cryo-freezers are used, including flighted freezers, spiral freezers, straight-belted freezers (standard carbon dioxide or liquid nitrogen tunnels, impingement tunnels, liquid nitrogen immersion tunnels), rotary drum freezers, triple-pass freezers, and cryo-mechanical systems. In the food industry and spice grinding, pre-coolers are typically straight-belt or spiral-type freezers, with spiral freezers requiring less labour, space, cost, technical expertise, and dwell time to operate. The pre-cooling device consists of a screw conveyor enclosed in an insulated barrel with a liquid nitrogen inlet. Mixing the raw material with liquid nitrogen enhances heat transfer. Key engineering considerations for the pre-cooler include screw design to match grinding capacity, a liquid nitrogen control mechanism for regulation, and the retention time of spices in the unit (in both liquid and gaseous states of nitrogen) (Wagner 1972; Singh and Goswami 1999a). Residence time of spice seeds in the liquid and gaseous state of nitrogen can be calculated by Modified Plank's equation.

9.4.3 Distribution System for Liquid Nitrogen

It consists of a distributor, a cryogenic delivery system, a cryogenic storage vessel, a conveyor-style heat exchanger connected to distribution lead tubes, etc. The cryogenic delivery system comprises vacuum-insulated supply hoses and valves, utilizing a redundant safety valve along with a proportional control valve. To distribute cryogen and create a pressure drop zone for enthalpic processes, the system includes one or more relief valves. Nozzles are used by the distributor to keep the cryogen distributed uniformly. The re-liquefier often recycles the gas to a different recovery process or reuses the cryogen or exhaust capture unit.

9.4.4 Grinder

Compression, cutting, shearing, impact, or combinations of these methods are used by size reduction machines. While tough material necessitates a combination of impact, cutting, and shearing. Cutting, shearing, and, in some cases, compression and impact can all degrade soft material (with a Mohs hardness of 1–4). During size reduction, such material typically behaves like a difficult liquid, dispersing, agglomerating, and homogenizing. The choice of the most suitable mechanism (individual or combined) for material reduction depends on the properties of the constituents and the desired outcomes. Selecting a proper size reduction machine needs knowledge of the material's size, shape, type, hardness, and other properties affecting the size reduction mechanisms and their handling.

There are three types of grinders: pin mills (Murthy and Bhattacharya 2008), attrition mills (Goswami and Singh 2003), and hammer mills (Singh and Goswami 1999a). For improved results, pin mills are typically utilized in contemporary spice grinding. The pin mill also can switch out the pins according to the physical properties of the raw materials and spices. To prevent overfeeding the grinder, which raises the temperature in the grinding zone, design considerations should balance the pre-cooler's capacity with the grinders. Nevertheless, there is a dearth of information on the design specifications for cryogenic spice grinders. Although it requires careful design, liquid nitrogen must be distributed efficiently to both the pre-cooler and the grinder at minimal cost. As the thermal and physical characteristics of the raw material are the primary consideration for a cryogenic or low-temperature grinder, specialized design processes and systems are necessary to ensure optimal liquid nitrogen utilization.

9.4.5 Sensors

Although temperature sensors operate on various principles, resistance temperature detectors (RTDs) are widely used for accurate measurement due to their high accuracy, stability, and repeatability. Their relative immunity to electrical noise makes RTDs ideal for the measurement of temperature in industrial environments, especially near high-voltage equipment, motors, and generators. Because the probe's resistance becomes unnoticeable at very low temperatures, a standard sensor is ineffective at cryogenic temperatures. Therefore, for cryogenic temperature measurements, specialized sensors including capacitance sensors, RTDs (carbon–glass, ruthenium oxide, rhodium–iron, platinum, or germanium), and diodes (silicon or GaAIAs) are employed. The most widely used RTDs, however, are platinum RTDs, which have a high sensitivity and follow an industry standard curve (−200 to 600 °C). These RTDs are cheap and require basic instrumentation.

9.5 Impact on Spice Quality and Preservation

9.5.1 Effects of Cryogenic Processing on Flavour and Aroma Retention

Enhancing shelf life, preserving food by rapid freezing, and grinding food are essential components of the food processing sector. With its low-temperature uses, cryogenics has proven to be a potent tool in the food business. Since the latter half of the twentieth century, many researchers have studied the application of cryogenics in freezing food products to enhance quality, shelf-life extension, and enhance production capacity (Cowley et al. 1962; Streeter and Spencer 1973; Arafa and Chen 1978). According to De Kock et al. (1995), foods frozen under cryogenic settings create higher-quality products because they quickly freeze the cellular starch as opposed to slowly freezing it. Awonorin (1997) used liquid nitrogen spray to assess several freezers. Later, Miller and Roberts (2001) investigated ways to reduce the cryogenic grinding system's beginning costs. Ramakrishnan (2004) used simulation modelling to assess grinding factors for precise control on freezers, improving economics by enhancing their low set-up costs, high flexibility, along with advantages in the food business.

Information on cryogenics in spice grinding has been reported in recent years (Goswami 2010; Balasubramanian et al. 2012). Bharat produces and consumes a lot of spices, which are essential components of many different dishes and cuisines. It is recommended to utilize ground spices for improved acceptability of prepared treats. The most common processed spice items nowadays are ground spice powders, either alone or in the right ratio with other spices. When it comes to ground spice powder marketing, quality and cleanliness are the main priorities. The industries that process spices can better investigate the advantages of cryogenic grinding. When spices are ground at cryotemperature, their original colour is preserved (Pesek and Wilson 1986). Additionally, using cryo-grinding equipment on a laboratory scale, it was found that enhanced black pepper product attributes were enhanced (Murthy et al. 1996). Compared to 1.15% in standard grinding technology, cryoground black pepper has substantially higher volatile oil (2.61%) (Jacob et al. 2000).

The comparison of cryogenic grinding technology with well-established methods was examined (Singh and Goswami 1999b). It was found that when cumin is ground conventionally, the product temperature rises to 95 °C, which significantly reduces volatile chemicals (Bera et al. 2001). The impact of grinding turmeric rhizomes at a cryogenic temperature on the size of the particles and thermally sensitive chemicals in turmeric oils was investigated (Manohar and Sridhar 2001). In the end, the grinding proved to be more effective than the previous conventional methods. Recent studies on the quality profile of spices and other products have been reported with significant retention of flavour, aroma compounds, and overall quality as compared to conventional pulverization technologies (Singh et al. 2022; Sardar et al. 2024; Yan et al. 2025).

9.5.2 Comparison with Traditional Spice Processing Method

The primary problem with spices is typically their fat content, but other factors to consider include yield, product particle size, homogeneity, economy, freedom from contamination, and operating dust. The resulting cryogenically ground product has double the flavour power per unit mass as the traditionally ground spice product. According to Pruthi and Misra (1963), during routine grinding, the product's temperature rose from 42 to 95 °C, while a considerable amount of its flavouring or volatile oil was lost. In conventional grinding mills, the rate of heat production is somewhat decreased by the use of cooling water, circulating air, etc. Cryogenic grinding, also called freezer grinding, freezer milling, or cryo-milling, involves grinding material near its glass transition temperature. This method allows heat-sensitive materials to be ground economically to extremely fine particle sizes. Recently, the use of cryogens such as liquid nitrogen or liquid carbon dioxide in grinding has gained attention in the milling industry for preserving food quality. Sossa et al. (2021), Ashok et al. (2021), and Qiu et al. (2022) reported the benefits of cryogenic grinding over conventional grinding techniques being used in the spice industry.

9.6 Challenges and Limitations of Cryogenic Grinding

Many advantages and benefits have been reported for cryogenic grinding of spices with better retention of aromatic compounds, flavour, and colour. However, sustainability has become a major concern in the industry, particularly regarding the reduction of cutting fluid consumption through lowest quantity lubrication, the use of cryogenics, and hybrid lubrication-cooling techniques, which still require thorough exploration. Another issue is the management of grinding swarf. Ongoing advancements demonstrate the potential of the cryogenic grinding process, indicating a promising and expanding future (Kishore et al. 2022). Moreover, other challenges still exist, such as higher initial investment costs, and it also requires proper handling mechanisms for cryogenic liquids. However, improved product quality, longer shelf life, enhanced flavour, and aroma retention outweigh these major concerns of the spice processor, particularly for high-value spices.

9.7 Conclusion

Cryogenic grinding utilizes extremely low temperatures of grinding, mostly using liquid nitrogen, to make spices with fine particles and superior quality. Cryogenic grinding retains flavours, volatile oils, vibrant colour, and improved sensory qualities. Also, cryogenic grinding extends shelf life by reducing oxidation and microbial growth, improves processing efficiency by avoiding clogging in oily or moist spices, and produces ultra-fine, uniform particles suitable for diverse culinary and industrial uses. However, cost effectiveness and sustainability are still major concerns of the spice processors and the cryogenic industry that need to be addressed so that spice processors in developing countries can easily afford and adopt this technology for size reduction and pulverization of spices and other food/non-food products.

References

Adejumobi, A.O., Anifowose, M.T., and Alabi, A.A. (2025). Evaluating the impact of grinding methods and frequency on iron chip contamination in ground pepper. *African Journal of Agricultural Science and Food Research* 19: 82–96.

Almela, L., Nieto-Sandoval, J.M., and Fernández López, J.A. (2002). Microbial inactivation of paprika by a high-temperature short-x time treatment. Influence on color properties. *Journal of Agricultural and Food Chemistry* 50: 1435–1440.

Andres, C. (1976). Grinding spices at cryogenic temperatures retains volatiles and oils. *Food Processing* 37: 52–53.

Arafa, A. and Chen, T. (1978). Liquid nitrogen exposure as an alternative means of chilling poultry. *Journal of Food Science* 43: 1036–1037.

Ashok, K.K., Chandra, B.H., Sunil, S., and Kulkarni, G.V. (2021). Application of cryogenics in grinding of spices for value addition: a review. *Natural Volatiles and Essential Oils* 8: 10580–11059.

Awonorin, S.O. (1997). An appraisal of the freezing capabilities of tunnel and spiral belt freezers using liquid nitrogen sprays. *Journal of Food Engineering* 34: 179–192.

Balasubramanian, S., Gupta, M.K., and Singh, K. (2012). Cryogenics and its application with reference to spice grinding: a review. *Critical Reviews in Food Science and Nutrition* 52: 781–794.

Balasubramanian, S., Roselin, P., Singh, K. et al. (2016). Postharvest processing and benefits of black pepper, coriander, cinnamon, fenugreek, and turmeric spices. *Critical Reviews in Food Science and Nutrition* 56: 1585–1607.

Banerjee, M. and Sarkar, P.K. (2003). Microbiological quality of some retail spices in India. *Food Research International* 36: 469–474.

Barnwal, P., Singh, K., Kumar, R., and Zachariah, T. (2013). Thermal properties of cryo-ground powder of black pepper (Panniyur 1). *Journal of Spices and Aromatic Crops* 22 (2).

Barnwal, P., Singh, K., Mohite, A. et al. (2015). Influence of cryogenic and ambient grinding on grinding characteristics of fenugreek powder: a comparative study. *Journal of Food Processing and Preservation* 39: 1243–1250.

Bera, M., Shrivastava, D., Singh, C. et al. (2001). Development of cold grinding process, packaging and storage of cumin powder. *Journal of Food Science and Technology-Mysore* 38: 257–259.

Bosland, P.W., Votava, E.J., and Votava, E.M. (2012). *Peppers: Vegetable and Spice Capsicums*. Cabi.

Chan, E., Menovsky, T., and Welch, A.J. (1996). Effects of cryogenic grinding on soft-tissue optical properties. *Applied Optics* 35: 4526–4532.

Cho, S.-H., Lee, C.-H., Jang, M.-R. et al. (2008). Aflatoxins contamination in spices and processed spice products commercialized in Korea. *Food Chemistry* 107: 1283–1288.

Choi, Y.-M., Hwang, S.-I., Park, G.-Y. et al. (2021). Microbial contamination levels of red pepper powder purchased in gyeonggi province and changes in characteristics according to the storage method. *Journal of Food Hygiene and Safety* 36: 382–391.

Condori, M., Echazu, R., and Saravia, L. (2001). Solar drying of sweet pepper and garlic using the tunnel greenhouse drier. *Renewable Energy* 22: 447–460.

Cowley, C., Timson, W., and Sawdye, J. (1962). A method for improving heat transfer to a cryogenic fluid. In: *Advances in Cryogenic Engineering: Proceedings of the 1961 Cryogenic Engineering Conference University of Michigan Ann Arbor, Michigan August 15–17, 1961*, 385–390. Springer.

Das, P. and Sarma, S. (2001). Drying of ginger using solar cabinet dryer. *Journal of Food Science and Technology-Mysore* 38: 619–621.

De Kock, S., Minnaar, A., Berry, D., and Taylor, J. (1995). The effect of freezing rate on the quality of cellular and non-cellular par-cooked starchy convenience foods. *LWT- Food Science and Technology* 28: 87–95.

Díaz-Maroto, M.C., Pérez-Coello, M.S., Gonzalez Vinas, M., and Cabezudo, M.D. (2003). Influence of drying on the flavor quality of spearmint (*Mentha spicata L.*). *Journal of Agricultural and Food Chemistry* 51: 1265–1269.

Duxbury, D. (1991). Cryogenically milled natural spices extend stability. *Food Processing* 52: 70.

Esmaeili, S., Berengi-Ardestani, S., Khanniri, E. et al. (2021). Effect of storage time on the microbial and physicochemical properties of gamma irradiated turmeric powder under various atmospheres of packaging. *Radiation Physics and Chemistry* 187: 109580.

Farkas, J. and Mohácsi-Farkas, C. (2011). History and future of food irradiation. *Trends in Food Science & Technology* 22: 121–126.

Goswami, T.K. (2010). Role of cryogenics in food processing and preservation. *International Journal of Food Engineering* 6.

Goswami, T. and Singh, M. (2003). Role of feed rate and temperature in attrition grinding of cumin. *Journal of Food Engineering* 59: 285–290.

Guclu, G., Keser, D., Kelebek, H. et al. (2021). Impact of production and drying methods on the volatile and phenolic characteristics of fresh and powdered sweet red peppers. *Food Chemistry* 338: 128129.

Hara-Kudo, Y., Ohtsuka, K., Onoue, Y. et al. (2006). *Salmonella* prevalence and total microbial and spore populations in spices imported to Japan. *Journal of Food Protection* 69: 2519–2523.

Iancu, C.V., Wright, E.R., Heymann, J.B., and Jensen, G.J. (2006). A comparison of liquid nitrogen and liquid helium as cryogens for electron cryotomography. *Journal of Structural Biology* 153: 231–240.

Ibrahim, H.M., Ragab, G., and Moharram, H. (1997). Paprika color quality inview of air and natural drying treatments. *Grasas y Aceites* 48: 200–206.

Jacob, S., Kasthurirengan, S., and Behra, U. (2000). Development of a cryogenic grinding plant for spices. *Paper presented at Cryogenic Engineering Conference and International Cryogenic Materials Conference, 45(B), Montreal, Quebe*, pp. 1731–1738.

Jung, H., Lee, Y.J., and Yoon, W.B. (2018). Effect of moisture content on the grinding process and powder properties in food: a review. *Processes* 6: 69.

Kamogawa, M.Y., Nogueira, A.R.A., Costa, L.C.M. et al. (2001). A new strategy for preparation of hair slurries using cryogenic grinding and water-soluble tertiary-amines medium. *Spectrochimica Acta Part B: Atomic Spectroscopy* 56: 1973–1980.

Karam, L., Salloum, T., El Hage, R. et al. (2021). How can packaging, source and food safety management system affect the microbiological quality of spices and dried herbs? The case of a developing country. *International Journal of Food Microbiology* 353: 109295.

Kishore, K., Sinha, M.K., Singh, A. et al. (2022). A comprehensive review on the grinding process: advancements, applications and challenges. *Proceedings of the Institution of Mechanical Engineers, Part C: Journal of Mechanical Engineering Science* 236: 10923–10952.

Klieber, A. and Bagnato, A.T. (1999). Colour stability of paprika and chilli powder.

Krajayklang, M., Klieber, A., and Dry, P.R. (2000). Colour at harvest and post-harvest behaviour influence paprika and chilli spice quality. *Postharvest Biology and Technology* 20: 269–278.

Liu, H., Zeng, F., Wang, Q. et al. (2013). The effect of cryogenic grinding and hammer milling on the flavour quality of ground pepper (*Piper nigrum L.*). *Food Chemistry* 141: 3402–3408.

Manikanta, J.E., Ambhore, N., and Nikhare, C. (2024). Application of sustainable techniques in grinding process for enhanced machinability: a review. *Journal of the Brazilian Society of Mechanical Sciences and Engineering* 46: 199.

Mankee, A., Ali, S., Chin, A.-L. et al. (2005). Microbial quality of "doubles" sold in Trinidad. *Food Microbiology* 22: 601–607.

Manohar, B. and Sridhar, B. (2001). Size and shape characterization of conventionally and cryogenically ground turmeric (*Curcuma domestica*) particles. *Powder Technology* 120: 292–297.

Marquardt, E., Le, J., and Radebaugh, R. (2002). Cryogenic material properties database. In: *Cryocoolers 11*. Springer.

Meghwal, M. and Goswami, T. (2014). Comparative study on ambient and cryogenic grinding of fenugreek and black pepper seeds using rotor, ball, hammer and Pin mill. *Powder Technology* 267: 245–255.

Miller, J. and Roberts, W. (2001). How to minimize startup costs. Process cooling and equipment.www.process–cooling.com.

Murthy, C. and Bhattacharya, S. (2008). Cryogenic grinding of black pepper. *Journal of Food Engineering* 85: 18–28.

Murthy, C., Krishnamurthy, N., Ramesh, T. et al. (1996). Effect of grinding methods on the retention of black pepper volatiles. *Journal of Food Science and Technology* 33: 299–301.

Oberoi, H., Ku, M., Kaur, J., and Baboo, B. (2005). Quality of red chilli variety as affected by different drying methods. *Journal of Food Science and Technology-Mysore* 42: 384–387.

Ostashko, F. (1963). On the cause of cold shock of spermatozoa, artificial insemination of agricultural animals. *Research Institute for Animal Breeding*, Kharkov, USSR, pp. 15–32.

Öztekin, S., Başçetinçelik, A., and Soysal, Y. (1999). Crop drying programme in Turkey. *Renewable Energy* 16: 789–794.

Pesek, C. and Wilson, L. (1986). Spice quality: effect of cryogenic and ambient grinding on color. *Journal of Food Science* 51: 1386–1386.

Peter, K., Indira, P., and Mini, C. (2003). The cultivation and processing of capsicum in India. In: *Capsicum*. CRC Press.

Pruthi, J.S. (1980). Spices and condiments: chemistry, microbiology, technology. *Advances in Food Research* 4: 1–449.

Pruthi, J.S. (2003). Advances in post-harvest processing technologies of capsicum. In: *Capsicum*, 195–233. CRC Press.

Pruthi, J.S. and Misra, B.D. (1963). Studies on the physico-chemical and microbiological changes in curry powders during their drying, milling and mixing operations. *Spices Bulletin* 3: 8–13.

Qiu, L., Zhang, M., Mujumdar, A.S., and Liu, Y. (2022). Recent developments in key processing techniques for oriental spices/herbs and condiments: a review. *Food Reviews International* 38: 1791–1811.

Rahman, M., Islam, M., Das, K.C. et al. (2021). Effect of gamma radiation on microbial load, physico-chemical and sensory characteristics of common spices for storage. *Journal of Food Science and Technology* 58: 3579–3588.

Ramakrishnan, S., Wysk, R.A., and Prabhu, V.V. (2004). Prediction of process parameters for intelligent control of tunnel freezers using simulation. *Journal of Food Engineering* 65: 23–31.

Ramesh, M., Wolf, W., Tevini, D., and Jung, G. (2001). Influence of processing parameters on the drying of spice paprika. *Journal of Food Engineering* 49: 63–72.

Sahar, N., Arif, S., Iqbal, S. et al. (2015). Moisture content and its impact on aflatoxin levels in ready-to-use red chillies. *Food Additives and Contaminants: Part B* 8: 67–72.

Sahay, K.M. and Singh, K.K. (1996). *Unit Operations of Agricultural Processing*. Vikas Publishing House Pvt. Ltd.

Sardar, N., Prasad, R., Modi, R., and Rathod, J. (2024). Optimizing volatile oil retention in ajwain (*Trachyspermum ammi L.*) through cryogenic grinding techniques. *Journal of Applied Horticulture* 26: 275–279.

Scharnow, R. (1986). Codiertes Handbuch der Güter des Seetransports. *VE Kombinat Seeverkehr und Hafenwirtschaft-Deutfracht/Seereederei-Ingenieurhochschule für Seefahrt Warnemünde/Wustrow, Rostock Bd,* 1.

Sieme, H., Harrison, R.A.P., and Petrunkina, A.M. (2008). Cryobiological determinants of frozen semen quality, with special reference to stallion. *Animal Reproduction Science* 107: 276–292.

Singh, K.K. and Goswami, T.K. (1999a). Design of a cryogenic grinding system for spices. *Journal of Food Engineering* 39: 359–368.

Singh, K.K. and Goswami, T.K. (1999b). Studies on cryogenic grinding of cumin seed. *Journal of Food Process Engineering* 22: 175–190.

Singh, B. and Solanki, R.K. (2015). Status of seed spices research and development in India. *The Indian Journal of Agricultural Sciences* 85: 151–156.

Singh, H., Meghwal, M., Prabhakar, P.K., and Kumar, N. (2022). Grinding characteristics and energy consumption in cryogenic and ambient grinding of ajwain seeds at varied moisture contents. *Powder Technology* 405: 117531.

Sossa, J.W., Pérez, P.J.P., Piedrahita, J.C.P., and Mesa, A.F.R. (2021). Trends in grinding of agroindustrial products-a literature review. *Recent Patents on Food, Nutrition & Agriculture* 12: 19–28.

Spice Board of India (2019). Spice statistics. Spice Board Cochin, Kerala, p. 281.

Steponkus, P.L., Myers, S.P., Lynch, D.V. et al. (1990). Cryopreservation of drosophila melanogaster embryos. *Nature* 345: 170–172.

Streeter, E.M. and Spencer, J.V. (1973). Cryogenic and conventional freezing of chicken. *Poultry Science* 52: 317–325.

Subbulakshmi, G. and Mridula Naik, M.N. (2002). Nutritive value and technology of spices: current status and future perspectives. *Journal of Food Science and Technology-Mysore* 39: 319–344.

Sweet, D. and Hildebrand, D. (1998). Recovery of DNA from human teeth by cryogenic grinding. *Journal of Forensic Sciences* 43: 1199–1202.

Tateo, F. and Bononi, M. (2006). Determination of ethylene chlorohydrin as marker of spices fumigation with ethylene oxide. *Journal of Food Composition and Analysis* 19: 83–87.

Tripathi, S. and Mishra, H. (2009). Studies on the efficacy of physical, chemical and biological aflatoxin B1 detoxification approaches in red chilli powder. *International Journal of Food Safety, Nutrition and Public Health* 2: 69–77.

Wagner, R.C. (1972). Engineering considerations for the design of a cryogenic food freezer. *Application of Cryogenic Technology* 4: 202–214.

Yan, J., Zhao, S., Liang, Z. et al. (2025). Application of cryogenic grinding combined with ultrasonic pretreatment to enhance the yield and quality of essential oils from nutmeg (*Myristica fragrans* Houtt.). *Sustainable Chemistry and Pharmacy* 44: 101979.

Zachariah, T.J., Leela, N.K., and Shamina, A. (2012). Methods of analysis of herbs and spices. In: *Handbook of Herbs and Spices*, 89–117. Woodhead Publishing.

10

Preserving Nature's Remedies: Cryogenic Applications in Herbal Medicine Processing

K. U. Pavitra Krishna

Department of Food Science and Processing Management, Subbalakshmi Lakshmipathy College of Science, Madurai, Tamil Nadu, India

10.1 Introduction

Cooling applications in the food industry are made possible by cryogenic technology. Using cryogenic technologies to preserve and treat food holds immense potential. The most effective method for ensuring the long-term survival of plant and animal species is cryopreservation. Because cryogenic preservation ensures food safety and quality, it has enormous promise for the nation's exports as well as local consumption. Evaporative air-cooling devices are used in most sectors; cryogenic preservation is less prevalent in this sector. The main quality factors when freezing foods, such as meat products, are shrinkage, loss of tenderness, toughening, shelf life, and microbiological activity, dehydration losses and drip loss. Cryogenic technology preservation will enhance the circumstance. Appropriate economic considerations, such as system life and payback period, should be made when selecting cryogenic applications. In India, cryogenic grinding of spices is a recent innovation that requires promotion and demonstration in spice pilot facilities. The ability to process food using indigenous cryogenic technology would guarantee the creation of higher-quality goods domestically and allow the exported goods to be sold abroad.

The investigation of the way materials behave at extremely low temperatures is known as cryogenics. As stated by the International Institute of Refrigeration (1975), cryogenics is a field of study that encompasses scientific and technological disciplines related to temperatures below −153 °C. An extremely frigid gas in the form of a liquid is typically used in cryogenic grinding and other related fields, which can cause the sample to become brittle when ground. Cryogens are substances such as liquid nitrogen, argon, oxygen, xenon, neon, krypton, helium, hydrogen, liquefied natural gas/methane, and carbon dioxide that reduce the heat produced during the grinding process. These fluids have a variety of applications, such as liquid oxygen (LO_2) and liquid hydrogen (LH_2) in rockets as cryogenic liquid propellants, liquid nitrogen (LN_2) and LO_2 in metallurgical and chemical processes, and LH_2

Cryogenic Processing: Advances in the Food Industry, First Edition. Edited by Sumit Sudhir Pathak, Rama Chandra Pradhan, and Sudarshan Ramanathan.

as a renewable energy source for vehicles. Cryogenics has been used in biology by a number of researchers, including those who have frozen living systems (Peter 1970), preserved embryos (Steponkus et al. 1990), ground soft tissues (Chan et al. 1996), ground teeth for DNA extraction (Sweet and Hildebrand 1998), and performed electron cryotomography (Cristina et al. 2006). To identify heavy metals, Kamogawa et al. (2001) used cryo-milling to grind up a hair sample. Cryopreservation has also been discussed in relation to germ plasm conservation of commercially significant, endangered plant and animal breeds. Choi et al. (2010) examined the thermal properties of biomaterials in the cryogenics regime for use in cryobiology. Sieme et al. (2009) reported on current cryopreservation techniques, cryobiological determinants, and mechanisms of cryo-injury and cryoprotectant action.

Enhancing shelf life, preserving food by rapid freezing, and grinding food are essential components of the food processing sector. With its low-temperature uses, cryogenics has proven to be a potent tool in the food business. Since the second half of the twentieth century, numerous researchers have worked on food products with cryogenic freezing to improve quality, extend shelf life, and increase manufacturing capacity (Hoeft et al. 1973; Arafa and Chen 1978). In his discussion on the use of cryogenics in the food sector, Davidge (1981) covered technological and control systems approaches to distribute LN_2 for the operation's economic viability. Kock et al. (1995) discovered that meals frozen under cryogenic settings rapidly freeze the cellular starch, as opposed to gradual freezing. Awonorin (1997) used liquid nitrogen spray to assess several freezers. Later et al. (2001) investigated ways to reduce the cryogenic grinding system's initial startup costs. By enhancing its cheap setup costs, high flexibility, and advantages in the food business, Ramakrishnan et al. (2004) used simulation modelling to assess grinding parameters for precise control on freezers, leading to improved economics. Recent years have seen the compilation of knowledge on cryogenics in spice grinding by Goswami (2010) and Balasubramanian et al. (2012a, b). Spices, which are vital ingredients in a wide variety of foods and cuisines, are produced and consumed in large quantities in India. For better prepared meal acceptability, ground spices are recommended. Ground spice powders are currently the most widely used processed spice products, either alone or in suitable amounts with other spices. The two most important factors in ground spice powder marketing are cleanliness and quality. The benefits of cryogenic grinding can be more thoroughly examined by the companies that handle spices. Food technology researchers began evaluating cryogenic grinding technology to enhance the quality of spices. When spices are ground at cryotemperature, their original colour is preserved (Pesek and Wilson 1986). Using a laboratory-scale cryogrinding equipment, Murthy et al. (1996) also discovered improved product qualities of black pepper. According to Jacob et al. (2000), the quantity of volatile oil in cryoground black pepper was substantially higher (2.61%) than in standard grinding technique (1.15%). Singh and Goswami (2000) have also examined the comparison of cryogenic grinding technology with well-established methods. Bera et al. (2001) found that when cumin is ground conventionally, the product temperature rises to 95 °C, which significantly reduces volatile chemicals. The impact of grinding turmeric rhizomes at a cryogenic temperature on the size of the particles and thermally sensitive chemicals in turmeric oils was investigated by Manohar and Sridhar (2001). Concurrently, Gouveia et al. (2004) homogenized breakfast cereals using cryogenic grinding technology. Compared to older traditional technologies, the grinding proved to be more efficient. Byadagi chilli was also prepared using cryogenic grinding technology by Mallappa et al. (2015). When compared to other conventional chilli pulverisers, they found that output from cryogenic grinders retained more of its colour, nutrients, and capsaicin concentration. But they expressed worry about the high initial and ongoing expenses associated with the cryogenic grinder.

10.2 Herbal Medicine Processing

In most of the world's civilizations, herbal medicine products have been utilized as the primary source of treatment for centuries (Motti 2021). These products, which are plant-based and derived from plants, contain an abundance of bioactive chemicals that potentially possess therapeutic potentials (Chrysant and Chrysant 2017). Herbal medicine is an integral part of many traditional healings, Ayurveda, and traditional medicine (Kabak and Dobson 2017). The belief in their natural origins and long-standing use are components that have led to their growing prominence. Herbal medicines have existed in many societies often supporting their health issues as well as maintaining the well-being of the collectivities (Alipour et al. 2022).

This tradition has been inherited over time and is foundational to our understanding of the therapeutic potential of much plant variety (Cho and Kim 2020). Herbal medicines have shown considerable promise in treating a variety of conditions such as respiratory, chronic pain, gastrointestinal, digestive, and immune system (Shaikh et al. 2020). Biologically active components, such as terpenes, alkaloids, polyphenols, flavonoids, and other bioactive compounds could have therapeutic effects, i.e. interact with biological systems (Brown 2017). Similarly, as interest in herbal preparations increases, so have concerns about their effectiveness, safety, and quality control (Tao et al. 2023). As herbal medicines make their way to market, their quality and standardization all contribute to ensuring herbal medicines are effective and safe (Rahath Kubra et al. 2016).

It is imperative to ensure health and wellness of consumers by applying stringent quality-control measures (Deutch et al. 2019). You can minimize any risk of prevalence of contamination, adulteration, and toxicity etc. within herbal products by incorporating stringent testing and quality-control processes (Giacometti et al. 2018). Poor quality control directly threatens consumer safety because contaminants, adulterants, and formulations when used in herbal products have the potential to result in toxicity, adverse reaction, or health complications (Cai et al. 2019).

Furthermore, varying levels of active pharmaceutical ingredients in herbal products due to insufficient quality control complicate prescribing and treatment management for healthcare providers (Liu et al. 2020). Varied levels of active ingredients can introduce challenges for healthcare practitioners prescribing and managing treatment for their patients, resulting in varied therapeutic outcomes (Ma et al. 2022). Yago and Pla (2020) identified crucial quality-control processes that mediate consistency of the potency and composition of herbal medicine products. If there are standardized testing practices and production practices in place, this can consider together the grades of active ingredients, which can be monitored and managed to achieve a product delivered to market to manage and achieve therapy objectives (Braga et al. 2021) that benefit customers, practitioners, and regulators. The aim of this chapter is to present a detailed discussion on the implications, preparation, and factors affecting the effectiveness, superiority, and security of herbal pharmaceutical outcomes.

Herbal medicines are created from plants. They use the medicinal qualities of plants, including flowers, leaves, stems, roots, or extricate, to ensure well-being and manage malady (Sinha, et al. 2015). Terpenes, alkaloids, flavonoids, and phenolic compounds are some of the key active ingredients that usually contribute to the therapeutic effects found in herbal pharmaceutical products (Ranasinghe et al. 2015). Products made from herbal medicine can be grouped in several ways. Practitioners in East Asian countries and China have used Traditional Chinese Medicine (TCM) and Ayurvedic herbal medicine as holistic healthcare systems for over 2,000 years (Prasad et al. 2021). These systems treat various illnesses and restore balance by using different medicinal plants, minerals, and natural elements (Wang et al. 2022). Many people have recognized these natural compounds for their medicinal and health advantages for thousands of years (Dina et al. 2022). Herbal medicine products are designed to meet individual health needs. For instance, there's *Echinacea*, a popular herb that helps the immune system; ginger, which is great for digestion; and *Ginkgo biloba*, renowned for its cognitive benefits (Sharifi-Rad et al. 2018). In addition, many traditional medicines also utilize minerals and mineral-rich substances to provide healing.

Traditional medicine recognizes calcium, magnesium, iron, sulphur, and zinc as a potential recommendation for various issues including bone health, correct muscle function, treating anaemia, increasing haemoglobin levels, skin issues such as eczema and acne, boosting the immune system, and assisting in regeneration (Song et al. 2021; Xiao et al. 2022). In East Asia, animal products have also been utilized to treat various issues in traditional medicine (Cheng et al. 2022). Bear gallbladder bile is used in some forms of orthodox Chinese medicine, but there are ethical issues of animal welfare; deer antler velvet is said to work as a source of energy and to increase strength and joint support (Yu et al. 2017). Other notable examples are *Cordyceps*, a parasitic fungus that grows on some insects that is noted for respiratory function and increased energy.

It should be noted, however, that even though these ancient practices have been performed for centuries, current scientific evidence may not always support the efficacy and safety of the aforementioned minerals, plants, and animal products (Yuan et al. 2016). Some of these drugs may have negative side effects or interfere with prescription drugs (Saini et al. 2022). It's really wise to get advice from experienced health professionals who know how to use these natural remedies safely.

There's been a growing interest in understanding orthodox medicine implementation and assessing the healing prospective of these products through scientific exploration (Jaiswal et al. 2016). With the vast array of Chinese herbal medicines available, it's crucial to maintain quality assurance in the herbal medicine industry to ensure that these products are both safe and effective (Kankanamalage et al. 2014). This is part of the goal of integrative medicine so that the best approach of orthodox and mainstream medicine can be combined to produce extensive and tailored healthcare services.

10.3 Quality Control for Phytomedicines

Quality check for herbal medicinal products can be seen as an ordered process composed of a multitude of operational components whose monitoring and management occur at various analytically definable points in the development, manufacture, and delivery of herbal products (Ding et al. 2008). Any business or organization needs specific measures for quality control because these procedures are critical in meeting specifications and requirements for products and services (Chen et al. 2023). Quality control effectively aims to monitor processes and recognize performance gaps. Improvements to a process can led to increased output with the same resources, or increased productivity with less resource utilization, by optimizing a process and removing blockages and unnecessary steps (Gong et al. 2023).

As outlined in Aleksieva and Yordanov (2018), the first step in quality control is to standardize and identify the herbs. Every herbal medicine product must have consistent and reliable active ingredient concentrations. This makes it easier to identify and standardize the herbs. Controlling the quality of each product and minimizing batch variations are the main goals (Ketai et al. 2000). Establishing reference standards can be aided by identifying the active components, performing quantitative assessments, and examining published work, traditional knowledge, and literature (Zhao et al. 2022). A quantitative measurement of the component or label is also crucial (An et al. 2024). Spectroscopy, chromatography, or suitable chemical tests can be used to obtain these measurements (Pan et al. 2018). In addition, reference standards that contain the specific concentrations of active ingredients should be referred (Xiong et al. 2022). Reference standards help ensure that different batches of herbal products have some consistency and can be used as reference points in quality-control tests (Jin et al. 2018). The standardization in herbal medicines within healthcare settings allows practitioners to provide guarantees of treatment, quality control, and treatment effectiveness at the same consistent level. Given the differences in safety and therapeutic action in phytomedicine species or parts of plants, authenticity testing is essential for internal quality control in herbal medicine, because it guarantees herb identification, and quality verification (Zhang et al. 2022). Plant or herb identification, as explained by Lo and Shaw (2019), is to classify the plant species or material which has been utilized for herbal products. Authenticity testing can be done using macroscopic and microscopic observation, high-performance liquid chromatography (HPLC), and thin-layer chromatography (TLC), profiling, and DNA analysis. Macroscopic observation concentrates on the colour, shape, size, and texture of the herbs or their parts (Ma et al. 2021).

Reference materials are important when calibrating markers and active components in the samples. These materials are essential in maintaining the traceability and consistency of different product lots and in evaluating the quality control system in biopharmaceutical testing (Jin et al. 2018). Currently, with the standardization of treatments included in the pharmacopoeias, and the licencing of Providence, healthcare practitioners have a well-defined framework for treatment prescriptions. This system enables us to control and evaluate the phytoproducts' quality and effectiveness. Different therapeutic and safety profiles exist among different species, and parts of a single plant; that is why systemic authentication is crucial for the quality assurance of herbal products. Zhang et al. (2022) stressed the requirement to confirm the herbal constituents that are offered in the formulations due to their increasing importance for the appropriate herbal postulations. Herb identification, as stated by Lo and Shaw (2019), is the accurate identification and authentication of species of plant material in herbal preparations. Authenticity testing methods include macroscopic and microscopic examination, TLC, HPLC, deoxyribonucleic acid (DNA) barcoding, and chemical profiling. Macroscopic examination consists of inspecting the physical characteristics of the herb including colour, shape, size, texture, and any others. This is useful for recognizing the medicinal plants and differentiating them from similar-looking plants to improve the imitation of herb identification and to understand the quality of herbal imitation. Microscopic examination on the other hand, assesses the tissues and characteristics of the herb with the aid of a microscope (Xiong et al. 2019). This can be useful for identifying and confirming specific parts of the plant, for example, the leaves, the stems, and the roots. We should recognize the herb's botanical identity can be methodical in separating evidence of the herb. TLC provides us with the opportunity to separate and differentiate the herb at the scientific level in order to evaluate the authenticity of its essence (i.e. its substance) and its quality. TLC is carried out with a thin layer of the medicinal plant essence on a solid support and with a suitable elution solvent.

In determining a single herb and investigating adulteration or contamination, the chromatogram can be further analysed more closely and compared with a reference standard (Del Bubba et al. 2013). HPLC is a strong analysis technique that effectively isolates and quantifies the many constituents within a medicinal plant (Sontag et al. 2019). Furthermore, it can also qualify the quality of the herbal data we can obtain through confirming presence or several specific active ingredients (Malherbe et al. 2012). Another effective way to determine and confirm species of plants is through DNA testing, DNA barcoding is one such example (Li et al. 2015). In a basic sense, with DNA barcoding is used to identify the specific barcode region of the plant's DNA that was extracted. After obtaining the sequence, it can be compared to a reference database to confirm the herb. This process will not only confirm the aiming correct species but also demonstrate any contamination or substitution (Senapati et al. 2022). DNA testing has the potential for genuine herb identification, especially for dried, processed, and powdered plant material. The second part of chemical fingerprinting is phasing using nuclear magnetic resonance, gas chromatography–mass spectrometry, or HPLC to identify the unique compounds within each herb (Li et al. 2022). When manufacturers are able to compare the chemical pattern of their herb sample with control samples using preset reference profiles, they can better determine the purity, consistency, and authenticity of their product. Chemical fingerprinting is also necessary for the identification of certain marker compounds or active ingredients so that herbal products can assure the correct strength and therapeutic action (Gainza et al. 2020). In combination, DNA testing and chemical fingerprinting represent the best quality control for natural health products, backed by independent evidence and science (Yin et al. 2022). This enables manufacturers to implement the required standards and provide a uniform product that meets the regulatory practices. This enables manufacturers to implement the required standards and provide a uniform product that meets the regulatory practices.

Integrating these steps as part of their quality assurance program will provide food companies with the opportunity to set realistic standards, maintain specifications, support compliance assurance, evaluate the genuineness of herbal substances, identify potential contamination or adulteration, and confirm the safety and consistency of herbal products. Moreover, it is important to recognize that a joined-up quality assurance system for herbal medicine should consider both macroscopic and microscopic assessments and include various analytical processes such as TLC, HPLC, DNA barcoding, or chemical profiling (Hu et al. 2021). Genuineness assessment is important for the quality control of the herb inputs to the product, but it is just as important for assuring the safety and efficacy of outputs.

10.3.1 Production Controls

Extraction methods are crucial within the manufacturing of herbal medicines because they facilitate the process of removing the active components from the herbs. The quality and effectiveness of the herbal products are directly related to the efficiency we use to extract the bioactive components, which is why optimizing our extraction methods is needed (Zhu et al. 2006). In developing these methods there are several aspects that need to be considered, like the correct solvent choice, processing the herb raw materials, correct solvent-to-herb ratio, controlling extraction time and temperature, multiple extraction steps, and pH adjustments, quality assurance protocols, sustainability and safety considerations, confirming and validating methods, repeatability, as well as adding co-solvents or enhancers (Boateng et al. 2023). Also, when considering an extraction method, some of the variables for consideration will include the particular herb chosen, the bioactive compounds to isolate, the required characteristics of the final product, and bulk production scale (Gollahon and Holt 2000). Extraction technique selection is critical for achieving extraction of bioactive compounds because every approach is accompanied by a specific set of benefits and limitations. It is extremely significant to choose the right extraction method because various methods have various advantages and disadvantages (Silva et al. 2022). The solvent chosen is the key to successful extraction (Brennan et al. 2009).

Some chemicals may be extracted using various solvents of varying polarities (Kalkan et al. 2019). Water, ethanol, methanol, or any combination of the three is usually used as solvents. When selecting any solvent, solubility of the target compound, toxicity, environmental issues, and regulation must be considered (Tasfiyati et al. 2022). A well-defined solvent-to-herb ratio is essential to attain more efficient extraction. Also, in order to extract bioactive compounds, maintain them stable, and minimize degradation through extraction, parameters like temperature, time, and pressure needs to be varied (Ma et al. 2021). These parameters will likely also depend on the herb and the designated compounds. It is essential to control temperature and extraction duration to retain heat labile compounds and avoid breakdown of components that are thermally unsteady (Feldmann and Bondemark 2006). Pretreatment considerations or methods (e.g. drying, size reduction; e.g. milling or grinding; or special treatments such as blanching) can have a significant impact on the extraction process (An et al. 2024). Pretreatments improve accessibility of bioactives to the extraction solvent through breaking apart the plant matrix to facilitate release of substances that are bioactive from its encapsulation (Dai et al. 2022). Upon selecting an extraction method, process validation is a vital step in guaranteeing consistent and repeatable results. Process validation constitutes process verification to ensure that the extraction process consistently provides results that meet parameters of quality, for example, yield, efficiency, and desired bioactive compounds present at harvest. For process validation, testing analytically for extracts, reference versus bioactive extract comparison, and performing statistical tests are all critical validation parameters (Onishi et al. 2018). Quality control in production is critical to make herbal medicine products safe, effective, and consistent (Fillingim et al. 2016).

It entails measurement and assessment of different factors during production (Sargeant et al. 2022). Let's proceed with the highlights when it comes to quality assessment while processing. The first step to establishing control is assessing the raw materials, herbs in this case (Liang et al. 2023). There are several tests and techniques for assessing and controlling quality, identity, authentication, and safety of the starting materials for herbal drugs, with authenticity, identity, purity, and quality enhancement being effective ways of assessing products (Benedetti et al. 2019). One of the steps we can take is the organoleptic assessment of raw material. This method employs our senses (site, smell, taste, and touch) in observing and describing the information of the raw material (Sridharan et al. 2016). At the organoleptic assessment stage, it is necessary to be aware of any unusual taste, smell, or visual presentation of the raw material. After the complete dried raw material is adequate identity, it can be seen in its entirety using a microscope to test for any potential adulteration, substitution, or extraneous matter (Saletnik et al. 2021). HPLC testing is frequently used to extract and separately identify the various chemical constituents of a sample while studying botanical herbal raw materials. Also typically included is the evaluation of markers or candidate materials (Sontag et al. 2019). Fourier transform infrared spectroscopy assists in amateur identification of functional groups within the raw materials, which assist with the finished product's overall chemical composition (Barnes et al. 2023).

It will also validate the safety of the raw materials through heavy metals testing using atomic absorption spectrophotometry (Ortegón et al. 2022). Active compounds in plant material can be identified by HPLC. The standard, originality, and protection of herbal ingredients in the product can be ensured by manufacturers securely by adopting these guidelines. These verifications are essential in meeting regulatory requirements and ensuring the product consistency and consumer protection in herbal products.

10.3.2 Good Manufacturing Practices for Herbal Products

Good Manufacturing Practices (GMP) regulation is very important in the herbal industry to provide a consistent assurance of product standard and ultimately for health of the consumer (Lesch et al. 2021). GMP for herbal medicines covers a vast array of regulations and standards to formulate the quality, safety, and constancy of herbal drugs, supplements, and other similar items (Castiglia et al. 2018). This was established from the availability of the growing number of herbal raw materials and herb based products at the time at the legislative/regulatory level, ranging from requirements on packaging and labelling to their manufacture, for instance facility and equipment requirements, standard operating procedures, batch records, quality-control test and standards, validation and testing procedures, stability testing, recall and complaint handling (Food and Drug Administration [FDA]), regulatory compliance, ongoing improvement and raw material sourcing and identification (Cundell et al. 2023). GMP retains a uniform quality control standard through documentation management, staff training of activities and proper procedures and practices, consistent and hygienic practices, and the use of quality control (Castiglia et al. 2018). It is ultimately necessary to utilize quality, authentic, and accurately identified herbal raw materials in accordance with GMP.

For ensuring uniformity and purity of the substances, it is essential to qualify and choose suppliers (Lesch et al. 2021). The operation, design, and maintenance of facilities should be governed by GMP. This means considering the controlled environmental conditions of manufacturing, cleanliness, and cross-contamination (Bai et al. 2022). GMP regulations have the mandate of developing and implementing valuable standard operating procedures for all of the important aspects of manufacturing (Shukla and Gottschalk 2013). They are important for assuring consistency and control of manufacture. Also, it is important to keep batch records and accessible records for all products manufactured (Mager et al. 2007). Records must include all raw materials, process information on manufacturing, quality-control test results, and all deviations and corrective actions compliant during manufacture (Smith 2020). GMP requires strict quality assurance of all raw materials, in-process samples, and finished products (Morgan et al. 2015). Quality assurance may also require the identification of herbal ingredients, such as testing for presence of active compounds and possible contaminants. Qualification of any tools and analytical methods can be important in establishing if they work well for the purpose of herbal products and to identify their effectiveness and integrity of production methods (Williams et al. 2012). Quality assurance can include conducting stability tests to establish shelf life and best storage conditions, which is important to herbal products. Quality assurance processes can ensure that the product can still maintain strength and quality over shelf life. Manufacturing procedures should be audited periodically, self-examined frequently, and examined from time to time to identify areas for improvement and ensure continuous adherence to GMP (Deagle et al. 2017). GMP in the herbal industry plays a role in setting the quality, safety, and consistency of products.

10.3.3 Process Control

In order to sustain product quality, it is critical to monitor and manipulate key parameters throughout the processing stages (Qu et al. 2021). Morgan et al. (2015) suggested monitoring framework such as pressure, temperature, mixing time, pH, and drying conditions. Continually monitoring and documenting these parameters is a productive activity that will assist with product consistency by monitoring expected values and assists on identifying deviations from product specification, allowing timely corrections or adjustments (Wasalathanthri et al. 2020). Temperature is pivotal to many processes as it can affect the rate of chemical reaction, phase changes, and it will affect microbial growth (Reischauer et al. 2009). Being able to monitor temperature will assist in ensuring that reactions occur as intended, eliminate or minimize the presence of contaminants or unwanted by-products for acceptable quality; and maintain appropriate conditions for preparation of the final product. Temperature is important in thermally driven processes, drug syntheses, and is important to prepare stable and effective pharmaceuticals. Pressure is a significant contributor in processes where gases or liquids may exist (Seo and Shin 2022), and both excessive and lack of pressure can modify reaction rates, solubility, production yields, and overall system efficiencies. Managing pressures allows for a safe operating conditions, reasonable operating windows, and reduces the chances of equipment failure in chemical reactions (Schmidt et al. 2023).

Although pH, measuring the acidic or alkaline trend of a product, is also a significant contributing factor to many products for stability and performance (Wright 2021), maintaining the pH level in the proper range is important for product preservation, flavour improvement, and the efficiency of products in the pharmaceutical, cosmetic, and food and beverage production industries. For instance, pH control is important in enzyme-based processes, as some enzymes are active at specific pH levels (Daniel et al. 2022). The time spent mixing or blending affects the uniformity and consistency of the resultant product (Hernandez and Perera 2022). When formulating products like pharmaceutical tablets, or food products, controlling mixing time so the ingredients are spread evenly is necessary (Adra et al. 2022). Drying is common in many industries that manufacture herbal medication-like products, pharmaceuticals, and chemicals. When drying products, especially with herbal materials, controlling the drying conditions including temperature, humidity, airflow, moisture content, texture, and stability can help improve product quality, shelf life and even keep the product fresh (da Cunha et al. 2021). Proper analysis is needed to quality test herbal medication products and confirm they are free of contaminants, pesticide residues, heavy metals, and microbiological safety.

10.3.4 Concept of Work-in-Progress Testing

Work-in-progress testing refers to performing quality checks to monitor the quality and consistency of a product at different stages of manufacturing (Silva et al. 2022). In-process testing involves a battery of tests including chemical analysis, microbiological analysis, attempts for dissolution, as well as physical measurements. In-process testing is important because it allows the manufacturer an opportunity to trouble shoot and identify any atypical or anomalous course of action that may occur during production and may jeopardize the resultant product quality (Chan and Choo 2013). Aspects of concern with any herbal medicine product would include risk of contamination by any number of constituents, such as pesticide residues, heavy metals, and microbiological safety, just to name a few (Omeje et al. 2021). For this reason, the quality of the herbal product relies on detecting pollutants, pesticides, counts of microbiological safety, and all of the levels of heavy metals present in the environment (Varol and Sünbül 2017). Documenting environmental pollutants, as noted from Simpson and McKelvie (2009), means taking time to check for harmful substances such as aflatoxins, mycotoxins, and residual solvents, and environmental contaminants such as heavy metals, pesticides, and industrial chemicals.

To detect these individual contaminants through validated analytical procedures in order to quantify the defined pollutants, there are a number of analytical techniques that can be utilized such as mass spectrometry, spectrophotometry, and chromatography (HPLC, GC) (Levy et al. 2018). Heavy metals such as mercury, lead, arsenic, and cadmium can pose serious health concerns when their presence occurs in herbal medicine products. Examples of infrastructure used to determine acute heavy metal concentrations would be atomic absorption spectroscopy (AAS) or inductively coupled plasma mass spectrometry (Lockwood et al. 2021). Pesticide residues could arise from agricultural systems or contamination during cultivation. There are also analytical methods for pesticide residue analysis that meet regulatory needs including gas chromatography or liquid chromatography with mass spectrometry to analyse pesticide residues (Lockwood et al. 2021).

Microbiological analysis is essential to assess the presence of pathogenic microorganisms. Levy et al. (2018) indicated microbiological testing should be done for certain pathogens, total microbial counts, and indicator organisms. Common tests are checks for *Escherichia coli, Staphylococcus aureus, Salmonella* spp. and microbiological plate counts (York et al. 2000). The methods described are used to assess total coliform counts, total viable counts, and total yeast and mould counts, as described in the guide by Hervert et al.

(2016) on microbiological testing for herbal medication products, with a few minor tweaks. These tests are vital to substantiate the claimed safety profile as well as ascertain whether the product meets the microbial limits set by regulatory authorities. Analytical methods are utilized to identify, quantify, and qualify active ingredients found within herbal medicines (Seo and Shin 2022). So far, HPLC methods have become the most commonly used technique to separate, identify, and measure individual chemical components, including active ingredients, in herbal extracts (Wright 2021).

This method helps researchers get precise measurements and supports standardization and quality control. For volatile compounds and essential oils found in herbal medicine products, researchers use gas chromatography–mass spectrometry (Silva et al. 2022). By examining retention time and mass spectra, researchers can identify and quantify specific compounds accurately. TLC is a rapid and inexpensive technique for the qualitative analysis and identification of ingredients present in herbal medicinal preparations (Del Bubba et al. 2013). TLC mainly separates constituents based on their migration through a thin layer of adsorbent medium. Several spectroscopic methods, such as UV–Vis spectroscopy, infrared spectroscopy, and nuclear magnetic resonance spectroscopy, are often employed to examine constituents of herbal medicine products.

These methods give information on functional groups, chemical structure, and levels. Through the use of quality tests and analysis methods, manufacturers have been able to improve the quality, safety, and efficiency of herbal drugs. When doing these analyses, they can ensure quality and safety by identifying potential contaminants, ensuring active constituents, and fulfilling regulating requirements.

10.3.5 Concept of Purity Management

Purity management is a key aspect of the herbal medicine business since it has direct impacts on product quality, safety, and environmental protection (Raeisossadati et al. 2016). It is the set of methodologies and processes designed to avoid, reduce, or eliminate unwanted or toxic material from products, process, and the environment (Simpson and McKelvie 2009). Based on the industry and the nature of the contamination, there exist various specific means of controlling it. Fundamental means of controlling contamination are sanitizing and cleaning; sterilization; disinfection; hand hygiene; clean room technology; air filtering; regulatory compliance; and environmental monitoring (Wan et al. 2019).

Control of containments involves routine washing and disinfection of surfaces, equipment, and facilities. Proper cleaning reduces the risk of cross-contamination by removing dirt, debris, and potential contaminants. In the pharmaceutical and medical fields, sterilization is a crucial procedure used to get rid of all living germs, such as bacteria, viruses, and spores. Specific methods are autoclaving, irradiation, and chemical sterilization, among others. Disinfection involves using chemical agents to reduce the total number of microorganisms on objects and surfaces. Disinfection is commonly used in healthcare and food production to limit the spread of microbial infections. Clean rooms are spaces designed to decrease airborne particles and microorganisms (Safra 2019). Clean rooms are widely used in assembly of electronics – for example semiconductor manufacturing, aerospace, and pharmaceuticals. Clean air is maintained through use of high-efficiency particulate air filters within a clean rooms' ventilation system to filter out contaminants, including bacteria.

For food production and healthcare delivery, pest control can have significant implications for contamination control. Many industries have developed formalized processes and procedures within the umbrella of contamination control (Panteghini et al. 2001).

Following governmental regulations enables a clean and safe environment, consumer health and safety, and protects the customer product quality. The environment is often controlled, monitored, and established through one or more of the following factors: temperature, humidity, and air quality can help pinpoint potential contamination sources and confirm that the right controls are in place (Omeje et al. 2021). According to Levy et al. (2018), the term 'contamination control methods' refers to a collection of techniques and strategies employed in the production of herbal medicine products to prevent or reduce contamination. The quality, safety, and effectiveness of these products and the public health and environment depend upon the means of contamination control applied.

The medicinal advantages of herbal remedies are increasingly recognized, but numerous barriers and inconsistencies in quality control persist. Variability in the chemical makeup and quality of herbal products is a major issue. The inconsistency stems from numerous factors, including poor control in different parts of the world, agricultural practices, post-harvest handling, and available local water and air. Assessment of standards of quality of herbal products is a complex task with this diversity. On the positive side, quality control has been revolutionized in most areas since new analysis technologies such as DNA barcoding and mass spectrometry have emerged. However, standardization of the use of these technologies to measure the efficacy of herbal medicines has been somewhat uneven. Molecular fingerprinting, such as metabolic profiling and DNA barcoding, is recommended to authenticate the medicinal properties of herbal medicines. We further encourage the use of Internet-of-Things devices and real-time monitoring across the herbal supply chain, allowing us to constantly capture environmental data to maintain the integrity of the plant and quality of the product.

Moreover, quality-assurance practices in the herbal sector should be based on data-driven predictive modelling; artificial intelligence (AI) and machine learning (ML) can be utilized to play a significant role in understanding how downstream processing practices affect the quality of the resulting herbal products that have been manufactured. Efficient end-to-end practices may improve the efficacy and safety of practices from herbal medical manufacturers and regulators, and therefore reassuring consumers and supporting responsibly growing the herbal medicine industry. Additionally, the plan provides a basis for Herbal remedies frequently incorporate centuries-old wisdom that needs to be carefully incorporated into contemporary quality assurance procedures. Many of the quality-control innovations proposed for herbal medicines are already well-established procedures in the context of biomedical medication products, which is important to keep in mind. Enhancing the findings of earlier studies, integrating state-of-the-art analytical methods, strategically adhering to international standards, and working together. There are some distinctions between the two fields, despite the fact that some elements of quality control for biomedical drugs can also be applied to herbal medicines.

Adopting various elements of biomedical drug quality control can help with the quality control of herbal medicines, despite the fact that there are unique considerations when using such techniques. We illustrate the potential creation of quality-control methods for herbs by contrasting the current procedures with the suggested improvements. Implementing these innovations can reform the herbal medicine industry and build trust with consumers who want to consume herbal medicines. We provide businesses with essential information for improving quality control of herbal medicines by bringing together an intimate understanding of the threats, risks, and opportunities of quality control in herbal medicine. The findings we include show an overall a systematic approach to quality control, the gaps and barriers, and recommendations for how to improve the quality control, including embracing modern practices, harmonizing regulations, using networks of support, encouraging innovation, and promoting continuous improvement. These companies will learn how to improve essential GMP components – cleanliness, training of personnel, documentation, and maintenance of equipment, while shrinking uncertainty about the quality of active ingredients, dosage forms, and potency. With this new-found information and knowledge, businesses will be able to quality-audit their products prior to releasing them into the marketplace. Based on the work of this text, which shows ahead of safety emphasizing quality control, this generic information on safety allows businesses to assess risk – mostly related to quality in the manufacture of herbal medicines. In today's ever-changing landscape, where continuous research, advanced technology, and shifting consumer preferences prevail, companies must frequently reassess and improve their quality control and quality assurance practices as necessary. By following the recommendations in this text, businesses will be able to appropriately respond to the challenges of producing herbal medicines, and their customers and consumers will have equitable access to herbal medicines.

10.4 Importance of Cryogenic Applications in Herbal Medicine Processing

Although the basic idea behind cryogenic grinding, or cryogrinding, for herbal medicine is comparable to that of grinding techniques for conventional materials, the compositions vary greatly, including highly volatile aromatics, oils, and fats that are readily oxidized. It is possible to grind all of these thermosensitive herbal medications below their brittle temperature by using liquid nitrogen or liquid air as the cryogen. The flavour and nutritional value of the medications will not be lost, nor will the colour and other characteristics of the cryogrinding products change.

Spices' high fat content causes issues with temperature rise and sieve clogging when grinding. Spices lose a large portion of their volatile oil or flavouring ingredients as a result of this temperature increase. In order to maintain the cryogenic temperature in the grinding zone and to cool the spices before putting them into the grinder, a cryogenic grinding system was created. The two primary parts of the cryogenic grinding system are pre-cooler and a grinder. A screw conveyor assembly, a compressor, a liquid nitrogen Dewar, and a power transmission unit make up the pre-cooler. For this, a commercially available grinder was used. Grinding cumin seed below −70 °C was successfully accomplished, according to tests done on the subject. Clogging of the sieve occurs above this temperature. Sieve blockage occurred above this temperature. The product's particle size and specific energy consumption increased significantly when the grinding temperature was raised from −160 to −70 °C. The volatile oil concentration varied between 3.30 and 3.26 mL/100 g as the temperature increased from −160 to −70 °C; however, this fluctuation was not significant at the 5% level.

The creation of the cryogenic superfine TCM reducing mechanism reveals a machine that uses chillness, a material-grinding agent, to turn traditional medicines into ultra-fine powder. A type of air slide dissolving mill is the subject of the innovation. The device is composed of three parts: a low-temperature gas flow grinding system, a coarse grinding system, a cyclone separator, a dust removing collector, and a gas source, a gas storage tank, a gas compressor, a cooling machine, and a heat exchanger. The low-temperature gas-flow grinding system is primarily composed of a compressor, a gas-flow grinding machine, a spiral pre-cooler, a sealing feeder, and a coarse grinding machine. The process of coarse grinding involves treating raw materials at a low temperature, separating gas and solid substances into gas and solid after coarse grinding, and then pre-cooling the coarsely ground materials once more. The materials are then fed into a gas-flow grinding machine, which grinds them finely in a fluidized bed. Thus, high grinding productivity, high granularity of ground materials, and high quality of powder are the benefits of the current invention. At the same time, cold gas is recycled in a complete hermetic system, which further lowers costs and saves energy.

Currently, most powder separators are significantly below normal temperature and pulverize; some materials can't even pulverize in this manner. Because liquid nitrogen refrigerating costs more, the output of some materials that are pulverized under normal temperature is lower, and some materials pulverize easily oxidizing under normal temperature. In order to address the aforementioned shortcomings, the invention offers a type of cryogenic superfine TCM reducing mechanism. As a result, the material and output that this mechanism can easily oxidize at a certain normal temperature are low in cost. The current technological scheme invention that resolves its technical issue and is adopted is: The refrigeration exchange system, coarse crushing system, and low-temperature airflow crushing system, are all part of this cryogenic superfine TCM-reducing mechanism.

10.5 Conclusion

In order to protect consumer health, create dependable treatment alternatives, adhere to legal requirements, and promote the expansion of the herbal medicine sector, quality control for the effectiveness and safety of herbal medication products is crucial. The amount and presence of active ingredients determine how well herbal medicines work. Standardized methods for sourcing, processing, and synthesizing herbal constituents should be the main goal of quality-control measures. There are ways that consumers can utilize measures to produce herbal medicines that are safe and effective. To clarify the safety and effectiveness of herbal medicines, there are many aspects to consider. The aspects involve tracking raw materials, verifying that active compounds are accurately used, use of GMP during manufacturing, and conscientious tracking of products once they have been sold. It is part of required processes to test raw material and finished products for contaminants. Quality-control processes are a major part of identifying safety or quality issue quickly and allowing you to rectify the supplant. The consistent and meticulous cataloguing of the genus and species of these plants in products can be problematic.

DNA barcoding has rapidly improved this area for standardization and accuracy. Quality and verification of efficacy, safety, and standardization must include procedures, use of identification for plants that are present in products, contaminant testing, concentrations of active components, and others, proving that compliance to regulations, stability, GMP, validation, post-market, and stakeholder involvement are essential facets to quality control for herbal medicine products. These data sets build the basis from which quality-control measures become effective for herbal medicines. This dialogue also draws attention to the importance of quality-management systems for herbal medicine, and the importance of maintaining safe and effective production for consumers. By implementing strict quality-control measures, we can create safe and effective herbal medicines for all.

References

Adra, H.J., Zhi, J., Luo, K., and Kim, Y.R. (2022). Facile preparation of highly uniform type 3 resistant starch nanoparticles. *Carbohydrate Polymers* 294: 119842. https://doi.org/10.1016/j.carbpol.2022.119842.

Aleksieva, K.I. and Yordanov, N.D. (2018). Various approaches in EPR identification of gamma-irradiated plant foodstuffs: a review. *Food Research International* 105: 1019–1028. https://doi.org/10.1016/j.foodres.2017.11.072.

Alipour, R., Marzabadi, L.R., Arjmand, B. et al. (2022). The effects of medicinal herbs on gut microbiota and metabolic factors in obesity models: a systematic review. *Diabetes and Metabolic Syndrome: Clinical Research and Reviews* 16: 102586. https://doi.org/10.1016/j.dsx.2022.102586.

An, Y.L., Wei, W.L., and Guo, D.A. (2024). Application of analytical technologies in the discrimination and authentication of herbs from fritillaria: a review. *Critical Reviews in Analytical Chemistry* 54: 1775–1796. https://doi.org/10.1080/10408347.2022.2132374.

Arafa, A.S. and Chen, T.C. (1978). Liquid nitrogen exposure as an alternative means of chilling poultry. *Journal of Food Science* 43: 1036–1037.

Awonorin, S.O. (1997). An appraisal of the freezing capabilities of tunnel and spiral belt freezers using liquid nitrogen sprays. *Journal of Food Engineering* 34 (2): 179–192.

Bai, R., Sun, Q., He, Y. et al. (2022). Ceramic toughening strategies for biomedical applications. *Frontiers in Bioengineering and Biotechnology* 10: 840372. https://doi.org/10.3389/fbioe.2022.840372.

Balasubramanian, S., Gupta, M.K., and Singh, K.K. (2012a). Cryogenics and its application with reference to spice grinding: a review. *Critical Reviews in Food Science and Nutrition* 52: 781–794.

Balasubramanian, S., Kumar, R., and Singh, K.K. (2012b). Influence of moisture content on physico-mechanical properties of coriander seeds. *International Agrophysics* 26 (4): 419–422.

Barnes, M., Sulé-Suso, J., Millett, J., and Roach, P. (2023). Fourier transform infrared spectroscopy as a non-destructive method for analysing herbarium specimens. *Biology Letters* 19: 20220546. https://doi.org/10.1098/rsbl.2022.0546.

Benedetti, A., Khoo, J., Sharma, S. et al. (2019). Data analytics on raw material properties to accelerate pharmaceutical drug development. *International Journal of Pharmaceutics* 563: 122–134. https://doi.org/10.1016/j.ijpharm.2019.04.002.

Bera, M.B., Shrivastava, D.C., Singh, C.J. et al. (2001). Development of cold grinding process, packaging and storage of cumin powder. *Journal of Food Science and Technology* 38: 257–259.

Boateng, I.D., Kuehnel, L., Daubert, C.R. et al. (2023). Updating the status quo on the extraction of bioactive compounds in agroproducts using a two-pot multivariate design. A comprehensive review. *Food & Function* 14: 569–601. https://doi.org/10.1039/d2fo02520e.

Braga, F., Pasqualetti, S., Aloisio, E., and Panteghini, M. (2021). The internal quality control in the traceability era. *Clinical Chemistry and Laboratory Medicine* 59: 291–300. https://doi.org/10.1515/cclm-2020-0371.

Brennan, A.A., You, J., and Lydy, M.J. (2009). Comparison of cleanup methods for fipronil and its degradation products in sediment extracts. *Talanta* 78: 1408–1413. https://doi.org/10.1016/j.Talanta.2009.02.034.

Brown, A.C. (2017). Liver toxicity related to herbs and dietary supplements: online table of case reports. Part 2 of 5 series. *Food and Chemical Toxicology* 107: 472–501. https://doi.org/10.1016/j.fct.2016.07.001.

Mallappa, Sharankumar, and Roopabai (2015). *Byadagichilli by Effect of milling methods on quality parameters of Byadagi chilli*. Lambert Academic Publishing ISBN: 978-3-659-16344-9.

Cai, Z., Wang, C., Zou, L. et al. (2019). Comparison of multiple bioactive constituents in the flower and the caulis of *Lonicera japonica* based on UFLC-QTRAP-MS/MS combined with multivariate statistical analysis. *Molecules* 24: 1936. https://doi.org/10.3390/molecules24101936.

Castiglia, S., Adamini, A., Rustichelli, D. et al. (2018). Cytokines induced killer cells produced in good manufacturing practices conditions: identification of the most advantageous and safest expansion method in terms of viability, cellular growth and identity. *Journal of Translational Medicine* 16: 237. https://doi.org/10.1186/s12967-018-1613-5.

Chan, E., Menovsky, M., and Ashley, J.W. (1996). Effect of cryogenic grinding on soft tissues optical properties. *Applied Optics* 25 (22): 4526–4532.

Chan, S.Y. and Choo, W.S. (2013). Effect of extraction conditions on the yield and chemical properties of pectin from cocoa husks. *Food Chemistry* 141: 3752–3758. https://doi.org/10.1016/j.foodchem.2013.06.097.

Chen, R., Liu, F., Zhang, C. et al. (2023). Trends in digital detection for the quality and safety of herbs using infrared and Raman spectroscopy. *Frontiers in Plant Science* 14: 1128300. https://doi.org/10.3389/fpls.2023.1128300.

Cheng, W.J., Yang, H.T., Chiang, C.C. et al. (2022). Deer velvet antler extracts exert anti-inflammatory and anti-arthritic effects on human rheumatoid arthritis fibroblast-like synoviocytes and distinct mouse arthritis. *The American Journal of Chinese Medicine* 50: 1617–1643. https://doi.org/10.1142/S0192415X22500689.

Cho, E.C. and Kim, K. (2020). A comprehensive review of biochemical factors in herbs and their constituent compounds in experimental studies on alopecia. *Journal of Ethnopharmacology* 258.

Choi, J.A., John, C., and Bischof, A.B.C. (2010). Review of biomaterial thermal property measurements in the cryogenic regime and their use for prediction of equilibrium and non-equilibrium freezing applications in cryobiology. *Cryobiology* 60: 52–70.

Chrysant, S.G. and Chrysant, G.S. (2017). Herbs used for the treatment of hypertension and their mechanism of action. *Current Hypertension Reports* 19: 77. https://doi.org/10.1007/s11906-017-0775-5.

Cristina, V.L., Elizabet, R.W., Heymann, B., and Jensen, G.J. (2006). A comparison of LN_2 and LHe as cryogens for electron cryptomography. *Journal of Structural Biology* 153: 231–240.

Cundell, T., Atkins, J.W., and Lau, A.F. (2023). Sterility testing for hematopoietic stem cells. *Journal of Clinical Microbiology* 61: e0165422. https://doi.org/10.1128/jcm.01654-22.

da Cunha, R.S., Amorim, K.S., Gercina, A.C. et al. (2021). Herbal medicines as anxiolytics prior to third molar surgical extraction. A randomized controlled clinical trial. *Clinical Oral Investigations* 25: 1579–1586. https://doi.org/10.1007/s00784-020-03468-1.

Dai, Y.L., Li, Y., Wang, Q. et al. (2022). Chamomile: a review of its traditional uses, chemical constituents, pharmacological activities and quality control studies. *Molecules* 28: 133. https://doi.org/10.3390/molecules28010.33.

Daniel, S., Kis, Z., Kontoravdi, C., and Shah, N. (2022). Quality by design for enabling RNA platform production processes. *Trends in Biotechnology* 40: 1213–1228. https://doi.org/10.1016/j.tibtech.2022.03.012.

Davidge, H. (1981). Cryogenics in the food industry. *Cryogenics* 21 (5): 287–290.

Deagle, R.C., Wee, T.E., and Brown, C.M. (2017). Reproducibility in light microscopy: maintenance, standards and SOPS. *The International Journal of Biochemistry & Cell Biology* 89: 120–124. https://doi.org/10.1016/j.biocel.2017.06.008.

Del Bubba, M., Checchini, L., and Lepri, L. (2013). Thin-layer chromatography enantioseparations on chiral stationary phases: a review. *Analytical and Bioanalytical Chemistry* 405: 533–554. https://doi.org/10.1007/s00216-012-6514-5.

Deutch, M.R., Grimm, D., Wehland, M. et al. (2019). Bioactive candy: effects of licorice on the cardiovascular system. *Food* 8: 495. https://doi.org/10.3390/foods8100495.

Dina, E., Vontzalidou, A., Cheilari, A. et al. (2022). Sustainable use of Greek herbs by-products, as an alternative source of biologically active ingredients for innovative products. *Frontiers in Nutrition* 9: 867666. https://doi.org/10.3389/fnut.2022.867666.

Ding, L., Luo, X.B., Tang, F. et al. (2008). Quality control of medicinal herbs Fructus gardeniae, common Andrographis herb and their preparations for their active constituents by high-performance liquid chromatographyphotodiode array detection-electrospray mass spectrometry. *Talanta* 74: 1344–1349. https://doi.org/10.1016/j.Talanta.2007.09.001.

Feldmann, I. and Bondemark, L. (2006). Orthodontic anchorage: a systematic review. *The Angle Orthodontist* 76 (3): 493–501.

Fillingim, R.B., Loeser, J.D., Baron, R., and Edwards, R.R. (2016). Assessment of chronic pain: domains, methods, and mechanisms. *The Journal of Pain* 17: T10–T20. https://doi.org/10.1016/j.jpain.2015.08.010.

Gainza, P., Sverrisson, F., Monti, F. et al. (2020). Deciphering interaction fingerprints from protein molecular surfaces using geometric deep learning. *Nature Methods* 17: 184–192. https://doi.org/10.1038/s41592-019-0666-6.

Giacometti, J., Bursać Kovačević, D., Putnik, P. et al. (2018). Extraction of bioactive compounds and essential oils from Mediterranean herbs by conventional and green innovative techniques: a review. *Food Research International* 113: 245–262. https://doi.org/10.1016/j.foodres.2018.06.036.

Gollahon, L.S. and Holt, S.E. (2000). Alternative methods of extracting telomerase activity from human tumor samples. *Cancer Letters* 159: 141–149. https://doi.org/10.1016/s0304-3835(00)00544-9.

Gong, L., Xie, J.B., Luo, Y. et al. (2023). Research progress of quality control for the seed of Ziziphus jujuba var. spinosa (Bunge) Hu ex HF Chow (Suan-Zao-Ren) and its proprietary Chinese medicines. *Journal of Ethnopharmacology* 307: 116204. https://doi.org/10.1016/j.jep.2023.116204.

Goswami, T.K. (2010). Role of cryogenics in food processing and preservation. *International Journal of Food Engineering* 6 (1): 2.

Gouveia, S.T., Lopes, G.S., Orlando, F.F. et al. (2004). Homogenization of breakfast cereals using cryogenic grinding. *Journal of Food Engineering* 51: 36–59.

Hernandez, G.A. and Perera, R.M. (2022). Autophagy in cancer cell remodeling and quality control. *Molecular Cell* 82: 1514–1527. https://doi.org/10.1016/j.molcel.2022.03.023.

Hervert, C.J., Alles, A.S., Martin, N.H. et al. (2016). Evaluation of different methods to detect microbial hygiene indicators relevant in the dairy industry. *Journal of Dairy Science* 99: 7033–7042. https://doi.org/10.3168/jds.2016-11074.

Hoeft, R., Bates, R.P., and Ahmed, E.M. (1973). Cryogenic freezing of tomato slices. *Journal of Food Science* 38 (2): 362.

Hu, G., Li, X., Zhang, J. et al. (2021). An integrated strategy for the identification and screening of anti-allergy components from natural products based on calcium fluctuations and cell extraction coupled with HPLC-Q-TOF-MS. *Analytical and Bioanalytical Chemistry* 413: 6253–6266. https://doi.org/10.1007/s00216-021-03580-5.

International Institute of Refrigeration (1975). *New International Dictionary of Refrigeration*. Paris, France: IIF/IIR.

Jacob, S., Kasthurirengan, S., and Behra, U. (2000). Development of a cryogenic grinding plant for spices. *Paper presented at Cryogenic Engineering Conference and International Cryogenic Materials Conference*, 45(B), Montreal, Quebec, Canada, pp. 1731–1738.

Jaiswal, Y., Liang, Z., and Zhao, Z. (2016). Botanical drugs in Ayurveda and traditional Chinese medicine. *Journal of Ethnopharmacology* 194: 245–259. https://doi.org/10.1016/j.jep.2016.06.052.

Jin, Y., Ma, Y., Xie, W. et al. (2018). UHPLC-Q-TOF-MS/MS-oriented characteristic components dataset and multivariate statistical techniques for the holistic quality control of Usnea. *RSC Advances* 8: 15487–15500. https://doi.org/10.1039/c8ra00081f.

Kabak, B. and Dobson, A.D. (2017). Mycotoxins in spices and herbs-an update. *Critical Reviews in Food Science and Nutrition* 57: 18–34. https://doi.org/10.1080/10408398.2013.772891.

Kalkan, E., Maness, N.O., and Chrz, D.R. (2019). Partial propane extraction of aromatic compounds from dehydrated basils (Ocimum Lamiaceae). *Journal of the Science of Food and Agriculture* 99: 3776–3784. https://doi.org/10.1002/jsfa.9592.

Kamogawa, M.Y., Rita, A., Nogueira, A. et al. (2001). A new strategy for preparation of hair slurries using cryogenic grinding and water-soluble tertiary-amines medium. *Spectrochimica Acta Part B* 56: 1973–1980.

Ketai, W., Huitao, L., Yunkun, Z. et al. (2000). Separation and determination of alantolactone and isoalantolactone in traditional Chinese herbs by capillary electrophoresis. *Talanta* 52: 1001–1005. https://doi.org/10.1016/s0039-9140(00)00467-7.

Kock, S.D., Minnaar, A., Berry, D., and Taylor, J.R.N. (1995). The effect of freezing rate on the quality of cellular and non-cellular par-cooked starchy convenience foods. *LWT - Food Science and Technology* 28: 87–95.

Kankanamalage, T.N.M., Dharmadasa, R.M., Abeysinghe, D.C., and Wijesekara, R.G.S. (2014). A survey on medicinal materials used in traditional systems of medicine in Sri Lanka. *Journal of Ethnopharmacology* 155: 679–691. https://doi.org/10.1016/j.jep.2014.06.016.

Later, Miller and Roberts (2001) How to minimize start up costs. Process cooling and equipment. www.process- cooling.com.

Lesch, H.P., Valonen, P., and Karhinen, M. (2021). Evaluation of the single-use fixedbe bioreactors in scalable virus production. *Biotechnology Journal* 16: e2000020. https: //doi.org/10.1002/biot.202000020.

Levy, J.H., Neal, M.D., and Herman, J.H. (2018). Bacterial contamination of platelets for transfusion: strategies for prevention. *Critical Care* 22: 271. https://doi.org/10.1186/s13054-018-2212-9.

Li, M., Xu, J., Zheng, Q. et al. (2022). Chemical-based surface plasmon resonance imaging of fingerprints. *Analytical Chemistry* 94: 7238–7245. https://doi.org/10.1021/acs.analchem.2c00389.

Li, X., Yang, Y., Henry, R.J. et al. (2015). Plant DNA barcoding: from gene to genome. *Biological Reviews of the Cambridge Philosophical Society* 90: 157–166. https://doi.org/10.1111/brv.12104.

Liang, X., Dang, W., Yang, G., and Zhang, Y. (2023). Environmental feasibility evaluation of cement co-production using classified domestic waste as alternative raw material and fuel: a life cycle perspective. *Journal of Environmental Management* 326: 116726. https://doi.org/10.1016/j.jenvman.2022.116726.

Liu, H., Lu, X., Hu, Y., and Fan, X. (2020). Chemical constituents of Panax ginseng and Panax notoginseng explain why they differ in therapeutic efficacy. *Pharmacological Research* 161: 105263. https://doi.org/10.1016/j.phrs.2020.105263.

Lo, Y.T. and Shaw, P.C. (2019). Application of next-generation sequencing for the identification of herbal products. *Biotechnology Advances* 37: 107450. https://doi.org/10.1016/j.biotechadv.2019.107450.

Lockwood, T.E., Westerhausen, M.T., and Doble, P.A. (2021). Pew(2): open-source imaging software for laser ablation-inductively coupled plasma-mass spectrometry. *Analytical Chemistry* 93: 10418–10423.

Ma, K., Wu, Z.Z., Wang, G.L., and Yang, X.P. (2021). Separation and purification of pyrroloquinoline quinone from Gluconobacter oxydans fermentation broth using supramolecular solvent complex extraction. *Food Chemistry* 361: 130067. https://doi.org/10.1016/j.foodchem.2021.130067.

Ma, L., Yang, Z., Kong, Q., and Wang, L. (2017). Extraction and determination of arsenic species in leafy vegetables: method development and application. *Food Chemistry* 217: 524–530. https://doi.org/10.1016/j.foodchem.2016.09.015.

Ma, P.Y., Li, X.Y., Wang, Y.L. et al. (2022). Natural bioactive constituents from herbs and nutraceuticals promote browning of white adipose tissue. *Pharmacological Research* 178: 106175. https://doi.org/10.1016/j.phrs.2022.10.175.

Mager, S.R., Oomen, M.H., Morente, M.M. et al. (2007). Standard operating procedure for the collection of fresh frozen tissue samples. *European Journal of Cancer* 43: 828–834. https://doi.org/10.1016/j.ejca.2007.01.002.

Malherbe, C.J., De Beer, D., and Joubert, E. (2012). Development of on-line high performance liquid chromatography (HPLC)-biochemical detection methods as tools in the identification of bioactives. *International Journal of Molecular Sciences* 13: 3101–3133. https://doi.org/10.3390/ijms13033101.

Manohar, B. and Sridhar, B.S. (2001). Size and shape characterization of conventionally and cryogenically ground turmeric (*Curcuma domestica*) particles. *Powder Technology* 120: 292–297.

Morgan, L., New, S., Robertson, E. et al. (2015). Effectiveness of facilitated introduction of a standard operating procedure into routine processes in the operating theatre: a controlled interrupted time series. *BMJ Quality and Safety* 24: 120–127. https://doi.org/10.1136/bmjqs-2014-003158.

Motti, R. (2021). Wild plants used as herbs and spices in Italy: an ethnobotanical review. *Plants* 10: 563. https://doi.org/10.3390/plants10.30563.

Murthy, C.T., Krishnamurthy, N., Ramesh, T., and Srinivasa, R.P.N. (1996). Effect of grinding methods on the retention of black pepper volatiles. *Journal of Food Science and Technology* 33 (4): 299–301.

Omeje, K.O., Ezema, B.O., Okonkwo, F. et al. (2021). Quantification of heavy metals and pesticide residues in widely consumed Nigerian food crops using atomic absorption spectroscopy (AAS) and gas chromatography (GC). *Toxins (Basel)* 13: 870. https://doi.org/10.3390/toxins13120870.

Onishi, T., Kadohira, T., and Watanabe, I. (2018). Relation extraction with weakly supervised learning based on process-structure-property-performance reciprocity. *Science and Technology of Advanced Materials* 19: 649–659. https://doi.org/10.1080/14686996.2018.1500852.

Ortegón, S., Peñaranda, P.A., Rodríguez, C.F. et al. (2022). Magnetic torus microreactor as a novel device for sample treatment via solid-phase microextraction coupled to graphite furnace atomic absorption spectroscopy: a route for arsenic pre-concentration. *Molecules* 27: 6198. https://doi.org/10.3390/molecules27196198.

Pan, M., Pei, W., Yao, Y. et al. (2018). Rapid and integrated quality assessment of organic-inorganic composite herbs by FTIR spectroscopy-global chemical fingerprints identification and multiple marker components quantification of indigo naturalis (qing dai). *Molecules* 23: 2743. https://doi.org/10.3390/molecules23112743.

Panteghini, M., Ceriotti, F., Schumann, G., and Siekmann, L. (2001). Establishing a reference system in clinical enzymology. *Clinical Chemistry and Laboratory Medicine* 39: 795–800. https://doi.org/10.1515/CCLM.2001.131.

Pesek, C.A. and Wilson, S.A. (1986). Spice quality: effect of cryogenic and ambient grinding on color. *Journal of Food Science* 51 (5): 1386–1388.

Peter, M. (1970). Cryobiology: the freezing of biological systems. *Science* 138 (3934): 939–949.

Prasad, S., Kulshreshtha, A., Lall, R., and Gupta, S.C. (2021). Inflammation and ROS in arthritis: management by ayurvedic medicinal plants. *Food & Function* 12: 8227–8247. https://doi.org/10.1039/d1fo01078f.

Qu, C.C., Sun, X.Y., Sun, W.X. et al. (2021). Flexible wearables for plants. *Small* 17: e2104482. https://doi.org/10.1002/smll.202104482.

Raeisossadati, M.J., Danesh, N.M., Borna, F. et al. (2016). Lateral flow based immunobiosensors for detection of food contaminants. *Biosensors & Bioelectronics* 86: 235–246. https://doi.org/10.1016/j.bios.2016.06.061.

Rahath Kubra, I., Kumar, D., and Jagan Mohan Rao, L. (2016). Emerging trends in microwave processing of spices and herbs. *Critical Reviews in Food Science and Nutrition* 56: 2160–2173. https://doi.org/10.1080/10408398.2013.818933.

Ramakrishnan, S., Wysk, R.A., and Prabhu, V.V. (2004). Prediction of process parameters for intelligent control of tunnel freezers using simulation. *Journal of Food Engineering* 65 (1): 23–31.

Ranasinghe, S., Ansumana, R., Lamin, J.M. et al. (2015). Herbs and herbal combinations used to treat suspected malaria in Bo, Sierra Leone. *Journal of Ethnopharmacology* 166: 200–204. https://doi.org/10.1016/j.jep.2015.03.028.

Reischauer, C., Staempfli, P., Jaermann, T., and Boesiger, P. (2009). Construction of a temperature-controlled diffusion phantom for quality control of diffusion measurements. *Journal of Magnetic Resonance Imaging* 29: 692–698. https://doi.org/10.1002/jmri.21665.

Safra, P. (2019). Importance of photographic standardization. *American Journal of Orthodontics and Dentofacial Orthopedics* 155: 756–757. https://doi.org/10.1016/j.ajodo.2019.03.014.

Saini, R., Sharma, N., Oladeji, O.S. et al. (2022). Traditional uses, bioactive composition, pharmacology, and toxicology of Phyllanthus emblica fruits: a comprehensive review. *Journal of Ethnopharmacology* 282: 114570. https://doi.org/10.1016/j.jep.2021.114570.

Saletnik, A., Saletnik, B., and Puchalski, C. (2021). Overview of popular techniques of Raman spectroscopy and their potential in the study of plant tissues. *Molecules* 26: 1537. https://doi.org/10.3390/molecules26061537.

Sargeant, J.M., Brennan, M.L., and O'Connor, A.M. (2022). Levels of evidence, quality assessment, and risk of bias: evaluating the internal validity of primary research. *Frontiers in Veterinary Science* 9: 960957. https://doi.org/10.3389/fvets.2022.960957.

Schmidt, R.L., Moore, R.A., Walker, B.S., and Rudolf, J.W. (2023). Precision quality control: a dynamic model for risk-based analysis of analytical quality. *Clinical Chemistry and Laboratory Medicine* 61: 679–687. https://doi.org/10.1515/cclm-2022-1094.

Senapati, A., Basak, S., and Rangan, L. (2022). A review on application of DNA barcoding technology for rapid molecular diagnostics of adulterants in herbal medicine. *Drug Safety* 45: 193–213. https://doi.org/10.1007/s40264-021-01133-4.

Seo, C.S. and Shin, H.K. (2022). Simultaneous analysis of 19 marker components for quality control of oncheong-eum using HPLC-DAD. *Molecules* 27: 2992. https://doi.org/10.3390/molecules27092992.

Shaikh, A.S., Thomas, A.B., and Chitlange, S.S. (2020). Herb-drug interaction studies of herbs used in treatment of cardiovascular disorders-A narrative review of preclinical and clinical studies. *Phytotherapy Research* 34: 1008–1026. https://doi.org/10.1002/ptr.6585.

Sharifi-Rad, M., Mnayer, D., Morais-Braga, M. et al. (2018). Echinacea plants as antioxidant and antibacterial agents: from traditional medicine to biotechnological applications. *Phytotherapy Research* 32: 1653–1663. https://doi.org/10.1002/ptr.6101.

Shukla, A.A. and Gottschalk, U. (2013). Single-use disposable technologies for biopharmaceutical manufacturing. *Trends in Biotechnology* 31: 147–154. https://doi.org/10.1016/j.tibtech.2012.10.004.

Sieme, H., Harrison, R., and Petrunkina, A. (2009). Cryobiological determinants of frozen semen quality, with special reference to stallion. *Animal Reproduction Science* 107 (3): 276–292.

Silva, D., Silva, M.S., Coelho, T. et al. (2022). Combining high intensity ultrasound and experimental design to improve carotenoid extraction efficiency from Buriti (Mauritia flexuosa). *Ultrasonics Sonochemistry* 88: 106076. https://doi.org/10.1016/j.ultsonch.2022.106076.

Simpson, M.J. and McKelvie, J.R. (2009). Environmental metabolomics: new insights into earthworm ecotoxicity and contaminant bioavailability in soil. *The Analyst* 394: 137–149.

Singh, K.K. and Goswami, T.K. (2000). Cryogenic grinding of cloves. *Journal of Food Processing and Preservation* 24: 57–71.

Sinha, R., Abnet, C.C., White, O. et al. (2015). The microbiome quality control project: baseline study design and future directions. *Genome Biology* 16: 276. https://doi.org/10.1186/s13059-015-0841-8.

Smith, W.S. (2020). Standardizing brain death globally. *JAMA Neurology* 77: 1353–1354. https://doi.org/10.1001/jamaneurol.2020.1243.

Song, P., Zhou, F., Li, F. et al. (2021). Superfine pulverisation pretreatment to enhance crystallinity of cellulose from Lycium barbarum L. leaves. *Carbohydrate Polymers* 253: 117207. https://doi.org/10.1016/j.carbpol.2020.117207.

Sontag, G., Pinto, M.I., Noronha, J.P., and Burrows, H.D. (2019). Analysis of food by high performance liquid chromatography coupled with coulometric detection and related techniques: a review. *Journal of Agricultural and Food Chemistry* 67: 4113–4144. https://doi.org/10.1021/acs.jafc.9b00003.

Sridharan, B., Mohan, N., Berkland, C.J., and Detamore, M.S. (2016). Material characterization ofmicrosphere-based scaffolds with encapsulated raw materials. *Materials Science & Engineering. C, Materials for Biological Applications* 63: 422–428. https://doi.org/10.1016/j.msec.2016.

Steponkus, P.L., Myers, S.P., Lynch, D.V. et al. (1990). Cryopreservation of Drosophila melanogaster embryos. *Nature* 345: 170–172.

Sweet, D.J. and Hildebrand, D.P. (1998). Recovery of DNA from human teeth by cryogenic grinding. *Journal of Forensic Science* 43 (6): 1199–1202.

Tao, Y., Bao, J., Zhu, F. et al. (2023). Ethnopharmacology of rubus idaeus linnaeus: a critical review on ethnobotany, processing methods, phytochemicals, pharmacology and quality control. *Journal of Ethnopharmacology* 302: 115870. https://doi.org/10.1016/j.jep.2022.115870.

Tasfiyati, A.N., Antika, L.D., Dewi, R.T. et al. (2022). An experimental design approach for the optimization of scopoletin extraction from Morinda citrifolia L. using accelerated solvent extraction. *Talanta* 238: 123010. https://doi.org/10.1016/j.Talanta.2021.123010.

Varol, M. and Sünbül, M.R. (2017). Organochlorine pesticide, antibiotic and heavy metal residues in mussel, crayfish and fish species from a reservoir on the Euphrates River, Turkey. *Environmental Pollution* 230: 311–319. https://doi.org/10.1016/j.envpol.2017.06.066.

Wan, Q., Yiner-Lee Leemaqz, S., Pederson, S.M. et al. (2019). Quality controlmeasures for placental sample purity in DNA methylation array analyses. *Placenta* 88: 8–11. https://doi.org/10.1016/j.placenta.2019.09.006.

Wang, Y.Y., Sun, Y.P., Yang, B.Y. et al. (2022). Application of metabolomics and network analysis to reveal the ameliorating effect of four typical "hot" property herbs on hypothyroidism rats. *Frontiers in Pharmacology* 13: 955905. https://doi.org/10.3389/fphar.2022.955905.

Wasalathanthri, D.P., Rehmann, M.S., Song, Y. et al. (2020). Technology outlook for real-time quality attribute and process parametermonitoring in biopharmaceutical development-a review. *Biotechnology and Bioengineering* 117: 3182–3198. https://doi.org/10.1002/bit.27461.

Williams, D.J., Thomas, R.J., Hourd, P.C. et al. (2012). Precision manufacturing for clinical-quality regenerative medicines. *Philosophical Transactions. Series A, Mathematical, Physical, and Engineering Sciences* 370: 3924–3949. https://doi.org/10.1098/rsta.2011.0049.

Wright, J.F. (2021). Quality control testing, characterization and critical quality attributes of adeno-associated virus vectors used for human gene therapy. *Biotechnology Journal* 16: e2000022. https://doi.org/10.1002/biot.202000022.

Xiao, Z., Yang, L., Chen, C. et al. (2022). Redox reaction between solid-phase humins and Fe(III) compounds: toward a further understanding of the redox properties of humin and its possible environmental effects. *Journal of Environmental Management* 310: 114793. https://doi.org/10.1016/j.jenvman.2022.114793.

Xiong, C., Huang, C.H., Wu, L. et al. (2022). Identification of Andrographis Herba and its common products using mini-barcode. *Chinese Journal of Natural Medicines* 20: 393–400. https://doi.org/10.1016/S1875-5364(22)60157-2.

Xiong, R., Li, Y., Zheng, K. et al. (2019). Er Shen Wan extract alleviates polyuria and regulates AQP 2 and AVPR 2 in a rat model of spleen kidney Yang deficiency-induced diarrhea. *Biomedicine & Pharmacotherapy* 110: 302–311. https://doi.org/10.1016/j.biopha.2018.11.147.

Yago, M. and Pla, C. (2020). Reference-mean-centered statistical quality control. *Clinical Chemistry and Laboratory Medicine* 58: 1517–1523. https://doi.org/10.1515/cclm-2019-1034.

Yin, F., Song, Z., He, Z. et al. (2022). Chemical fingerprinting and characterization of spilled oils and burnt soot particles - a case study on the Sanchi oil tanker collision in the east China sea. *Science of the Total Environment* 824: 153896. https://doi.org/10.1016/j.scitotenv.2022.153896.

York, M.K., Baron, E.J., Clarridge, J.E. et al. (2000). Multilaboratory validation of rapid spot tests for identification of Escherichia coli. *Journal of Clinical Microbiology* 38: 3394–3398. https://doi.org/10.1128/JCM.38.9.3394-3398.2000.

Yu, Y., Jin, Y., Wang, F. et al. (2017). Protein digestomic analysis reveals the bioactivity of deer antler velvet in simulated gastrointestinal digestion. *Food Research International* 96: 182–190. https://doi.org/10.1016/j.foodres.2017.04.002.

Yuan, H., Ma, Q., Ye, L., and Piao, G. (2016). The traditional medicine and modern medicine from natural products. *Molecules* 21: 559. https://doi.org/10.3390/molecules21050559.

Zhang, D.Y., Peng, R.Q., Wang, X. et al. (2022). A network pharmacology-based study on the quality control markers of antithrombotic herbs: using *Salvia miltiorrhiza - Ligusticum* chuanxiong as an example. *Journal of Ethnopharmacology* 292: 115197. https://doi.org/10.1016/j.jep.2022.115197.

Zhao, N., Liu, D., Wang, Y. et al. (2022). Screening and identification of anti-acetylcholinesterase ingredients from Tianzhi granule based on ultrafiltration combined with ultra-performance liquid chromatography-mass spectrometry and in silico analysis. *Journal of Ethnopharmacology* 298: 115641. https://doi.org/10.1016/j.jep.2022.115641.

Zhu, X., Su, Q., Cai, J., and Yang, J. (2006). Optimization of microwave-assisted solvent extraction for volatile organic acids in tobacco and its comparison with conventional extraction methods. *Analytica Chimica Acta* 579: 88–94. https://doi.org/10.1016/j.aca.2006.07.005.

11

Vegan Chills: Cryogenic Technologies for Vegan-Based Products

Dhiraj Kumar Yadav[1], Khalid Bashir[2], Pintu Choudhary[3], and Shweta Suri[4]

[1] *Department of Agricultural Engineering, North Eastern Regional Institute of Sciences and Technology, Nirjuli, Arunachal Pradesh, India*
[2] *Department of Food Technology, School of Interdisciplinary Sciences and Technology, Jamia Hamdard, New Delhi, India*
[3] *Department of Food Technology, College of Agricultural Engineering and Technology, Dr. Rajendra Prasad Central Agricultural University, Samastipur, Bihar, India*
[4] *School of Health Sciences & Technology, University of Petroleum and Energy Studies (UPES), Dehradun, Uttarakhand, India*

11.1 Introduction

Freezing is a well-established method of food preservation, effectively maintaining the quality as well as nutritional value of a wide range of food items over an extended duration. This process encompasses reducing the temperature of food to below the freezing point, typically at or below 0 °C (32° F) (James et al. 2015). One of the key benefits of freezing as a cold preservation technique is its ability to allow for long-term storage while better preserving essential nutrients and sensory attributes – such as taste, aroma, and texture – compared to many alternative preservation methods, including canning and drying (Barbosa-Cánovas et al. 2005). During freezing, the water activity in food decreases, which plays a critical role in inhibiting various biochemical processes. This reduction in water activity leads to a significant decrease in enzymatic reactions, effectively slowing down the spoilage and deterioration of the food. Additionally, the reduced moisture content helps mitigate microbial growth, thus lowering the risk of spoilage or foodborne illness. Consequently, freezing successfully delays cellular metabolic reactions in the food, resulting in a considerably longer shelf life (Jha et al. 2022).

The phase transition of water into ice during freezing induces physical changes and chemical modifications that can substantially impact the freshness of fresh fruit. This procedure frequently leads to irreversible alterations in quality, including textural degradation and chemical modifications (Jha et al. 2022). Fruit tissues are typically more susceptible to turgidity loss due to freezing than vegetables. This increased susceptibility arises from their fragile composition, elevated moisture levels, and decreased fibre content (Charoenrein 2018). Previously, the negative effects of freezing can be mitigated by

Cryogenic Processing: Advances in the Food Industry, First Edition. Edited by Sumit Sudhir Pathak, Rama Chandra Pradhan, and Sudarshan Ramanathan.

employing circumstances that enhance the freezing rate or by utilizing pretreatment techniques prior to freezing (Jha et al. 2022).

The use of cryogenic liquids for freezing is a major breakthrough in food preservation, providing exceptionally fast freezing speeds. Cryogenically frozen fruits, vegetables, and plant-based products, *viz.*: 'Vegan Chills', deliver significant advantages, including the preservation of maximum nutrients, aroma, and flavour. Their pre-prepared nature, which often involves peeling, dicing, or cutting, ensures unparalleled convenience for consumers (Barbosa-Cánovas et al. 2005). This groundbreaking technique leads to the creation of many tiny ice crystals in both intercellular and intracellular areas, significantly minimizing the adverse effects of dehydration and shrinkage. As a result, this approach is crucial for preserving the overall quality of frozen foods.

Cryogenic technology operates at low temperatures, typically between 0 and −150 °C, to safeguard food quality. By quickly freezing food items using cryogenic substances such as liquid nitrogen or carbon dioxide, this method guarantees the preservation of both the food's nutritional content and its shelf life. Embracing this technology can lead to superior food products that meet consumer demands for quality and storage life. Freezing with cryogenic fluids occurs at a significantly faster rate compared to most conventional freezing systems. This quick freezing is due to the significant temperature gap present for heat exchange. Although the heat transfer coefficient is relatively low, the greater temperature difference (ΔT) facilitates efficient heat transfer. Additionally, fans that improve convective heat transfer use very little energy. In comparison, traditional blast freezers employ air cooled below −40 °C. To make up for the smaller ΔT in these systems, they rely on generating a high heat transfer coefficient through air blast freezing. Despite the strong convective forces produced, freezing time in blast freezers can take several hours, while cryogenic freezing typically results in freezing time measured in mere minutes for similar products (Goswami 2010). Moreover, cryogenic processing significantly affects the quality of vegan products, including fruits, vegetables, and plant-based foods. Studies have shown that it has substantial effects on various quality parameters, including texture, moisture retention, structural integrity, flavour preservation, drip loss, microbial, and spoilage considerations.

This chapter offers an in-depth overview of cryogenic agents utilized in the food sector, emerging cryogenic technologies (such as cryogenic freezing, pulse freezing, cryo-milling, ultra cryo-milling, cryogenic cooling, and cryo-concentration), their influence on the quality characteristics of vegan products, consumer perceptions of vegan products treated with cryogenic methods, and the future prospects of cryogenic technology within the food industry.

11.2 Concept of Vegan Chills

Plant-based foods, such as fruits and vegetables, aid as a crucial repository of micronutrients (vitamins and minerals), dietary fibre, and other vital elements that significantly influence human nutrition. These foods typically have a perishable nature and require adequate temperature-controlled set-up throughout storage, transportation, and marketing to ensure food quality and reduce the loss of nutrients, thus ensuring that the available food satisfies consumers' quality prospects. With the increasing accessibility of food markets and supply networks, ensuring the quality of food has emerged as a significant issue. Several situations can arise where vegan products are subjected to unsuitable storage conditions, causing unwanted alterations in sensory and nutritional characteristics, which ultimately results in a decline in food quality. A constant decline in food quality across one or more quality factors ultimately makes the product unacceptable. Furthermore, inadequate food storage directly and adversely affects the economic value of food items, resulting in the wastage of resources utilized in production, including land, water, energy, and additional inputs. Therefore, preserving food quality throughout the food supply chain would enhance the effectiveness of the food distribution system and boost consumer access (Parfitt et al. 2010; You et al. 2021).

Furthermore, to preserve the quality of food for long durations, freezing could be one option. Freezing is typically considered a better technique for the long-term preservation of perishable vegan products than other preservation techniques in terms of organoleptic qualities and nutritional benefits. Still, the effects of freezing and thawing continue to pose a major issue for the quality of food items. The process of freezing can lead to irreversible harm to tissue structures due to the crystallization of water (James et al. 2015). Freezing also alters the sensory characteristics of foods that are eaten after being delivered and stored.

Nowadays, cryogenic freezing is utilized to preserve vegan products owing to its merits over conventional methods. Hence, vegan and plant-based products processed using cryogenic processing technology are referred to here as 'Vegan Chills'. In-depth, the cryogenic freezing technology utilizes cooling refrigerant straight on food and is categorized into three primary methods: (1) vaporizing the refrigerant to blow it over the foods, (2) immersing the foods in the refrigerant, or (3) spraying the refrigerant directly onto the food. The third method (3) is the most frequently utilized approach; the refrigerants employed in food applications are liquid nitrogen or carbon dioxide. The process is quite efficient because of the high heat transfer rates and extremely low freezing temperatures linked with refrigerants; however, the expensive nature of refrigerants has restricted this technology's use to high-end products (Gates 2013; You et al. 2021).

11.3 Concept of Cryogenic Techniques

Cryogenic freezing is an advanced technique that utilizes refrigerants, for instance, liquid nitrogen or solid carbon dioxide. When these refrigerants interact with the product, they boil off, producing a significant cooling effect. This process effectively uses both the latent heat absorbed during the boiling phase and the sensible heat derived from the resultant cold gas (Fennema 1978). The majority of cryogenic systems use total-loss refrigerants, which are intended to be discharged into the atmosphere rather than recovered. The alternatives are mainly limited to ambient air and its constituents since total-loss refrigerants must be easily accessible and safe due to economic and environmental concerns.

In the 1930s, liquid air was used for the first commercial use of cryogenic freezing (Bald 1991). Although a significant amount of liquid oxygen, a strong oxidizing agent, is present in liquid air, the gas can potentially be created locally, avoiding the need to buy and store it. Safer substitutes like liquid nitrogen and liquid or solid carbon dioxide have generally replaced liquid air, despite advertising its use.

11.3.1 Importance of Cryogenics in Food Preservation

Cryogenic technology provides low-temperature solutions for the food sector. The application of this technology for food preservation has a lot of potential. The physical characteristics of cryogenic gases serve as a vital resource for the food industry to enhance plant automation, flexibility, efficiency, and production costs (Zohuri and Zohuri 2018). Cryogenics is a useful way of food preservation because it helps maintain the quality of food and extend its shelf life in several ways:

- In cryogenic freezing, a thin layer of frozen crust forms quickly on the surface of food material. This layer can act as a barrier to lessen environmental contamination, for instance, salt diffusion from the saline solution employed in the freezing process that follows (Lan and Farid 2004). With cryogenic freezing, water loss and the development of big ice crystals that could weaken the food's structure are avoided.
- Cryogenic freezing also reduces food material deterioration by preventing moisture loss, which is a major issue typically linked to other traditional freezing techniques. It leads to uniform freezing of the product. Additionally, it reduces the chances of freeze burns in the product by forming smaller ice crystals.
- Cryogenic freezing creates an environment that's hostile to microorganisms and biochemical reactions, which slows spoilage and increases the storage life of food.
- Cryogenic freezing can freeze large quantities of food at once. It is considered a quick method for preserving foods. Besides, it requires minimum manpower, and low installation and maintenance costs, thereby making it a cost-effective freezing method.
- Cryogenic freezing prevents the loss of volatile compounds that contribute to the taste and aroma of the food. This is especially important for specialty foods.
- The physical integrity of food products can be enhanced through this freezing method. Additionally, it offers the benefit of reducing the total freezing time (Lan and Farid 2004).

11.4 Cryogenic Principle and Mechanism

11.4.1 Definition, Temperature Range, and Principle

The word cryogenics originates from the Greek term for icy cold, '*Kryos*', and when paired with the suffix '*genics*', it translates to 'appropriate for creation in icy cold environments'. In summary, cryogenics is the application of extremely low-temperature techniques to induce physical transformations in gases, and liquids through liquefaction, or solids. It is the study of materials at very low temperatures (Zohuri 2017). Additionally, cryogenics refers to the study and application of producing and utilizing artificial cold at very low temperatures. For many years, the boundaries of cryogenics remained undefined until the 13th International Congress of Refrigeration, which took place in Washington D.C. in 1971. This congress established a universal definition for 'cryogenics' and 'cryogenic' by setting a lower limit of 120 K, or −153 °C, to distinguish these terms from conventional refrigeration (Popov et al. 2019).

The temperature range for cryogenic freezing is generally considered to be from around −150 °C (−238 °F) down to absolute zero, which is −273 °C (−459 °F) (Goswami 2010). Cryogenic freezing in food preservation involves quickly reducing the temperature of a food item to very low levels, often utilizing cryogens such as liquid nitrogen or carbon dioxide. This process effectively slows down or entirely stops the action of enzymes and microorganisms that contribute to spoilage, thereby maintaining the food's quality for longer durations. The rapid freezing process created by cryogens is key to minimizing ice crystal formation, which can damage cell structures and affect the texture of the food. Furthermore, the latent heat of vaporization is obtained from food in the cryogenic freezing process done by cryogens.

11.4.2 Cryogenic Agents

Cryogenic agents such as liquid nitrogen, hydrogen, helium, oxygen, inert gases, NH_4, CO_2 (in solid or liquid forms), air, etc. are used as cryogens on the industrial scale. Essentially, cryogenic agents or cryogenic liquids, referred to as cryogens, exist as gases under standard temperatures and pressures. Nonetheless, at lower temperatures, they exist in their liquid form. The boiling points of these liquids are below −150 °C (238 °F), making them extremely cold. Extremely cold gases and vapours are even released from cryogenic liquids. They frequently compress the moisture in the air, producing a very noticeable vapour. Various cryogens liquefy under distinct temperature and pressure conditions, yet they share two characteristics: they are extremely cold, and even minimal quantities of liquid can transform into vast volumes of gas (Zohuri and Zohuri 2018).

Among different cryogens, majorly liquid nitrogen and carbon dioxide (both solid and liquid forms) are widely used in food preservation. Table 11.1 presents the properties of cryogenic agents. In cryogenic freezers for food usage, the predominant cryogen (refrigerant) employed is liquid nitrogen, which is a colourless, as well as odourless liquefied gas with a boiling point of −196 °C (or −195.6 °C) and a cooling capacity of approximately 400 kJ/kg at an exhaust temperature of 0 °C (Goswami 2010; Barnwal and Singh 2012).

Table 11.1 Properties of the liquid cryogens used in food industry.

Properties	Liquid nitrogen	Liquid carbon dioxide
Boiling point (°C)	−195.60	−78.50
Density (kg/m^3)	808.00	464.00
Specific heat of liquid (kJ/kgK)	2.05	2.26
Thermal conductivity (W/mK)	0.14	0.19
Latent heat of evaporation (kJ/kg)	199.00	352.00
Total usable refrigeration effect (kJ/kg)	690.00	565.00

Source: Adapted from Barnwal and Singh (2012).

Table 11.2 Comparison of properties of cryogenic freezing vs. conventional freezing.

Property	Cryogenic freezing	Conventional freezing
Types of freezers used in the process	Immersion-type freezers, spray freezers, etc.	Air blast type freezers, contact freezers, etc.
Dehydration loss	Approximately lower than 1%	Between 3% and 10%
Flavour, texture, and aroma of food products	Retained after cryogenic freezing	Lost after conventional freezing
Oxidative rancidity	Minimum	Maximum
Drip loss	Minimum	Maximum

11.4.3 Advantages and Disadvantages of Cryogenic Freezing and Conventional Freezing

Cryogenic freezing is superior to traditional conventional methods for maintaining the quality of food items after they are frozen and throughout their time in frozen storage. Table 11.2 presents a comparison of cryogenic freezing and conventional freezing. Usually, the high-quality preservation is apparent through different quality aspects, such as the upkeep of the original microstructural properties of food, enhanced retention of texture, decreased loss of exudate, reduced dehydration losses, diminished lipid oxidation, and smaller changes in the colour of food products (Jha et al. 2022).

Studies indicate that cryogenic freezing improves the product's mechanical strength and serves as a moisture-retaining barrier. This happens because when food is exposed to cryogens, a layer of ice forms on the surface of the product. The ability of this crust to retain moisture is probably what accounts for the lesser dehydration seen after cryogenic freezing in contrast to traditional freezing, both during the freezing phase and while being stored. Additionally, the creation of this crust could help to enclose and protect the food's taste, providing an extra benefit regarding flavour retention (Agnelli and Mascheroni 2002). In addition to these advantages, cryogenic freezing also has the potential to improve the microbiological safety of food (Jha et al. 2022). Furthermore, because of the significant cooling capacity and temperature difference, the time needed for freezing in cryogenic processes is minimal and nearly instantaneous (Goswami 2010).

Additionally, ultra-fast freezing, enabled by cryogenic liquids, offers remarkably high freezing speeds. This process creates many tiny ice crystals in both intercellular and intracellular spaces, thus reducing the harmful impacts of dehydration and damage from shrinkage (Jha et al. 2022).

11.5 Cryogenic Techniques for Vegan Products

11.5.1 Emerging Cryogenic Technologies

11.5.1.1 Cryogenic Freezing

Utilizing cryogenic liquids like carbon dioxide or nitrogen, cryogenic freezing is accomplished and typically occurs at temperatures below −70 °C (Goswami 2017). Cryogenic freezing serves as an effective technique for preserving the quality and freshness of perishable fruits and vegetables. During the process, fresh produce is sorted and examined to guarantee consistency in size and quality, with any substandard fruits discarded to uphold high product standards. The cryogenic freezer is set to achieve a steady-state condition at an internal temperature of −120 °C. This low temperature ensures rapid freezing and minimal ice crystal formation, which is critical for maintaining the texture and nutritional quotient of the items. The freezing materials are introduced into the freezer tunnel at a controlled rate of 5 kg/min, depending on the freezer specification, ensuring consistent exposure to the freezing environment. The feeding mechanism allows for the even distribution of fruits within the tunnel, optimizing the freezing efficiency. The freezing process involves continuous monitoring of the temperatures of the freezing material and surrounding air using Type K thermocouples, which are connected to data loggers that record readings at set intervals. The recorded data are analysed to evaluate the thermal profile of the fruit and the air inside the tunnel, ensuring consistent freezing conditions throughout the process.

The frozen items are collected at the exit of the freezing tunnel. Rapid freezing within the tunnel averts the development of large ice crystals, preserving the structural integrity and quality of the fruit. The frozen fruits are immediately packaged in rigid polyethylene boxes, each with a capacity of ½ kg, providing adequate protection against physical damage and contamination during subsequent storage and transportation. The packaged dates are then directly transferred to ultra-freezers set at −40 °C. This secondary freezing ensures the long-term preservation of the product's quality. This refrigeration method is different from other techniques, such as plate contact systems, gas contact systems, and immersion and liquid contact systems. It does not use outside sources to create low temperatures. Low temperatures are generated by the phase transition of cryogenic liquids that come into contact with food in freezing cabinets. The process is exceedingly swift, although concurrently somewhat costly. In certain instances, cryogenic freezing is integrated with air contact freezing as a dual-phase method to enhance freezing rates while minimizing process expenses. This procedure, referred to as the cryo-mechanical process, generates a cold, hard crust on the material through cryogenic freezing, prior to transferring it to a cold chamber to complete the solidification process (Brennan and Grandison 2012).

A research study investigated the quality and stability of date fruits (cv. *Barhi*) at the Khalal maturity stage, utilizing two distinct freezing techniques: cryogenic freezing with liquid nitrogen and traditional deep freezing. Evaluations were performed every three months during a nine-month storage period at temperatures of −20 °C for conventional freezing and − 40 °C for cryogenic freezing (Alhamdan et al. 2018). The colour characteristics of the frozen dates were affected by both the freezing method and the length of frozen storage. Significant disparities exist in the quality of frozen fruits, favouring cryogenic freezing over normal gradual freezing methods. The results demonstrated a significant discrepancy in the freezing times of the two approaches. The cryogenic freezing process required 10 minutes of freezing time, but the standard gradual freezing approach required 1,800 minutes (Alhamdan et al. 2018).

11.5.1.2 Pulse Freezing

Pulse freezing is an innovative cryo-processing technique developed to enhance the freezing of food products by rapidly alternating cooling and holding phases during the freezing process. This technique aims to enhance the quality and structural integrity of frozen foods by minimizing the formation of large ice crystals, which can harm cellular structures and reduce the sensory and nutritional qualities of the product.

The fundamental concept of pulse freezing involves controlling the speed and consistency of heat transfer during the freezing process. A more even freezing is achieved by decreasing the temperature difference within the food item through quick, intense cooling pulses interspersed with short holding periods. This pulsing technique facilitates the creation of smaller, more evenly distributed ice crystals in comparison to traditional freezing methods, such as blast freezing or slow freezing. Pulse freezing has considerable potential in the food industry, especially in addressing consumer needs for high-quality frozen items with reduced sensory and nutritional losses.

11.5.1.3 Cryo-milling

The term 'cryo' originates from Greek, meaning very low temperatures. As a result, 'cryo-milling' refers to the mechanical process of milling or grinding at very low temperatures, ideally below 123 K or −150 °C. According to Katiyar et al. (2020), cryo-milling is a sort of top-down technology that uses liquid nitrogen (LN_2) or liquid argon (LAr) to grind at or below −150 °C. It is environmentally friendly and may be employed to create all forms of nanomaterials in huge amounts. Any material can be effectively made brittle and friable by freezing it before grinding. Therefore, reducing cold welding or powder agglomeration and speeding up fracturing are beneficial. There are two methods to accomplish this. In its most basic form, the powder is ground together with the milling media, which is then combined with cryogenic liquid to create a cryogenic slurry. After that, ball milling is done. However, when the powder and the cryogenic liquid come into direct contact, this could contaminate the powder. In the second method, a jacket of cryogenic liquid is placed around the vial to externally cool the milling media and powder without exposing the powder to the liquid. Thus, the likelihood of contamination is negligible. However, it affects the efficiency of cooling the powder due to inadequate heat transfer over the thick layer of the vial (Katiyar et al. 2020).

Cryo-milling can be divided into two categories based on cooling: wet milling, which involves mixing materials with cryogenic liquid, and dry milling, which involves cooling the milling chamber externally. The most basic method involves combining the powder to be ground and the milling medium with a cryogenic liquid to create a cryogenic slurry, which is then mechanically ground. It is commonly referred to as mechanical grinding at very low temperatures or the pulverization process. This is described as wet cryo-milling. Due to the cooling liquid's prolonged direct contact with the milling materials, wet milling may result in contamination from the liquid. Sometimes, in addition to the LN_2, ice flakes may also enter the grinding chamber. As a result, when the ice flakes turn to water due to heating after the ground powder is removed from the mill, the materials may become oxidized. Dry milling, on the other hand, can completely remove or significantly reduce this issue. The second method involves enclosing the vial in a jacket of cryogenic liquid to externally cool the milling media and powder. This prevents the powder from coming into touch with the cryogenic liquid and reduces the possibility of contamination. This is called dry cryo-milling. Theoretically, a variety of cryogenic liquids can be used, such as liquid helium (BP ~ 4 K), liquid oxygen (BP ~ 90 K), liquid neon (BP ~ 27 K), liquid argon (BP ~ 87 K), and liquid nitrogen (BP ~ 77 K). Nevertheless, due to their accessibility, safety, ease of handling, and affordability, liquid argon (LAr) and liquid nitrogen (LN_2) are also widely used. In general, pre-cooled spices are ground into a powder using a hammer mill. This permits rapid refinement of the spices, keeping of scent, as well as maintenance of health and cleanliness of the spices. Sophisticated mills like fluid energy impact mills (such as fluidized bed jet mills and spiral jet mills) are used in pharmaceutical cryo-milling to produce nanoparticles with a limited size distribution. The market offers a variety of cryo-mills, each with its own capacity and operating principle.

Low temperatures during cryo-grinding can effectively restrict or even inhibit recovery and recrystallization processes, hence facilitating early grain refinement. The continual milling can create wear and tear on the balls and vials, leading to the integration of these materials into the milled powder. Longer milling, in general, employed to generate nanocrystalline particles for ductile materials, might cause a high level of contamination. In cryo-milling, rapid nanocrystallization occurs, as discussed previously, requiring significantly shorter milling time and consequently less contamination. The fact that cryo-milling eliminates the need for a process control agent (PCA) further reduces the possibility of contamination, which is the subject of another benefit. It has been demonstrated that obtaining a finer scale just requires quick grinding at cryogenic temperature.

Cryo-milling has also been utilized to produce enhanced dispersion of carbon nanotubes (CNTs), carbon blacks, graphene, or even particles in polymeric matrices without affecting the characteristics of their surface by using harmful chemicals (Mittal et al. 2016). Despite the simplicity of cryo-milling simplicity, improvement could concentrate on eliminating labour-intensive aspects and contamination problems. The use of closed systems to reduce contamination, automated cryo-milling systems, and milling parameter optimization for increased efficiency are some of the opportunities. Advances in cryo-milling technology could boost its dependability and accessibility.

Numerous studies are concentrating on new techniques for mechanical particle reduction. In some situations, these techniques draw ideas from frequently used procedures, but in many cases, they provide a whole new approach to particle nanonization. Figure 11.1 illustrates a comparison between conventional and cryo-ball milling, emphasizing their difference in temperature range, time required, and the contamination risks of products.

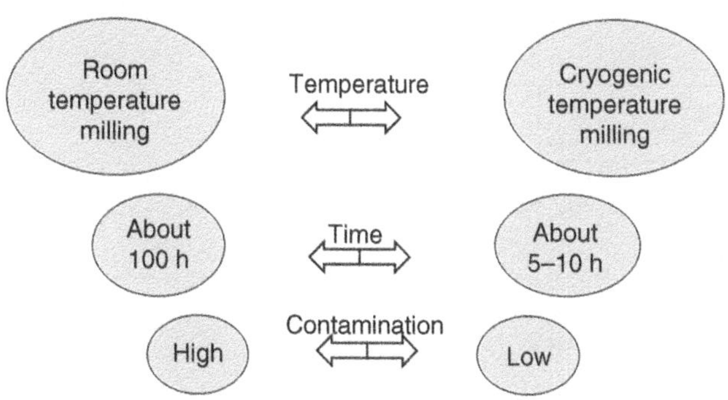

Figure 11.1 Comparison of cryo-ball milling vs. conventional room temperature milling.

11.5.1.4 Ultra Cryo-milling

Ultra cryo-milling is an advanced wet bead milling technique that utilizes cryogenic liquids, typically liquid nitrogen, instead of conventional aqueous milling media. This substitution offers several advantages. Liquid nitrogen, with its low surface tension and viscosity, effectively minimizes particle aggregation by forming a barrier between particles. Additionally, it is cost-effective, non-toxic, non-flammable, and has negligible solubility for most pharmaceutical compounds. A significant benefit of using liquid nitrogen is its spontaneous vaporization, which eliminates the need for a drying step and avoids associated challenges. Moreover, the cryogenic temperatures make drug particles more brittle, facilitating their breakdown into smaller nanocrystals. Zirconia beads are commonly employed as milling media in this process. Ultra cryo-milling has been shown to achieve finer particle sizes than traditional jet milling, in both dry and redispersed forms. The particles produced through this method also exhibit a larger specific surface area compared to jet-milled and untreated particles, potentially enhancing dissolution rates and bioavailability. Incorporating a stabilizing excipient during the milling process further improves particle stability and dissolution rates. For example, polyvinylpyrrolidone (PVP) has been used effectively as a co-grinding agent, resulting in slightly larger particle sizes compared to the drug alone. However, cryo-milled particles co-ground with PVP demonstrate significantly improved dissolution rates, outperforming both the original drug suspension and jet-milled samples. Other stabilizing agents, such as Eudragit L 100 and hydroxypropyl methylcellulose (HPMC), have also been employed for similar benefits. This innovative approach enhances drug performance by optimizing particle size, stability, and bioavailability.

11.5.1.5 Cryogenic Cooling

Cryogenic cooling utilizes liquid N_2 or solid CO_2 (dry ice) as the refrigerant to accomplish fast freezing. Figure 11.2 displays a cryogenic cooling system. Liquid N_2, having a boiling point of −196 °C at atmospheric pressure and a latent heat of evaporation of 199 kJ/kg, is often employed in cryogenic freezing systems. In this method, liquid nitrogen kept under pressure is sprayed over food carried through a tunnel. The extremely low temperature ensures quick freezing, while the cold nitrogen vapours precool incoming items before they encounter the liquid nitrogen spray. Typically, around 2.5 kg of liquid nitrogen is required to freeze 1 kg of food.

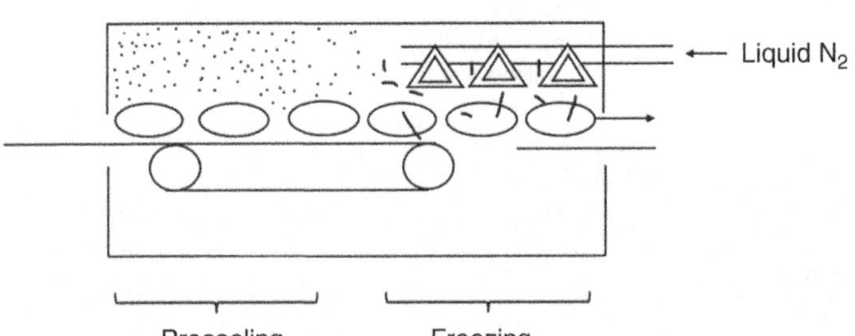

Figure 11.2 Cryogenic freezing cooling system.

Similarly, liquid carbon dioxide kept under pressure is employed in cryogenic systems. It transforms into a mixture of gaseous and solid CO_2 crystals when released through spray nozzles. These crystals sublimate upon contact with the meal, cooling it through a process known as evaporative cooling. Liquid CO_2 has an evaporation temperature of −78.5 °C and a latent heat of evaporation of 571 kJ/kg, making it an effective refrigerant for rapid freezing applications. Both technologies are very efficient, harnessing the unique features of cryogenic materials to protect food quality through ultra-rapid freezing.

11.5.1.6 Cryo-concentration

Cryo-concentration is a natural phenomenon that occurs during ice thawing (Aider and De Halleux 2009). From the original solution, a more concentrated phase is subsequently extracted. A widely used technique in the food industry, cryo-concentration concentrates a variety of liquid meals and recovers heat-sensitive biological constituents such as proteins, polyphenols, water-soluble vitamins, and aromatics. Cryo-concentration can be achieved by various techniques as summarized here.

Cryo-concentration technology based on the suspension crystallization method offers a unique approach to liquid food concentration but faces significant challenges due to its complexity and cost. To achieve effective concentration, the method requires a highly complex system comprising several components:

- Scraped Surface Heat Exchanger: Used for generating seed ice crystals
- Recrystallization Vessel: Facilitates the growth of ice crystals
- Washing Tower: Essential for separating ice crystals from the concentrated solution

A critical aspect of this method is the growth of ice crystals through the Ostwald ripening mechanism, which enhances separation efficiency but requires precise control. Despite its ability to preserve flavours and nutritional qualities, the high costs and complicated set-up have limited its practical application in the food industry, although it has been successfully employed for products like coffee extracts and fruit juices. Future studies on design and cost reduction are essential for wider adoption of this promising technology.

Progressive cryo-concentration technology operates on a fundamentally different principle compared to the suspension crystallization method. Instead of forming numerous small ice crystals, this method produces a single large ice crystal on a cooling surface within a crystallization vessel. This design simplifies the process by making the separation of the ice crystal from the mother solution much easier, thereby reducing system complexity and overall costs. Originating as a melt crystallization technique primarily used for purifying organic materials, progressive cryo-concentration has also been applied to small-scale concentration processes for both aqueous and organic solutions. Although historically used for analytical purposes, such as concentrating impurities in organic solvents, systematic studies on its mechanism and separation efficiency remain limited. This method holds potential for broader applications, requiring further research and optimization.

Partial-block cryo-concentration is a novel and efficient technology for concentrating liquid food solutions, as described by Burdo et al. (2007). In this method, a liquid solution is partially frozen in a crystallizer chamber, where cooling is applied through a central pipe. As the ice block grows, the remaining solution becomes progressively more concentrated. This technique significantly reduces energy consumption, with optimized cryo-separators achieving power expenses as low as 0.3–0.4 MJ/kg of ice. Additionally, utilizing energy stored in the ice block for refrigerant supercooling can further decrease energy consumption by 30%. Gravitational separation of the ice block minimizes the loss of dry substances in the solution to just 0.1–0.5% of the ice mass. The separation efficiency largely depends on the packing density of ice crystals, with higher densities improving separation. However, a balance must be struck between thermal and diffusion resistances in the boundary layer to avoid extreme regimes, such as undirected volumetric crystallization or equal solute concentration in the ice and initial solution. This method has been successfully applied to concentrate milk whey, fruit juices, coffee extracts, and aromatic compounds from grape seeds, demonstrating its versatility and potential in the food industry.

Complete block cryo-concentration is an advanced and efficient technique for producing concentrated liquid foods with superior nutritional and sensory properties (Aider and de Halleux 2008a). The process involves fully freezing a liquid food solution, ensuring the temperature at the product's centre drops significantly below the freezing point. Gravitational thawing is subsequently used to separate the concentrated fraction from the ice fraction after the entire frozen solution has thawed, often with the use of additional methods to increase separation efficiency (Aider and de Halleux 2008a,b). During thawing, the concentrated liquid flows through the ice block, which acts as a solid framework. By carefully controlling the thawing temperature, this method can achieve efficiency levels exceeding 90%, minimizing the solute trapped within the ice crystals.

Eutectic cryo-concentration technology is a process designed to separate inorganic salts from aqueous solutions through the simultaneous crystallization of both solute and ice. This occurs when the solution is cooled below its eutectic temperature – the lowest melting point achievable for a specific composition where all constituents crystallize simultaneously. At this eutectic point, the solution produces ice and salt crystals, which can then be separated under the influence of gravity due to differences in their densities. The process yields high-purity products with relatively low energy consumption, making it a promising technique for industrial applications. The crystal size distribution plays a crucial role in determining the efficiency of downstream separation and product quality, requiring precise modelling and control for optimal performance. While earlier studies explored its potential, recent research has highlighted its advantages, particularly for achieving steady-state crystallization and efficient separation under controlled conditions. Hence, high nutritional, biological, and organoleptic value goods can be obtained in the bio-food business through the application of cryo-concentration technology.

11.6 Effect of Cryogenic Processing on Quality Characteristics of Vegan Products

Cryogenic processing has a significant impact on the quality attributes of vegan products such as fruits, vegetables, and plant-based foods. This section discusses the effects on various quality parameters including texture, moisture retention, structural integrity, flavour preservation, drip loss, microbial, and spoilage consideration.

11.6.1 Packaging Materials for Frozen Foods

A variety of packaging materials, including plastics, metals, and paperboards, are utilized to preserve the quality of frozen vegan products. Common materials like low-density polyethylene (LDPE) and high-density polyethylene (HDPE) are widely used for packaging individual quick-frozen (IQF) fruits, vegetables, and plant-based products due to their cost-effectiveness and ease of sealing. However, they provide limited oxygen barrier protection. More advanced materials, such as polyester terephthalate (PET), offer excellent thermal stability but are expensive and brittle at low temperatures. Proper packaging minimizes moisture loss and prevents freezer burn, a critical factor in maintaining product quality during cryogenic storage.

11.6.2 Concentration Effects

During freezing, solutes in vegan products concentrate, leading to potential quality deterioration:

- Gritty or sandy texture from solute precipitation
- Protein denaturation due to salting-out effects and pH changes
- Supersaturation of dissolved gases, potentially releasing gases and affecting quality

11.6.3 Freeze Cracking

Certain vegan products, such as mango slices, are prone to cracking under high freezing rates, particularly in cryogenic processes using liquid nitrogen. Precooling the product can mitigate freeze cracking and maintain structural integrity.

11.6.4 Moisture Loss and Freezer Burn

Moisture loss during freezing, or sublimation of surface ice crystals, can cause freezer burn, leading to off-flavours and texture degradation. Utilizing moisture-proof packaging materials effectively prevents freezer burn and preserves sensory quality.

11.6.5 Rancid Flavour Development

Oxidative reactions between frozen products and residual air in the packaging can lead to rancid flavours. This issue can be mitigated by using airtight packaging materials and removing excess air from storage containers.

11.6.6 Ensuring High-Quality Frozen Products

Using nutrient-rich and tender raw materials ensures better sensory and nutritional outcomes. For example, sweet and tender pea varieties with low starch content are preferred for freezing. Rapid freezing immediately after harvesting prevents undesirable metabolic changes, such as sugar-to-starch conversion in peas, which can negatively affect texture, sweetness, and overall quality.

11.6.7 Texture

Cryofreezing significantly affects the texture of food products due to the impact of ice crystal formation on the food matrix. The size and distribution of ice crystals play a crucial role in determining the extent of texture degradation. Smaller ice crystals, typically formed under controlled freezing conditions such as microwave-assisted freezing (MAF), cause less structural damage compared to larger crystals formed during conventional freezing methods. Studies revealed that freezing and thawing substantially reduce the firmness of food products like apples and potatoes. However, texture retention is improved under optimized cryofreezing techniques. For instance, pulsed MAF at higher power levels resulted in better firmness retention due to smaller ice crystal sizes, leading to reduced mechanical and biochemical damage. Texture measurements, such as firmness and Young's modulus, confirmed that samples subjected to advanced cryofreezing methods exhibit less degradation, preserving structural integrity and mechanical properties more effectively than conventional freezing methods.

11.6.8 Drip Loss

Cryofreezing influences the water-holding capacity of food, with advanced freezing methods such as MAF showing notable improvements over conventional techniques. Drip loss, an indicator of moisture retention, was significantly reduced under MAF conditions. For instance, pulsed MAF resulted in lower drip loss (P1MAF: $2.03 \pm 0.88\%$, P2MAF: $1.62 \pm 0.70\%$) compared to conventional freezing ($2.67 \pm 1.12\%$). The reduction in drip loss is attributed to the smaller and more uniformly distributed ice crystals formed under MAF, which cause less mechanical damage to cell structures during freezing and thawing.

11.6.9 Colour

The impact of cryofreezing on colour varies depending on the pretreatment and freezing protocol. For blanched potatoes, neither freezing nor thawing significantly altered the colour characteristics, as blanching effectively inactivated enzymes responsible for browning. In contrast, unblanched samples exhibited colour changes due to enzymatic browning during thawing. Lightness (L*) remained stable, while blanching reduced yellowness (b*) and imparted a slight greenish tint (negative a*), with these characteristics preserved post-freezing.

11.6.10 Suppressed Microbial Growth and Reduced Activity of Spoilage Bacteria

Cryogenic processing, characterized by extremely low temperatures, is highly effective in suppressing microbial growth and mitigating spoilage in food products. The process primarily inhibits the metabolic activities of microorganisms, preventing their reproduction and reducing the risk of foodborne illnesses. Despite its efficacy, cryogenic conditions cannot entirely eliminate microorganisms, particularly cryogenic microorganisms such as psychrophilic, psychrotolerant, and psychotropic species. These microorganisms exhibit unique adaptations to low temperatures, enabling them to survive and even thrive under refrigeration and freezing conditions (Xu et al. 2024).

Microorganisms that may endure in refrigerators are categorized into three categories: psychrophilic (capable of growing at 0 °C or lower, with a maximum temperature of 20 °C), psychrotolerant (able to grow at 0 °C and temperatures surpassing 20 °C), and psychotropic (growing up to a maximum of 20 °C). Each group demonstrates unique molecular processes for adaptation to low temperatures and has varied ecological ranges. Psychrophilic bacteria flourish in nutrient-rich settings at 0 °C, demonstrating increased vitality, while psychrotolerant microorganisms display a broader adaptability range. However, due to the problems in exactly categorizing specific strains as psychrophilic, psychrotolerant, or psychotropic, these groupings are collectively referred to as 'cryobiotic microorganisms'.

The principal sources of cryogenic microorganisms in refrigerators include microorganisms that survive on the interior surfaces of refrigerators for extended durations and those originating from food or human interaction. Cryogenic microorganisms have been identified in various refrigerator components, such as inner walls, shelves, vegetable compartments, and even dishcloths used for cleaning, where they can contaminate food through air movement or direct contact (Macías-Rodríguez et al. 2013). The contamination risks vary across different places, with microorganisms on refrigerator walls posing a lesser danger (39%) compared to those at the bottom (61%), which shelters the majority of microbial flora. Factors such as limited storage space, the diversity of stored goods, insufficient classification, and mixing of raw and cooked foodstuffs contribute to high cross-contamination concerns. Additionally, long-term storage of chilled or frozen foods provides an excellent setting for microbial development.

The presence of cryogenic microorganisms in residential refrigerators is a substantial contributor to food spoiling and foodborne diseases, underlining the significance of effective prevention measures. Improper handling, mixing of raw and cooked foods, and long-term storage in cryogenic conditions can lead to microbial persistence on food surfaces or within refrigerator compartments. This persistence can result in food spoilage and increased risk of illness if effective hygiene and temperature control measures are not maintained. For instance, poorly managed refrigerator temperatures often observed in older units or due to frequent door openings can create micro-environments where cryogenic microorganisms flourish. Thus, while cryogenic processing is instrumental in enhancing food safety and shelf life, maintaining strict temperature controls and implementing hygienic practices is critical to minimizing spoilage and microbial risks (Xu et al. 2024).

Additionally, cryogenic processing has emerged as a promising technique for manipulating the structural and functional properties of materials, particularly in food and material sciences. In a recent study, poly (ethylene oxide) (PEO) nanofiber mats underwent cryogenic milling to investigate their potential as oleo gel scaffolds. The original PEO nanofiber mats were removed from aluminium foil, cut into pieces not exceeding 1 cm^2, and cryo-processed using liquid nitrogen and a mixer mill. Two different milling durations were evaluated: 8 and 14 minutes, with intermediate cooling steps in liquid nitrogen. The temperatures during milling were maintained below the glass transition temperature (Tg) of PEO (ranging between −55.5 and − 57.2 °C) (Valoppi et al. 2023).

Transitioning PEO from a rubbery to a glassy state increased its rigidity and fragility, enabling fragmentation during milling (Balani et al. 2015). However, despite successful fragmentation, the resulting cryo-milled nanofiber powders did not exhibit oil-structuring ability when dispersed in refined sunflower oil (RSO) at a 10% concentration. Polarized light micrographs revealed that the microstructure was characterized by large, compact crystalline aggregates dispersed in oil, lacking internal morphology indicative of long fibres. This absence of the porous structure observed in the original PEO nanofiber mat likely contributed to the inability of the cryo-milled material to form oleo gels.

Further experiments confirmed this hypothesis. When cryo-milled PEO powder, sourced directly from the manufacturer, was added to RSO under similar conditions, it also failed to form an oleo gel. The resulting compact particles, although smaller than those derived from cryo-milled nanofibers, shared a similar compact structure. The importance of maintaining a porous scaffold for oleo gel formation has been previously demonstrated (Tanti et al. 2016; Manzocco et al. 2017; Plazzotta et al. 2020). Porous structures facilitate oil absorption and retention, which are critical for the formation of oleo gels. Consequently, preserving the original porous structure of nanofiber mats during cryogenic processing is essential for achieving the desired functional properties.

11.7 Applications of Cryogenically Treated Vegan Products

By using liquid nitrogen or another refrigerant, food can be frozen in a matter of seconds or minutes, minimizing cellular damage and preventing the production of massive ice crystals. Food items that are brittle or have a tendency to cling to one another can also be preserved using cryogenic freezing. The goal of cryogenic food processing is to partially defrost food while hardening its exterior. This keeps things from deforming and is helpful in a number of procedures, such as making ice cream or cutting or moving things.

It helps minimize the growth of microbes while preserving the low temperatures of fresh goods that could be harmed by freezing, such as fruits, vegetables, dairy, meat products, etc., during storage or transit. In addition, cryogenic treatment has shown great potential in the development and enhancement of vegan products by preserving sensory attributes and nutritional qualities during processing. The cryogenic approach, particularly in food concentration and preservation, reduces thermal degradation, maintains volatile aromas, and preserves bioactive compounds that are otherwise lost in traditional heat- or vacuum-based methods (Aider and De Halleux 2009).

For vegan beverages such as plant-based juices, cryogenic processing ensures flavour retention, colour stability, and minimal nutrient loss. Similar to the findings for pineapple and tomato juice, the cryogenic concentration of vegan juices can result in products with enhanced natural flavour profiles and superior quality compared to those processed using evaporation techniques. This makes the method ideal for producing premium plant-based juices and functional beverages. In vegan protein concentrates, such as those derived from soy, almond, or oat milk, cryogenic treatment could concentrate protein fractions effectively while minimizing the loss of essential amino acids or denaturing sensitive proteins. This mirrors the success seen in cryo-concentration of cheese whey and provides opportunities for producing high-quality vegan protein supplements (Aider et al. 2007).

Cryogenic methods also support the development of vegan spreads, marmalades, and sauces, where flavour retention and textural integrity are critical. The ability to freeze and defrost at controlled stages, as seen in studies on whey and tomato juice, could optimize concentration efficiency and product uniformity in plant-based formulations (Aider and De Halleux 2009). Additionally, the process could enhance vegan dessert production, particularly for frozen desserts like plant-based ice creams, where low-temperature processing ensures creaminess and preserves flavour volatiles, aligning with consumer preferences for fresh and indulgent products.

11.8 Consumer Attitudes Towards Frozen Vegan Products and Cryogenic Processing

In both domestic and international markets, there is always a significant demand for fresh fruit. Even at higher costs than other readily available sources, consumers seek out fresh fruit. However, fruits have a limited shelf life because they are perishable. The primary cause of the fruits' perishable character is their physiological and biochemical processes, which begin as soon as they are removed from the mother plant. One of the simplest, most practical, and most widely used commercial methods of product preservation is freezing. Harvested plant products' living cells continuously breathe, absorbing available oxygen (O_2) from their surroundings and releasing carbon dioxide (CO_2).

Food quality is preserved through freezing at a competitive price. The initial condition of the food and the storage temperature have an impact on the quality of frozen food. In general, foods that are properly frozen retain more of their inherent nutrients and colour, flavour, and texture than foods that are preserved in other methods. The molecular mobility is limited during the freezing process, which slows down chemical and biological events including oxidation and enzyme activity. Unlike heat treatment, freezing lowers the activity of the microorganisms without killing them. Despite its microbiological stability, the product is nevertheless susceptible to physical and biochemical processes, which could have an impact on its organoleptic qualities.

Consumer attitudes towards frozen vegan products are influenced by perceptions of safety, convenience, and nutritional value. However, a significant gap exists in consumer awareness regarding the potential risks of consuming frozen vegetables and vegan products without proper cooking. Surveys reveal that many consumers mistakenly believe that freezing eliminates harmful pathogens such as *Listeria monocytogenes, Salmonella*, and *Escherichia coli*, despite research demonstrating that freezing only inhibits bacterial growth but does not destroy it (Chaves et al. 2011; Kataoka et al. 2017). Furthermore, a considerable portion of consumers either neglect to read cooking instructions or lack the equipment necessary to follow them, increasing the risk of foodborne illnesses (Marshall 2022). This underscores the critical need for clear labelling, public education on safe handling practices, and stringent preventive measures during the production of frozen vegan foods (Madad et al. 2023).

Cryogenic processing has emerged as a promising solution for enhancing the safety and quality of frozen vegan products. This technology enables rapid freezing at extremely low temperatures, which minimizes the formation of large ice crystals that can damage the texture and quality of food (Hoppu et al. 2021; Canning et al. 2025). Moreover, cryogenic processing aligns with consumer demand for minimally processed foods by maintaining nutrients and sensory characteristics better than conventional freezing methods. Despite these benefits, its application requires stringent controls to mitigate contamination risks during production and distribution. Studies indicate that the inclusion of preventive measures, such as proper blanching, sanitization, and efficient cold-chain management, are essential to ensuring consumer safety (Willis et al. 2020; Fay et al. 2024). By integrating advanced freezing techniques and addressing gaps in consumer awareness, the frozen vegan food industry can meet the dual challenges of maintaining high product quality and ensuring safety.

11.9 Future Perspectives and Challenges Associated with Production of Vegan Products

Cryogenic technologies hold immense potential for revolutionizing the production and preservation of vegan products, offering unique advantages in terms of quality retention and flavour preservation. However, their widespread adoption faces several challenges, particularly regarding cost implications and energy requirements.

11.9.1 Challenges and Cost Implications

When constructing a mill for operation at extremely low temperatures, several essential factors must be considered. The materials for the balls, vials, and accessories must resist cryogenic conditions. Tungsten carbide and stainless steel are widely utilized as they resist wear and bear shock loads at high temperatures. Planetary cryo-mills face fabrication issues due to the necessity for additional components to feed cryogenic liquid, and their revolving design limits the handling of liquid coolant. Other key design elements include maintaining consistent temperature during dry milling, guaranteeing an inert gas environment to prevent oxidation, and minimizing ice flake development produced by sporadic rather than continuous liquid nitrogen flow. Each cryo-mill type differs in energy transfer efficiency based on its design and the hardness difference between the milling powder and ball-vial materials. Thus, selecting a right cryo-mill is vital for specific items.

Despite the advantages, the high initial capital investment and energy costs associated with cryogenic technologies remain significant barriers. Energy consumption in cryogenic processes is heavily influenced by the temperature gradient between the evaporator and condenser. Lowering this temperature difference, such as through cascade crystallization, can improve energy efficiency but requires advanced system designs and precise operational control. For instance, energy consumption during wine cryo-concentration was reported to decrease significantly when multiple cascade stages were employed. However, the initial set-up costs and maintenance of these systems could be prohibitive, particularly for small-scale producers of vegan products.

Moreover, the energy demand for cryogenic systems is substantial. Even with heat pump integration, the production of ice and the associated cooling requirements result in higher energy usage compared to traditional heat evaporation methods. While the long-term energy savings and improved product quality may offset these costs, they remain a critical consideration for producers seeking to implement this technology.

11.9.2 Sustainability and Scalability

Future research must focus on developing cost-effective and energy-efficient cryogenic technologies tailored for vegan food production. This includes optimizing the integration of renewable energy sources, improving system designs for higher efficiency, and exploring hybrid approaches that combine cryogenic and conventional methods. Cryogenic technology holds the potential to transform the vegan food industry by enabling the production of superior-quality products. However, addressing the challenges of high energy demands and cost implications is essential for its large-scale adoption. As advancements continue, these technologies could become more accessible and sustainable, paving the way for a new era of innovation in vegan product manufacturing.

11.9.3 Future Perspectives

The adoption of cryogenic technologies in vegan product development is anticipated to grow as the demand for high-quality, minimally processed, and nutrient-rich vegan foods increases. Innovations such as cryo-concentration and freeze-drying can be instrumental in producing premium vegan beverages, protein concentrates, and plant-based desserts. The ability of cryogenic methods to preserve flavour volatiles, minimize nutrient loss, and maintain sensory attributes aligns perfectly with consumer preferences for fresh, natural, and wholesome plant-based products.

Recent advancements, such as the integration of heat pump technology into cryogenic systems, have demonstrated significant energy conservation potential. For instance, freeze concentration systems (FCS) with heat pumps have shown a Coefficient of Performance (COP) of 14.3, indicating high energy efficiency when used to pre-concentrate solutions like cane juice. These innovations could be adapted for vegan product manufacturing, reducing overall energy costs while maintaining superior product quality. The cascading crystallization design presents another promising avenue for improving energy efficiency. The heat of crystallization is recovered at temperatures above the freezing point of the treated solution; this method could significantly reduce energy consumption. For liquid vegan foods, such as plant-based juices or protein solutions, this approach could optimize the process while maintaining product integrity.

11.10 Conclusion

The present chapter on 'Vegan Chill' explores innovative advancements in chilling and freezing technologies tailored to the production of high-quality vegan food products. These technologies prioritize the preservation of nutritional value, sensory attributes, and product integrity, addressing the rising consumer demand for sustainable and minimally processed plant-based foods. Emphasis is placed on energy-efficient and cost-effective cooling solutions, which are critical for maintaining the freshness and quality of vegan products throughout processing and storage. State-of-the-art methods, including cryogenic freezing, cryo-milling, cryo-concentration, and novel cooling systems, are discussed as transformative tools for enhancing the shelf life and safety of plant-based foods. The chapter also highlights the importance of parameters that align with the unique challenges of vegan food production, such as texture preservation, colour retention, and minimizing moisture loss. Overall, the application of cryogenic techniques in vegan food processing offers a promising avenue for creating high-quality products with superior flavour, nutrient retention, and sensory appeal, meeting the increasing demand for plant-based options in global markets.

References

Agnelli, M.E. and Mascheroni, R.H. (2002). Quality evaluation of foodstuffs frozen in a cryomechanical freezer. *Journal of Food Engineering 52* (3): 257–263.

Aider, M. and De Halleux, D. (2008a). Passive and microwave-assisted thawing in maple sap cryoconcentration technology. *Journal of Food Engineering 85* (1): 65–72.

Aider, M. and De Halleux, D. (2008b). Production of concentrated cherry and apricot juices by cryoconcentration technology. *LWT- Food Science and Technology 41* (10): 1768–1775.

Aider, M. and De Halleux, D. (2009). Cryoconcentration technology in the bio-food industry: principles and applications. *LWT- Food Science and Technology 42* (3): 679–685.

Aider, M., De Halleux, D., and Akbache, A. (2007). Whey cryoconcentration and impact on its composition. *Journal of Food Engineering 82* (1): 92–102.

Alhamdan, A., Hassan, B., Alkahtani, H. et al. (2018). Cryogenic freezing of fresh date fruits for quality preservation during frozen storage. *Journal of the Saudi Society of Agricultural Sciences 17* (1): 9–16.

Balani, K., Verma, V., Agarwal, A., and Narayan, R. (2015). Physical, thermal, and mechanical properties of polymers. In: *Biosurfaces: A Materials Science and Engineering Perspective* (ed. K. Balani, V. Verma, A. Agarwal, and R. Narayan), 329. John Wiley & Sons.

Bald, W.B. (1991). Ice crystal growth in idealised freezing systems. In: *Food freezing: Today and tomorrow* (ed. W.B. Bald), 67–80. London: Springer.

Barbosa-Cánovas, G.V., Altunakar, B., and Mejia-Lorio, D.J. (2005). *Freezing of Fruits and Vegetables: An Agribusiness Alternative for Rural and Semi-rural Areas*, vol. 158. Food & Agriculture Org.

Barnwal, P. and Singh, K.K. (2012). Application of cryogenic technology in food processing and preservation. *Indian Dairy Engineers Association (IDEA) ICAR—National Dairy Research Institute (NDRI)*, Karnal, Haryana, India.

Brennan, J.G. and Grandison, A.S. (ed.) (2012). *Food Processing Handbook*. Weinheim, Germany: Wiley-Vch.

Burdo, O.G., Kovalenko, E.A. and Kharenko, D.A. (2007). Development of freeze concentration device. *Communication of the Odessa State Academy of Food Technologies*, Odessa, Ukraine.

Canning, M., Ablan, M., Crawford, T.N. et al. (2025). Preparation methods and perceived risk of foodborne illness among consumers of prepackaged frozen vegetables–United States, September 2022. *Journal of Food Protection* 100440.

Charoenrein, S. (2018). Microstructural changes and their relationship with quality and stability of frozen foods. In: *Food Microstructure and Its Relationship with Quality and Stability* (ed. S. Devahastin), 123–138. Woodhead Publishing.

Chaves, B.D., Han, I.Y., Dawson, P.L., and Northcutt, J.K. (2011). Survival of artificially inoculated *Escherichia coli* and *Salmonella Typhimurium* on the surface of raw poultry products subjected to crust freezing. *Poultry Science 90* (12): 2874–2878.

Fay, M.L., Salazar, J.K., Stewart, D.S. et al. (2024). Survival of *Listeria monocytogenes* on frozen vegetables during long-term storage at– 18 and– 10° C. *Journal of Food Protection 87* (3): 100224.

Fennema, O. (1978). Cryogenic freezing of foods. In: *Advances in Cryogenic Engineering*, 712–720. Boston, MA: Springer US.

Gates, K.W. (2013). *Advances in Food Science and Technology, Volume 1* (ed. P.M. Visakh, S. Thomas, L.B. Iturriaga, and P.D. Ribotta), 324. Salem, MA: Scrivener Publishing and Hoboken, NJ, USA, John Wiley and Sons.

Goswami, T.K. (2010). Role of cryogenics in food processing and preservation. *International Journal of Food Engineering 6* (1).

Goswami, T.K. (2017). Recent trends of application of cryogenics in food processing and preservation. *Journal of Food, Nutrition and Population Health 1* (3): 27.

Hoppu, U., Puputti, S., and Sandell, M. (2021). Factors related to sensory properties and consumer acceptance of vegetables. *Critical Reviews in Food Science and Nutrition 61* (10): 1751–1761.

James, C., Purnell, G., and James, S.J. (2015). A review of novel and innovative food freezing technologies. *Food and Bioprocess Technology 8*: 1616–1634.

Jha, P.K., Le-Bail, A., and Jun, S. (2022). Recent advances in freezing processes: an overview. In: *Food Engineering Innovations Across the Food Supply Chain* (ed. P. Juliano, R. Buckow, M.H. Nguyen, et al.), 187–206. Academic Press.

Kataoka, A., Wang, H., Elliott, P.H. et al. (2017). Growth of *Listeria monocytogenes* in thawed frozen foods. *Journal of Food Protection 80* (3): 447–453.

Katiyar, A., Kumar, N., Shukla, R.K., and Srivastava, A. (2020). Substrate free ultrasonic-assisted hydrothermal growth of ZnO nanoflowers at low temperature. *SN applied sciences* 2 (8): 1386.

Lan, S.H. and Farid, M.M. (2004). Experimental analysis of cryogenic freezing of food. *Journal of Chemical Engineering of Japan 37* (2): 304–309.

Macías-Rodríguez, M.E., Navarro-Hidalgo, V., Linares-Morales, J.R. et al. (2013). Microbiological safety of domestic refrigerators and the dishcloths used to clean them in Guadalajara, Jalisco, Mexico. *Journal of Food Protection* 76 (6): 984–990.

Madad, A., Marshall, K.E., Blessington, T. et al. (2023). Investigation of a multistate outbreak of *Listeria monocytogenes* infections linked to frozen vegetables produced at individually quick-frozen vegetable manufacturing facilities. *Journal of Food Protection 86* (8): 100117.

Manzocco, L., Valoppi, F., Calligaris, S. et al. (2017). Exploitation of κ-carrageenan aerogels as template for edible oleogel preparation. *Food Hydrocolloids 71*: 68–75.

Marshall, K.E. (2022). Appliances used by consumers to prepare frozen stuffed chicken products—United States, May–July 2022. *MMWR. Morbidity and Mortality Weekly Report* 71.

Mittal, G., Rhee, K.Y., and Park, S.J. (2016). The effects of cryomilling CNTs on the thermal and electrical properties of CNT/PMMA composites. *Polymers 8* (5): 169.

Parfitt, J., Barthel, M., and Macnaughton, S. (2010). Food waste within food supply chains: quantification and potential for change to 2050. *Philosophical Transactions of the Royal Society, B: Biological Sciences 365* (1554): 3065–3081.

Plazzotta, S., Calligaris, S., and Manzocco, L. (2020). Structural characterization of oleogels from whey protein aerogel particles. *Food Research International 132*: 109099.

Popov, D., Fikiin, K., Stankov, B. et al. (2019). Cryogenic heat exchangers for process cooling and renewable energy storage: a review. *Applied Thermal Engineering 153*: 275–290.

Tanti, R., Barbut, S., and Marangoni, A.G. (2016). Oil stabilization of natural peanut butter using food grade polymers. *Food Hydrocolloids 61*: 399–408.

Valoppi, F., Schavikin, J., Lassila, P. et al. (2023). Formation and characterization of oleogels obtained via direct dispersion of ultrasound-enhanced electrospun nanofibers and cold milling. *Food Structure 37*: 100338.

Willis, C., McLauchlin, J., Aird, H. et al. (2020). Occurrence of *Listeria* and *Escherichia coli* in frozen fruit and vegetables collected from retail and catering premises in England 2018–2019. *International Journal of Food Microbiology 334*: 108849.

Xu, Z., Li, Y., Liu, Z. et al. (2024). Current knowledge on cryogenic microorganisms and food safety in refrigerators. *Trends in Food Science & Technology* 104382.

You, Y., Kang, T., and Jun, S. (2021). Control of ice nucleation for subzero food preservation. *Food Engineering Reviews 13*: 15–35.

Zohuri, B. (2017). *Physics of Cryogenics: An Ultralow Temperature Phenomenon*, 387–473. Elsevier.

Zohuri, B. and Zohuri, B., 2018. Cryogenic technologies. Bahman Zohuri, *Hybrid Energy Systems: Driving Reliable Renewable Sources of Energy Storage*, Springer Cham, pp. 39–88.

12

Impact of Cryogenics on Microbial Populations and Food Safety

Pratiksha Wankhede[1], Jessica Hannah[2], Sumit Sudhir Pathak[3], Surendra Babu[4], and Bhushan Chaudhari[1]

[1] *Kavayitri Bahinabai Chaudhari North Maharashtra University, Jalgaon, Maharashtra, India*
[2] *Department of Food Processing and Preservation Technology, School of Engineering, Avinashilingam Institute for Home Science and Higher Education for Women, Coimbatore, Tamil Nadu, India*
[3] *Division of Food Processing Technology, School of Engineering and Technology, Karunya Institute of Technology and Sciences, Coimbatore, Tamil Nadu, India*
[4] *Rudhra Consultancy and Services, Bangalore, Karnataka, India*

12.1 Introduction

With advancements in technology over the past few decades, the different methods for creating low-temperature environments are commonly known as cryogenics. Cryogenics encompasses the fields of physics and engineering, deriving its name from the Greek words '*cryos*', meaning frost or cold, and '*gen*', which signifies to generate. Cryogenics encompasses the scientific and technological fields that focus on temperatures that fall below −153 °C. The initial milestone in the field of cryogenics is the liquefaction of gases. In 1845, Michael Faraday endeavoured to liquefy gases by immersing them in a bath of ether and dry ice. In the year 1877, Louis Cailletet of France and Raoul Pictet from Switzerland achieved the remarkable feat of creating the first droplets of liquid air. The gases O_2, H_2, N_2, CO_2, CH_4, and N_2O, which demonstrated resistance to liquefaction, were often subjected to attempts for liquefaction. In 1883, Von Wroblewski successfully generated a quantifiable amount of liquid oxygen and liquid nitrogen. During the investigation, oxygen was transformed into a liquid state at −183 °C, whereas nitrogen reached its liquid form at −196 °C. James Dewar achieved the remarkable feat of liquefying hydrogen, attaining an astonishing temperature of −259 °C. In the course of exploring these inventions, the year 1894 saw the discovery of argon, alongside the identification of its impurities in LN_2, as well as krypton and xenon during the process of fractional distillation. A summary of the development of cryogenics can be found in Table 12.1.

Nowadays cryogenics is associated with production and study of low-temperature environment. Cryogenics is a multidisciplinary science, and its technologies have wide

Cryogenic Processing: Advances in the Food Industry, First Edition. Edited by Sumit Sudhir Pathak, Rama Chandra Pradhan, and Sudarshan Ramanathan.

Table 12.1 Development in cryogenics.

Year	Events occurred
1877	Cailletet and Pictet achieved the liquefaction of O_2. Cryogenics developed as a distinct field apart from refrigeration.
1884	Wroblewski from Kracow University in Poland was the first to successfully liquefy hydrogen, achieving it as a mist.
1892	Sir James Dewar from England created the vessel designed to store cryogenic fluids with vacuum insulation.
1895	H.K. Onnes from Holland founded the Leiden Cryogenic Laboratory. Karl von Linde from Germany secured a fundamental patent for the process of air liquefaction.
1898	At the Royal Institute of London, James Dewar successfully produced LH_2 in large quantities.
1902	Georges Claude pioneered the initial air-liquefaction system by employing an expansion engine.
1908	H.K. Onnes was the first to achieve liquefaction. He was previously regarded as a permanent gas.
1911	H.K. Onnes made the groundbreaking discovery of superconductivity.
1916	The initial commercial facility for American-produced air liquefaction was set up.
1922	The initial commercial production of Ne was set up in the United States.
1926	Robert Goddard achieved the groundbreaking milestone of the first successful flight utilizing a LO_2-powered rocket alongside non-cryogenic gasoline propellant.
1933	Magnetic cooling has been utilized to achieve temperatures lower than −272 °C
1934	Peter Kapitza constructed the inaugural expansion engine for a helium liquefier.
1939	The inaugural vacuum-insulated railway tank car designed for the transportation of liquid oxygen.
1945	The advancement of H_2 propulsion technology for powering the upper stages of NASA's Saturn V rocket was achieved successfully.
1947	The Collins cryostat was developed for the liquefaction of helium.
1952	The cryogenic engineering laboratory was established by the National Bureau of Standards in Boulder, Colorado.
1957	The Atlas ICBM, fuelled by LOX/RP-1, underwent a test firing. The fundamental theory of superconductivity, known as the Bardeen–Cooper–Schrieffer theory, is presented.
1958	Multilayer insulation (MLI) emerged as a significant advancement.
1961	The Saturn launch vehicle powered by LO_2 and LH_2 underwent a test firing.
1966	The He3/He4 dilution refrigerator was created.
1969	A 3,250 hp DC superconducting motor constructed for ship drive application
1986	Georg Bednorz and Alex Muller unveil a remarkable high-transition-temperature ceramic superconductor, achieving a Tc of approximately −243 °C.

Source: Flynn (2004) / Taylor and Francis.

applications in chemical, medicinal, automobile, metallurgy, power plant, biological preservations and food processing industries. Cryogenic technology is commonly used to preserve high-value frozen items such as fruits and vegetables, livestock, and marine products. Cryogenic technology was also used for food grinding at low temperatures, which maintained the ideal look, medicinal, and aromatic qualities in spices, herbs, and other food commodities while delivering higher quality. Cryogenics is limitedly applicable in peeling of fruits and vegetables and homogenization of biological tissues (Balasubramanian et al. 2012).

12.2 Overview of Cryogenics in Food Microbiology

As we know, food is the basic need of life of all living entities in the world. Once the food enters into the food chain, it is metabolized on each tropic level to generate the energy. The formed energy helps for internal metabolism and regulation of body functions and regulates the body temperature in homeotherms. The gimmicks of taste and aroma of food always depend on the processing by which it is produced. Food materials are not sustainable in nature for extended periods of time due to their high water content, which includes fat or oil, protein, carbohydrate, and other water-soluble and insoluble elements. Spoilage of food material is caused by chemical, metabolic, and microbiological changes. As per reports of FAO (Food and Agriculture Organization of the United Nations), the larger (nearly one-third) quantity of food is wasted by spoilage before it reaches to consumer. The most of the spoilage is caused by microbial load present in food. Microbial contamination occurs during harvesting, storage, packaging, and transporting. Some of the inherent commensals may also be responsible for food spoilage. In recent years, crop production has not kept pace with the rapid population expansion, leading to the potential for disastrous hunger amid impending overpopulation. The global food supply must increase to maintain equilibrium with population expansion, necessitating the effective utilization of land and other resources. The growing food demand of expanding people necessitates not just production but also preservation. The advancement in agriculture necessitates the secure storing of surplus products. Microbiological standards have been established retrospectively to inhibit or slow the natural process of food rotting, and several techniques have been devised for this objective.

Historically, the methods of preserving food have relied heavily on reducing water activity through techniques like solar drying, salting, using concentrated sugar solutions (preserves), or employing fire-smoking methods. Refrigeration stands out as a highly efficient and cost-effective approach. Refrigeration has become a common practice in India, utilized for the preservation of milk and its products, vegetables and fruits, eggs, fish, and meat, among others. Canning, while a popular method of preservation, presents clear challenges due to the significant expense associated with tin cans. The cost of tin cans frequently surpasses 50% of the overall production expenses. Aside from items such as milk powder, mashed potatoes, and powdered coffee, the commercial success of preservation through dehydration has been limited, especially when dealing with solid pieces. Food is preserved without altering its texture, flavour, and aroma through cryogenic freezing methods. These methods utilize freezers that employ either liquid nitrogen (−196 °C) or solid carbon dioxide (−78 °C), which are applied directly to the food product to facilitate temperature reduction (Goswami 2010).

12.2.1 Microbial Populations in Food Products

The Japanese market is well known for its probiotic functional foods, such as Yakult, Meiji yoghurt and yoghurt drinks, Morinaga bifidus yoghurt, and Calpis Ameal S120. Activia creamy yoghurt, Danone GYoPlus, Bluebunny-Sedona yoghurt ice cream, chocolate sweet scoops-frozen yoghurt, and Yovation-Pierre's probiotic ice cream all have a foothold in the US market. Famous probiotic products in the United Kingdom include Vita-Yo creamy probiotic yogurt, Yeo Valley bio-live yoghurt, Yeo-Valley natural fat-free yoghurt, and Unilever's flora proactive cholesterol. In Canada, prominent items include Bio Best plant steroid probiotic yogurt, Kraft live active cheddar cheese and chocolate raspberry snacks, Liberte yoghurt, and Olympic natural, no-fat probiotic yoghurt. In India, commercially available functional foods include Nesvita probiotic yoghurt, Amul Prolife probiotic ice cream, B-Activ probiotic curd, and Heritage probiotic curd. In Italy, various commercial functional foods are available on the market, such as Ganeden BC30® probiotic low-fat yogurt, probiotic dairy drinks latteria sociale merano, Danacol fermented dairy beverage, and Yo-live frozen yogurt.

Despite the availability of numerous commercial options globally, the preference for homemade probiotic preparations continues to thrive. Many individuals favour items made using traditional methods; nonetheless, these items exhibit a remarkable variety globally. LAB play a crucial role as the primary microflora in the fermentation of dairy products, vegetables, and sourdough. Specifically, lactobacilli and pediococci are integral components of the starter cultures utilized in meat fermentation, contributing to the development of desirable acids and flavour compounds. The effective management of fermentation processes in an industrial setting demands current insights into the physiology, metabolism, and genetic characteristics of these microorganisms. Understanding their influence on food quality, safety, and shelf life is just as crucial. The nature of food fermentation, on the other hand, is quite intricate. Fermentation of agricultural products, whether from animals or plants, is carried out using either the natural microflora present or a specific starter culture. This process enhances or achieves various qualities such as shelf life, nutritional value, health benefits, flavour, or texture. In contrast to the Western world, Asia has cultivated a variety of foods that utilize vegetable proteins derived from fungi, frequently employing a solid-state fermentation method. In Europe and the United States, the primary emphasis of food fermentation has been on preservation through acid fermentation. In contrast, aspects like flavour, nutritional benefits, and health impacts take precedence in Asian fermentation, particularly those involving fungi.

The processes of food production, distribution, and storage depend on strategically applied harmful stresses or obstacles that effectively hinder or deactivate contaminating microorganisms within food systems. Nonetheless, the manufacture of fermented foods does not take place in an environment free of contaminating microbes. For instance, dairy items can harbour yeasts, moulds, and various types of bacteria whose metabolic processes compromise quality. Once fermentation is finished in cheese, only bacteria that can withstand acidic conditions are able to thrive. Nevertheless, should the acid development proceed at a sluggish pace or if the pH fails to drop adequately, contaminants that would typically be suppressed might find the opportunity to proliferate. The harmful microorganisms present in food can lead to foodborne illnesses. The pathogenic microbial flora encompasses various organisms, including *Clostridium botulinum* (type A and proteolytic types B), *Bacillus cereus, Campylobacter jejuni,* as well as *C. botulinum* (type E and non-proteolytic types B and F). Additionally, it includes *Clostridium perfringens, Listeria monocytogenes, Salmonella* spp., *Shigella* spp., *and Staphylococcus aureus,* among others.

The growth of microbial populations in food is influenced by a range of factors. The food showcases a rich variety of microbial life forms. The substances that fight against microbes found in food serve to restrict the growth of microbial populations on those foods. Some plant species are recognized for their essential oils, which exhibit antimicrobial properties. Eugenol in cloves, allicin in garlic, cinnamic aldehyde and eugenol in cinnamon, allyl isothiocyanate in mustard, eugenol and thymol in sage, and carvacrol (isothymol) and thymol in oregano (Shelef 1983) are among the compounds. Eggs, like milk, contain lysozyme, and this enzyme, together with conalbumin, provides a very powerful antibacterial system. Hydroxycinnamic acid derivatives, such as p-coumaric, ferulic, caffeic, and chlorogenic acids, are found in a variety of plant sources, including fruits, vegetables, tea, and molasses, and have antibacterial and antifungal activities.

12.3 Role of Temperature in Microbial Survival and Proliferation

The growth of microorganisms in food leads to different levels of alteration in the food's characteristics due to their metabolic processes. Certain transformations, such as those occurring during fermentation, are beneficial, whereas others, leading to food spoilage and foodborne illness, are detrimental (Preetha and Narayanan 2020). The inherent elements such as nutrient composition, pH levels, oxidation-reduction potential, water activity, and the biological structure of food significantly impact the microflora present in food. Various external factors, such as humidity and storage temperature, also impact the development of microbial flora. The presence or absence of various microorganism groups is influenced by the temperature of the environment. A kind of microbial strain known as psychrophiles thrives at 0 °C, with an ideal growth temperature of 15 °C or below. For example, *Bacillus psychrophilus* and *Chlamydomonas nivalis* flourish in cold conditions, but psychrotrophs may grow at temperatures ranging from 0 to 7 °C, with optimum development happening between 20 and 30 °C and a maximum tolerance around 35 °C (*L. monocytogenes; Pseudomonas fluorescens*). *Escherichia coli, Neisseria gonorrhoeae, and Trichomonas vaginalis* are notable examples that thrive abundantly within a moderate temperature range of 20–45 °C, categorized as mesophiles. Thermophiles represent a category of microbes capable of thriving at temperatures of 55 °C or above, with their ideal conditions typically falling between 55 and 65 °C, as seen in species such as *Bacillus stearothermophilus, Thermus aquaticus, and Cyanidium.* In contrast, hyperthermophiles prefer even more extreme environments, with optimal temperature ranges spanning from 80 to approximately 113 °C, exemplified by organisms like *Sulfolobus, Pyrococcus, and Pyrodictium.* The ideal temperature range for specific bacteria facilitates the production of primary metabolites and effectively demonstrates metabolic processes. The enzymes involved in metabolism necessitate a specific temperature range for bacteria to effectively metabolize food sources and reproduce, thereby increasing their population.

Microorganisms thrive across various temperatures, each species exhibiting a specific range for growth that includes a minimum, maximum, and optimal temperature. The temperature significantly influences the time it takes for an organism to develop and the duration of its initial lag phase. Microorganisms can be categorized into one of four distinct groups based on their ideal growth conditions. Increased amounts of unsaturated fatty acids will maintain their fluidity and functionality even at reduced temperatures. At lower temperatures, the fluidity of the cytoplasmic membrane of mesophilic and thermoduric bacteria diminishes, resulting in less food transfer across it. Growth temperature influences the regulation of virulence genes in particular foodborne pathogens. Protein expression levels driven by the *Yersinia enterocolitica* virulence plasmid are considerably higher at 37 °C (99 °F), lower at 22 °C (72 °F), and virtually absent at 4 °C (39 °F). The temperature at which an organism develops has a substantial impact on its susceptibility to thermal fluctuations.

When *L.* monocytogenes is maintained at 48 °C (118 °F) in inoculated sausages, its D value at 64 °C (147 °F) experiences a 2.4-fold increase. The duration of the lag phase and the growth rate of a microorganism are affected by temperature, along with various other internal and external factors. For example, the rate at which *C. perfringens* grows is notably reduced at pH 5.8 compared to pH 7.2 over a broad spectrum of temperatures. The generation of toxins tends to be sluggish when temperatures drop below approximately 20 °C. In laboratory medium maintained at pH 7, the time needed to produce detectable levels of enterotoxin ranged from 78 to 98 hours at 19 °C (66 °F) to 14 to 16 hours at 26 °C (79 °F). Less favourable conditions, like a decrease in pH, further hindered the production of enterotoxins.

Elevated temperatures lead to harm in membranes, depletion of vital nutrients and ions, clustering of ribosomes, breaks in DNA strands, deactivation of crucial enzymes, and coagulation of proteins. Nearly all cellular structures experience some degree of impact from increased temperatures. Every microorganism possesses a lipid–fatty-acyl composition that is tailored to suit its specific growth temperature range. As the temperature drops, the fluidity of the membrane will diminish, leading to a slowdown in metabolic processes associated with the membrane, which are facilitated by enzymes, cytochromes, and permeases. These occurrences will initiate adjustments in fatty-acyl composition to enhance the fluidity of the membrane. Provided that the temperature variation remains within the typical growth temperature range, the alterations in fatty-acyl composition will generally offset the kinetic reduction in activity. When the temperature is altered to or slightly exceeds the lower or upper threshold, cold shock or heat shock will ensue, respectively. The impact of low temperatures on microbes encompasses the inhibition of protein synthesis initiation, alteration of ribosome structure, development of secondary structures, and heightened negative supercoiling of DNA, all of which provoke a cellular response from the microbe (Preetha and Narayanan 2020).

12.4 Mechanisms of Microbial Inactivation Through Cryogenic Freezing

To understand the relevance of cryoprotective agents (CPAs), first consider how subzero temperatures affect normally healthy tissue. Without cryoprotectants, cells often die at temperatures below 0 °C. Because water accounts for about 80% of tissue mass, freezing it, whether within or outside cells, has a substantial influence on detrimental biochemical and structural modifications, which are thought to cause unprotected freezing damage.

There are two distinct theories that attempt to explain the negative effects of freezing on cells: (1) the formation of ice crystals causes mechanical disruption of cellular membranes, making it impossible to recover structurally intact cells post-thaw; and (2) as ice crystals form intracellularly during cooling, the remaining liquid phase experiences a perilous rise in solute concentration. Regardless of whether the mechanical or osmotic effects of freezing take priority, the end is the same: unprotected chilling and thawing of cells is a process that cannot support life. To mitigate these impacts, two critical procedures must be implemented: the use of a cryoprotectant and the careful selection of appropriate chilling and thawing rates.

Essentially, cryopreservation involves employing extremely low temperatures to maintain the integrity of living cells and tissues. The cooling effect is a significant phenomenon, primarily influenced by the freezing of water, leading to the concentration of solutes present in the remaining liquid phase. The occurrence of freezing injury has raised significant concerns, leading to a prevailing belief that the formation of ice within cells is detrimental, while freezing outside of cells is viewed as relatively benign. Reports indicate that ice crystals penetrate or interact with the cells, leading to their destruction either directly or through alterations in the composition of the liquid phase (Pegg 2007). The rates at which cooling occurs are crucial for the survival of cells. The formation of ice outside the cells can occasionally lead to detrimental effects, as tightly packed cells are susceptible to harm from mechanical stresses in the channels where they are confined. It is, nonetheless, recommended to ensure the preservation of cell viability and to prevent harm to the extracellular complex architecture. While vitrification can help mitigate ice formation to a degree, the primary issue remains the associated toxicity.

Given these considerations, there is a continual necessity to enhance and refine preservation protocols, as no single method can be universally applied to all types of microbes. It has been noted that various strains of fungi exhibit reduced growth when cultivated on solid nutrient media. Under severe circumstances, it may lead to complete devastation. The process of cryoprotection entails the application of cryoprotectant solutes to the cells or tissues intended for preservation. In varying concentrations, these solutes act as the driving force behind the movement of water through osmosis and the movement of solutes via diffusion. Various obstacles hinder the unrestricted movement of solutes, leading to temporary and occasionally stable alterations in compartment volumes, which can prove fatal. Consequently, the phenomena of diffusion and osmosis are significant and serve crucial functions in cryoprotection, with extensive applications in the field of biology (Kumar et al. 2017).

12.5 Comparison of Cryogenic Freezing with Conventional Freezing Methods

Cryogenic freezing provides reduced freezing durations in contrast to traditional air freezing, due to the significant temperature disparities between the cryogen and the product surface, along with the elevated rate of surface heat transfer that occurs from the boiling of the cryogen. Cryogenic freezing operates without the need for mechanical refrigeration systems; it only necessitates a cryogen tank and appropriate spray apparatus. Nonetheless, the cryogenic process could potentially lead to some distortion in the shape of the product, which may affect its commercial application. Additionally, the process of cryogenic freezing involves significant refrigerant usage, exceeding 1 kg of nitrogen for every kilogram of product processed, and incurs substantial operational expenses. This renders cryogenic freezing a feasible option solely for high-end items like seafood or premium fruits.

12.5.1 Advantages in Using LN_2 as Refrigerant

Advantages in use of LN_2 as refrigerant, in freezers in particular, over conventionally refrigerated freezers:

1) The loss due to dehydration is under 1%, compared to the 3–10% observed in traditional freezing methods.
2) The absence of O_2 reduces oxidative rancidity while freezing.
3) Least amount of liquid loss.
4) Enhanced characteristics regarding taste, consistency, and visual appeal.
5) Equipment that occupies less floor space.
6) Minimized losses during product handling.
7) Adaptability.
8) Elevated efficiency.
9) Minimal financial outlay.
10) Lower upkeep expenses.
11) Essential workforce needs
12) Quick setup and space efficiency.

12.5.2 Impact of Cryogenic Processing on Microbial Viability

Pathogenic bacteria in foods can be controlled and rendered inactive with the use of safe and efficient methods like gamma irradiation and cryogenic freezing. This study examined the impact of gamma irradiation and cryogenic freezing on shrimp in order to inactivate *Salmonella* spp. Using a small-scale industrial liquid nitrogen freezer, we discovered that cryogenic freezing raw shrimp at −82 °C for three minutes reduced *Salmonella* spp. on whole shrimp by 1.27 logarithmic units. This decrease persisted for 12 weeks of frozen storage at −20 °C. When evaluating a particular microbiological medium for the recovery and counting of *Salmonella* spp., we discovered that Brilliant Green Sulfur Agar yielded findings that were identical to those of non-selective Tryptic Soy Agar. *Salmonella* spp. on frozen shrimp had radiation D_{10} values of around 0.56 kGy. A > 5-log decrease in *Salmonella* spp. was achieved by cryogenic freezing (−82 °C) and gamma irradiation (2.25 kGy), and this reduction persisted for 12 weeks of frozen storage (−20 °C). These findings suggest that *Salmonella* spp. on frozen shrimp is inactivated by both cryogenic freezing and gamma irradiation (Sommers et al. 2015).

12.6 Safety Considerations in Cryogenic Processing

Cryogenic processing in food is the use of extremely low temperatures, usually obtained with liquid nitrogen or carbon dioxide, to quickly freeze or cool food items. Compared to traditional freezing processes, this approach helps retain the texture, taste, colour, and nutritional content. It is commonly used for freezing meats, seafood, fruits, vegetables, and prepared meals. Cryogenic freezing reduces the size of ice crystals formed in the food, minimizing cellular damage and improving product quality. It also enhances food safety by quickly lowering temperatures to inhibit microbial growth, making it ideal for high-quality, long-shelf life frozen food products.

Temperature contributes to the stability of biosystems. Generally, reactions occur more quickly with rising temperatures. Lowering the temperature has the direct effect of reducing reaction rates or rates of change for many significant processes. When all other factors are equal, frozen storage should significantly increase storage life. A complication arises from the freezing process's ability to induce alterations that may harm a system's state, leading to negative impacts on quality due to the freeze–thaw cycle. This is distinct from the matter of frozen storage stability.

Cryogenic processing in the food industry involves the use of extremely low temperatures, typically achieved through cryogens such as liquid nitrogen or liquid carbon dioxide, to freeze or chill food products rapidly. This method offers numerous benefits, including better preservation of texture, flavour, and nutritional content, compared to conventional freezing techniques. However, the use of cryogenic substances introduces a range of safety considerations that must be rigorously addressed to ensure the protection of workers, consumers, and the environment. One of the primary safety concerns is the handling of cryogenic liquids themselves. These substances are stored and transported at temperatures below −150 °C, and direct contact can cause severe cold burns or frost bite. Workers must use proper personal protective equipment (PPE), such as insulated gloves, face shields, and protective clothes, to avoid skin or eye contact. In addition, all cryogenic equipment must be specifically designed to withstand extremely low temperatures and should be regularly maintained to prevent leaks or equipment failure. Another major hazard is the potential for asphyxiation in confined spaces.

Cryogenic liquids vaporize rapidly into large volumes of gas, which can displace oxygen in the air. This creates a serious risk of oxygen deficiency, especially in enclosed or poorly ventilated areas. Oxygen monitors and proper ventilation systems must be in place to detect and prevent hazardous atmospheres. Moreover, the rapid expansion of cryogenic gases can lead to pressure build-up in storage containers if not properly vented. Pressure relief valves and venting systems are essential to prevent explosions or container rupture. In addition, liquid nitrogen can condense atmospheric oxygen, increasing the risk of fire or explosion if flammable materials are present. Food safety is also a crucial consideration.

12.7 Case Studies on Cryogenic Processing and Microbial Safety

Since ancient times, spices have been prized for their unique flavour and distinctive scent. Grinding spices has long been done using various grinding tools that are appropriate for the situation. The need for high-quality processed spices has grown in the modern period. In the food industry, cryogenics plays a crucial role. Of all the cryogens available, liquid nitrogen is the one that is most recommended for achieving the required low temperature in food grinding. Superior-grade ground powder with improved cleanliness was generated by grinding spices at a very low temperature. Despite being a proven technology, the hefty initial investment is keeping it unpopular in emerging nations like India. The current review paper compiles the initial research conducted to achieve the physico-mechanical, thermal, and grinding properties, aroma profile. Microbial contamination of spices is common, especially from heat-resistant pathogens such as *C. perfringens* and *B. cereus*. Conventional grinding techniques produce heat, which can break down volatile substances and has no effect on the microbial load. A firm that produces spices used liquid nitrogen to chill the spices both before and during the milling process, a technique known as cryogenic grinding. This worked as a microbiological control measure in addition to preserving taste and aromatic oils. During grinding, microbial cells were mechanically damaged at temperatures as low as −100 °C, which decreased the overall number of viable cells (Ashok et al. 2021).

Green grapes, blueberries, and sea buckthorn fruits were used to test the effects of cryogenic drying pretreatments' performance. The fruits were immersed in liquid nitrogen for one to five 2-minute cycles of freezing and thawing. The control samples were unprocessed samples. A variety of drying experiments were performed on both treated and untreated berries, including hot-air drying at 50 °C and 1 m/s, vacuum-drying at 50 °C and 25 Hg vacuum, freeze-drying at 25 °C shelf temperature, and total pressure of 30 m Torr. Time-dependent weight-loss development of the consumables was measured. The epicarp was determined by microscopic (SEM and optical) means. Before and after dehydration, a visual examination was conducted, and colour changes and volume decreases were evaluated. Between 20% and 50%, the berries' epicarp thickness shrank (Pavlov et al. 2019).

Initial microbial tests indicated a significant reduction in overall viable counts of microorganisms, particularly yeasts and moulds, with decreases of 1.5–2 log CFU/g. While the product was not sterilized, the implementation of cryogenic freezing effectively halted the development of these microorganisms. This process also contributed to quality retention, as the berries maintained their colour, taste, and texture much better, providing end users with a more 'fresh-like' sensory experience upon thawing. Moreover, the shelf life of the frozen berries was extended from 6 to 12 months in a frozen environment without any noticeable deterioration in quality. Although the initial investment for cryogenic freezing equipment and liquid nitrogen was substantial, it resulted in a notable economic impact by reducing export losses, which had previously ranged from 20% to 30% (Pavlov et al. 2019). The chicken fryer halves were frozen using explosion of air, 'Freon', 'Nitreon', and nitrogen-in-liquid form spray techniques. After cooking, measurements of drip loss, cooking loss, shear force, visible bone darkening, and sensory assessments were performed on the halves. A Warner-Bratzler tenderometer was used to assess the shear of the pectoralis major muscle. A sensory panel was used to assess the degree of bone darkening. The least amount of drip loss was seen in liquid nitrogen frozen halves. The air-blast-frozen half had the largest average cooking loss, whereas the non-frozen halves had the lowest. Bone darkening scores, shear force values, and cooking loss were lowest for non-frozen halves. Shear values and bone darkening showed only slight variations between freezing techniques microbiological contamination, especially from *Campylobacter* and *Salmonella*. Because of their slower freezing rates, conventional blast-freezing procedures were insufficient, resulting in the development of ice crystals that caused tissue injury and facilitated the survival of germs.

Following the installation of a liquid nitrogen cryogenic freezing system, the business saw a notable increase in the quality and safety of its products. Because of the quick freezing, less intracellular ice formed, preserving structural integrity and reducing water activity, both of which prevent the development of microorganisms. When compared to items treated by traditional freezing, the average microbial counts in frozen products decreased by 1.5–2 log CFU/g.

Although the results of experimental research differ, it is widely acknowledged that items produced at quicker freezing rates are of greater quality than those produced at slower freezing rates. Carcasses that were quickly frozen at −40 °C had superior overall quality, including palatability, in comparison to those that were traditionally frozen at −12 °C. Turkey carcasses frozen in air blast at 29 °C and 40 °C were 'superior' to those frozen at 17.8 °C based on appearance. Better texture and colour retention were observed in sensory assessments, along with increased safety. By increasing shelf life by 25–30%, spoilage and waste were decreased. Additionally, the factory was able to boost throughput due to the quick, consistent freezing.

12.8 Challenges, Limitations, and Future Direction

Cryogenics is the common term for low-temperature engineering. This scientific discipline relies on the process of air liquefaction and the separation of various components of liquids referred to as cryogen. This type of cooling is not appropriate for a vapour compression system. Nowadays, several liquefaction cycles are widely used, contingent on the kind of cryogen generated and the machine parts involved. In this context, however, liquid nitrogen is regarded as a cryogen that results from the production of oxygen. Over the past 10 years, there has been a substantial amount of research on non-heated inactivation technologies because of the growing consumer demand for fresh, healthful foods with an appropriate shelf life and organoleptic quality. Investigated inactivation technologies include ionization radiation, high-pressure homogenization, high-voltage arc discharge, pulsed electrical fields, high-pressure homogenization, UV decontamination, high-power ultrasound, oscillating magnetic fields, high-voltage arc discharge, streamer plasma, pulsed high-intensity light, high-intensity laser, and pulsed white light (Bhadra et al. 2024).

Due to consumer demand for superior quality with less processing, convenient and non-thermal solutions and safe food products are being researched. Non-thermal methods extend shelf life without requiring additives or preservatives, yet maintaining the product's colour, flavour, texture, nutrients, and functionality. The following conditions must be fulfilled in order to submit a new or unique manufacturing method application to the Food and Drug Administration (FDA). The FDA should be consulted at every stage of the process design; the FDA should visit the pilot and production facilities; the proposed filing should be drafted and sent to the FDA; the resistant organisms that pose the greatest threat to public health and commercial viability should be identified (Jan et al. 2017).

Protecting biodiversity, or the diversity and variability of life on Earth, is crucial for both current and future generations. Breeding programs and the development of new cultivars depend on maintaining variety at the genetic and ecological levels. Many plant species, wild variants, and regional types of fruit and ornamental plants are currently in risk of going extinct. The best long-term preservation technique for plant genetic resources is thought to be cryopreservation, which involves keeping biological samples in tanks with liquid nitrogen. However, developing effective cryogenic processes is a challenging undertaking that calls for careful consideration of a number of variables. Of special relevance is how cryopreservation affects the homogeneity and stability of the samples that are kept (Mokbel et al. 2024).

Using incredibly low temperatures, cryogenic processing preserves product quality, texture, and structural integrity. Because of its capacity to preserve product quality, texture, and structural integrity, liquid nitrogen (often carbon dioxide) is being used more and more in the food, pharmaceutical, and biomedical sectors. However, because of its weak impact on microbial inactivation, it is challenging to ensure microbiological safety.

In the future, a number of studies and technical developments might improve cryogenic processing's safety profile, especially with regard to microbiological risks (Wang et al. 2023). One of the most potential future avenues is the combination of cryogenic processing and other antibacterial therapies. Combining freezing with complementing technologies, such as ultraviolet (UV-C) radiation, pulsed electric fields (PEF), and high-pressure processing (HPP), or natural antimicrobials, might have a synergistic impact because freezing by itself only slows the development of most germs rather than killing them. For example, it has been demonstrated that HPP and cryogenic freezing together may be able to harm microbial cell membranes more efficiently than either procedure alone. These multistep methods can make the procedure more deadly while yet preserving the product's intended organoleptic and nutritional properties.

Real-time microbial detection technology will also play an important role. Continuous surface hygiene and microbiological load monitoring may be achieved by integrating biosensors and Internet of Things (IoT) technology into cryogenic tunnels and storage units. Early alerts from these systems allow for prompt remedial action and lessen the need for postponed laboratory testing.

Further investigation into the behaviour of microorganisms in cryogenic environments is also urgently needed. Although certain viruses and bacteria can withstand extremely low temperatures, there is still a dearth of thorough information on many species' post-thaw pathogenicity, recovery patterns, and survival rates. To make hazard analysis and critical point of control (HACCP) planning, future research should concentrate on developing predictive models for microbial activity over a range of cryogenic settings and food matrices.

The incapacity of cryogenic processing to render the majority of microorganisms inactive is one of its primary drawbacks. Instead of killing pathogens, cryogenic freezing mostly slows down microbial metabolism, in contrast to heat pasteurization or irradiation. *B. cereus, C. botulinum*, and *L. monocytogenes* are examples of spore-forming and psychrotrophic microbes that can withstand extremely low temperatures and even develop again after thawing. Because of this, cryogenic processing is more of a preservation technique than a cleaning procedure, and if appropriate post-thaw handling is not maintained, there is a chance that microorganisms will proliferate (Havelaar et al. 2010).

Cryogenic processing in the food sector involves the application of extremely minimal temperatures, typically achieved through cryogens such as liquid nitrogen and other cryogens or carbon dioxide in liquid form, to freeze or chill food products rapidly. This method offers numerous benefits, including better preservation of texture, flavour, and nutritional content, compared to conventional freezing techniques. However, the use of cryogenic substances introduces a range of considering safety that need to be rigorously stated for the workers protection, consumers, and the surroundings. One of the primary safety concerns is the handling of cryogenic liquids themselves. These substances are stored and transported at temperatures below −150 °C, and direct contact can cause severe cold burns or frostbite. Workers must wear the proper PPE to avoid skin or eye contact, such as face shields, insulated gloves, and protective clothes.

In addition, all cryogenic equipment must be specifically designed to withstand extremely low temperatures and should be regularly maintained to prevent leaks or equipment failure. Another major hazard is the potential for asphyxiation in confined spaces. Cryogenic liquids vaporize rapidly into large volumes of gas, which can displace oxygen in the air. This creates a serious risk of oxygen deficiency, especially in enclosed or poorly ventilated areas. Oxygen monitors and proper ventilation systems must be in place to detect and prevent hazardous atmospheres. Moreover, the rapid expansion of cryogenic gases can lead to pressure build-up in storage containers if not properly vented. Pressure-relief valves and venting systems are essential to prevent explosions or container rupture. In addition, liquid nitrogen can condense atmospheric oxygen, increasing the risk of fire or explosion if flammable materials are present. Food safety is also a crucial consideration (Kica et al. 2012).

All cryogenic materials used in food processing must be food-grade and meet regulatory standards to ensure they do not contaminate the product. Equipment used in cryogenic freezing must be designed to prevent cross-contamination and should be cleaned and sanitized regularly. Proper procedures must be followed to ensure that food does not become excessively brittle or damaged during freezing, as this could affect product quality and safety. Furthermore, training and education of personnel are vital components of cryogenic safety. Workers must be trained in the proper handling and storage of cryogenic substances, emergency response procedures, and the use of PPE.

Emergency protocols should be established and regularly practiced, including procedures for dealing with spills, leaks, and exposure incidents. Regulatory compliance with agencies such as Occupational Safety and Health Administration (OSHA), Food and Drug Administration (FDA), and local health departments is essential, and facilities must be routinely inspected to ensure adherence to safety standards. Finally, environmental considerations should not be overlooked. Improper disposal of cryogenic materials or accidental releases can have harmful effects on the environment. Waste management practices should be in place to minimize any potential impact. In conclusion, while cryogenic processing offers significant advantages for food preservation and quality, it involves a complex array of safety considerations that must be addressed through comprehensive planning, rigorous safety protocols, proper equipment, and thorough training. Ensuring safety in cryogenic food processing protects not only workers and consumers but also supports the long-term sustainability of the industry (Kica et al. 2012).

It is crucial to carry out thorough risk assessments for every technology in addition to compliance. Producers may create strong procedures to successfully reduce these risks by recognizing possible risks and weaknesses. To continuously review the safety precautions in place, regular audits and inspections should be carried out, guaranteeing continuous adherence to safety rules. In addition to protecting customers, this proactive strategy enhances the food producers' reputation and fosters consumer confidence in their goods. Most all, it is imperative that food manufacturers remain current with the constantly changing standards and regulations pertaining to these technologies. To ensure food safety and consumer protection, adherence to national and international laws is essential. This entails being aware of the particular limitations and requirements pertaining to fermentation, cryogenics, and nanotechnology, and making certain that all integration initiatives comply with these guidelines.

Keeping lines of communication open with customers is equally important. Food manufacturers have to disclose the cutting-edge technology included into their goods in an understandable and factual manner. Transparency helps consumers make educated decisions and fosters trust. An important factor in this is consumer education. By informing customers about any worries or misunderstandings, these can be allayed thanks to the advantages of these technologies and the strict safety precautions put in place. In order to guarantee that customers comprehend the advances involved and feel secure about the safety of the goods they eat, this education can take many different forms, including product labelling, instructional campaigns, and informational website.

Temperature contributes to the stability of biosystems. Generally, reactions occur more quickly with rising temperatures. Lowering the temperature has the direct effect of reducing reaction rates or rates of change for many significant processes. When all other factors are equal, frozen storage should significantly increase storage life. A complication arises from the freezing process's ability to induce alterations that may harm a system's state, leading to negative impacts on quality due to the freeze-thaw cycle. This is distinct from the matter of frozen storage stability.

12.9 Conclusion

Cryogenics, the study and use of extremely low temperatures, has had a profound impact on many scientific domains, including medicine, engineering, and food preservation. Cryogenics is essential to food safety and microbiology because it prolongs shelf life, preserves nutritional value, and suppresses microbiological activity. This paper examines how cryogenics affects bacteria populations and assesses how it affects food safety. Materials are chilled to temperatures below −150 °C in cryogenics, often using liquid nitrogen (−196 °C) or carbon dioxide (−78.5 °C). The food industry uses cryogenic freezing to quickly reduce the temperature of perishable goods while preserving cellular integrity and reducing the development of ice crystals. Cryogenic technologies guarantee faster freezing and greater quality preservation than traditional freezing, which frequently produces big ice crystals that harm food texture. Because it alters cellular architecture and metabolic activities, this extremely low-temperature environment has a substantial impact on microbial populations as well. Cryogenics improves food safety and extends shelf life by focusing on bacteria, moulds, and yeasts – common causes of food contamination and spoiling. Cold resistance among microorganisms, including bacteria, fungi, and viruses, exhibits significant variability.

While mesophilic and thermophilic species are generally hindered or destroyed by subzero temperatures, psychrophilic organisms possess the ability to endure and even thrive in cold environments. The process of cryogenically freezing food effectively inhibits the viability of microorganisms through multiple mechanisms. Rapid freezing results in the formation of ice crystals within microbial cells, which lead to lysis and cellular death by compromising cell membranes. Additionally, the freezing process induces dehydration and osmotic shock due to elevated solute concentrations as water freezes both inside and outside the cells, potentially resulting in cell collapse and plasmolysis. Furthermore, cryogenic temperatures inhibit enzymatic activity, effectively halting the growth and metabolism of microorganisms. The freezing process reduces water activity, rendering less water available for microbial activity, which is crucial since most microorganisms depend on adequate water levels for their survival and proliferation. Cryogenics is important in current food safety since it inhibits microbiological development while protecting food quality. It efficiently breaks down microbial life cycles by ultra-low temperature freezing, increasing shelf life and lowering the risk of food borne illness. However, rather than being a stand-alone remedy, cryogenics should be viewed as a component of an integrated food safety system. Complementary technologies and strict post-thaw handling procedures are required because to its limitations, especially with regard to pathogen survival and operating expenses. Cryogenic preservation will become increasingly important as customer demand for fresh, minimally processed foods continues to rise. Future developments will probably concentrate on increasing cost effectiveness, energy efficiency, and the complementary use of cryogenics to other microbial control techniques. In this sense, cryogenics is an effective technique for more than only preservation.

References

Ashok, K.K., Chandra, B.H., Sunil, S., and Kulkarni, G.V. (2021). Application of cryogenics in grinding of spices for value addition: a review. *Natural Volatiles & Essential Oils* 8 (5): 10580–10593.

Balasubramanian, S., Gupta, M.K., and Singh, K.K. (2012). Cryogenics and its application with reference to spice grinding: a review. *Critical Reviews in Food Science and Nutrition* 52 (9): 781–794. https://doi.org/10.1080/10408398.2010.509552.

Bhadra, L., Dhiman, P., Srivastava, A. et al. (2024). Processed food microbiology and safety: risks, trends and future perspectives. In: *Industrial Microbiology and Biotechnology: An Insight into Current Trends*, 175–227. Singapore: Springer Nature Singapore.

Flynn, T. (2004). Cryogenic Engineering, Revised and Expanded (2nd ed.). CRC Press. https://doi.org/10.1201/9780203026991.

Goswami, T.K. (2010). Role of cryogenics in food processing and preservation. *International Journal of Food Engineering* 6 (1): https://doi.org/10.2202/1556-3758.1771.

Havelaar, A.H., Brul, S., De Jong, A. et al. (2010). Future challenges to microbial food safety. *International Journal of Food Microbiology* 139: S79–S94. Professional, London.

Jan, A., Sood, M., Sofi, S.A., and Norzom, T. (2017). Non-thermal processing in food applications: a review. *International Journal of Food Sciences and Nutrition* 2 (6): 171–180.

Kumar, V., Teotia, P., Tuteja, N. et al. (2017). Metabolomics-mediated characterization of endophytic species in recalcitrant tree species. In: *Modern Tools and Techniques to Understand Microbes*, 251–257. Cham: Springer International Publishing https://doi.org/10.1007/978-3-319-49197-4_16.

Mokbel, K., Kodresko, A., Ghazal, H. et al. (2024). Cryogenic media in biomedical applications: current advances, challenges, and future perspectives. *In Vivo* 38 (1): 1–39.

Pavlov, A.V., Verzhuk, V.G., Orlova, S.Y. et al. (2019). Cryopreservation as a method to preserve some fruit and berry crops and wild medicinal plants. *Problems of Cryobiology and Cryomedicine* 29 (1): 044–057.

Pegg, D.E. (2007). Principles of cryopreservation. In: *Cryopreservation and Freeze-drying Protocols*, Methods in Molecular Biology, 2e, vol. 368 (ed. J.G. Day and G.N. Stacey), 39–57. Totowa, NJ: Humana.

Preetha, S.S. and Narayanan, R. (2020). Factors influencing the development of microbes in food. *Shanlax International Journal of Arts, Science and Humanities* 7 (3): 57–77. https://doi.org/10.34293/sijash.v7i3.473.

Shelef, L.A. (1983). Antimicrobial effects of spices. *Journal of Food Safety* 6: 29–44.

Sommers, C.H., Rajkowski, K.T., and Charles Samer, S.S. (2015). The effect of cryogenic freezing followed by gamma radiation on the survival of *Salmonella* spp. on frozen shrimp. *Journal of Food Processing & Technology* s8: https://doi.org/10.4172/2157-7110.s8-001.

Wang, N., Wang, Y., Bai, L. et al. (2023). Advances in strategies to assure the microbial safety of food-associated ice. *Journal of Future Foods* 3 (2): 115–126.

13

Environmental Impact and Sustainability of Cryogenic Processing in Food Industries

Abhishek Kumar

CSIR-National Metallurgical Laboratory, Jamshedpur, Jharkhand, India

13.1 Introduction

In developing countries, the increasing global need for temperature-controlled food distribution has necessitated the establishment of integrated food supply chains. The need for refrigerated vehicles has significantly increased their prevalence globally (Mustafa et al. 2024). By 2015, the expected count of refrigerated cars will rise from around 3 million in 2013 to 15.5 million (Lin et al. 2025). Comparably, in the United Kingdom, by 2025, the estimated number of transport refrigeration units (TRUs) will increase from 84,000 to 97,000 (Grego et al. 2025). Driven by auxiliary diesel engines and depending on refrigerants as the working fluid, most of these chilled-transport trucks run utilizing vapour-compression refrigeration systems. By omitting grocery shopping, the commercial food transport industry contributes to over 12 million metric tons (MMT) of CO_2 equivalent ($MtCO_2e$) emissions yearly in the United Kingdom (Kumar and Gangawane 2022a; Marks 2024).

13.1.1 Environmental Impact of Vapour-Compression Refrigeration Systems

Energy and greenhouse gas (GHG) emissions from temperature-controlled food delivery have been studied extensively. A 2016 study estimated that the U.S. food supply chain utilized 4,660 trillion British thermal units (TBTU) of site energy and created 970 MMT of GHG emissions, mostly from on-farm production. A secondary research found that global agrifood cold chain GHG emissions rose from 0.55 Gt CO_2 equivalent in 2000 to 1.32 Gt CO_2 equivalent in 2022, with the household sector contributing the most at 0.55 Gt CO_2 equivalent. Research shows that the UK food sector's refrigeration resulted in 12.9 MMT CO_2 equivalent carbon emissions, representing for 3.5% of the country's annual territorial GHG emissions. These results demonstrate the environmental effect of temperature-regulated food distribution and the need for cold chain energy efficiency and emission reduction (Kumar and Gangawane 2022b; Alan et al. 2023; Armstrong et al. 2025; Flammini et al. 2024).

Cryogenic Processing: Advances in the Food Industry, First Edition. Edited by Sumit Sudhir Pathak, Rama Chandra Pradhan, and Sudarshan Ramanathan.

13.1.2 Cryogenic Refrigeration as a Sustainable Alternative

Growing concerns about the environmental impact of TRUs have attracted a lot of attention to alternative refrigeration techniques. Among these alternatives, cryogenic TR systems using liquid nitrogen (LN_2) or liquid carbon dioxide (LCO_2) have become quite promising. Reducing reliance on diesel fuel and high-GWP refrigerants, these systems provide an ecologically friendly alternative to traditional vapour-compression systems. Limited research has been conducted on the environmental impact of cryogenic TRUs despite their prospective benefits. A limited number of case studies were provided by the United Nations Environment Programme (UNEP) to compare vapour-compression refrigeration systems with LCO_2 and LN_2-based cryogenic food TRUs (Protocol and Layer 2022; Saeed et al. 2023).

13.1.3 Comparative Analysis of Cryogenic and Vapour-Compression TRUs

The Californian Air Protection Agency has evaluated the well-to-wheel (WTW) GHG emissions of several TR substitutes, including cryogenic systems. The chapter proposes that because of the significant energy needs for manufacturing cryogenic liquids, the well-to-tank (WTT) emissions of cryogenic systems are almost double those of diesel. Assuming zero emissions throughout the operating period of cryogenic fluids, total WTW emissions for cryogenic systems were judged to be 50–60% lower than those of conventional diesel-powered TRUs (Rai and Tassou 2017b). Previous comparison studies primarily focused on GHG emissions during the operating phase of TRUs without including emissions related to the manufacturing of refrigerants and cryogenic fluids (Dong et al. 2021). This study seeks to evaluate the environmental impact of diesel-powered vapour-compression refrigeration systems and cryogenic TRUs, focusing on emissions produced during both the production and operational phases. This study aims to improve understanding of the sustainability of cryogenic refrigeration in food distribution, promoting the use of environmentally friendly cooling technologies (Kumar et al. 2022; Wu et al. 2022).

Various variables, including duty needs, weight, noise, maintenance, environmental concerns, and fuel taxes, determine the compressor drive technique of choice in vapour-compression TRUs. Comprising over 90% of the market, the two most often utilized compressor drive systems are auxiliary diesel engines. One approach runs the compressor and fans directly from a direct drive, while another utilizes an auxiliary diesel engine to operate a generator electrically driving the compressor and fans. Depending on the size of the unit, these engines usually run between 1 and 5 L/h in fuel consumption. Furthermore, some TRU systems operate the compressor using the main engine power of the vehicle either via a direct belt drive or an alternator unit. Still, these systems have a small market share, especially in long-distance travel uses (UNEP Technical Options Committee 2019; Kumar and Gangawane 2023a).

Operating on similar ideas, cryogenic transport refrigeration systems either mostly employ LN_2 or LCO_2. Usually having a storage capacity between 420 and 700 kg, these systems use a big vacuum-insulated tank installed under the chassis. The thermophysical characteristics of the cryogens define their storage pressure; LN_2 is held at 3 bar while LCO_2 is kept at 8.6 bar. At filling stations, however, these fluids are maintained at much greater pressures and lower temperatures; LN_2 is held at 18 bar and −196 °C and LCO_2 at around 22 bar and −57 °C. Direct, indirect, and hybrid systems are the three categories into which cryogenic refrigeration systems fall (Tassou et al. 2009).

Under direct systems, the cryogenic fluid is sprayed straight into the cargo area, and then it is discharged into the atmosphere when the cargo doors open. LN_2 has a boiling temperature of −185.24 °C; LCO_2's is −44.07 °C at storage pressure (Clauss 2008). The liquid cryogen rapidly phases to a gaseous form when it comes into touch with the warmer air within the trailer, therefore lowering the temperature from an ambient temperature of 30 °C in less than 30 minutes. This technique guarantees quick and effective cooling, but because of oxygen depletion in the cargo area, it raises safety concerns. Modern direct system designs have many safety procedures, including oxygen level monitoring and access limitation when oxygen concentrations fall below 19.5%, therefore addressing this problem (Maiorino et al. 2021). Expanding the cryogen in a heat exchanger within the cargo compartment before releasing it into the atmosphere helps indirect cryogenic refrigeration systems reduce the safety hazards associated with direct systems. Although the boiling temperatures of LCO_2 and LN_2 stay the same as in direct systems, air circulation enabled by a fan over the heat exchanger moves the cooling effect to the cargo area (Meneghetti and Ceschia 2020). These devices reach a 0.101 kWh cooling capacity per kilogramme of cryogenic fluid. Still, the pulldown rate of indirect systems is somewhat slower than that of straight (Engineering Guidelines & Minimum System Standards for Implementation in Cold Chain Components 2025) systems.

The regulated and sustainable cooling mechanism is guaranteed by the vented gas exits operating under ambient circumstances. Manufacturers' design of indirect systems differs depending on things such as cargo size, cooling capacity, and how other cooling units are integrated. In terms of lowered GHG emissions and enhanced sustainability, cryogenic refrigeration systems have clear benefits over traditional diesel-powered TRUs. These systems help to cut carbon footprints and lessen dependency on non-renewable energy sources by using cryogenic fluids instead of fossil fuel burning. To evaluate the viability of cryogenic refrigeration in the food sector, one must take into account elements such as cryogen supply, infrastructure for refuelling, and running expenses. Driven by developments in efficiency, safety, and environmental performance, cryogenic transport refrigeration is projected to be adopted as sustainability takes the front stage in food logistics (Rai 2019; Kumar 2023; Kumar et al. 2024b, d).

13.2 Technologies and Mechanisms

The selection of compressor drive techniques in vapour-compression TRUs depends on various factors, including duty requirements, weight, noise, maintenance, environmental impact, and fuel taxation. Over 90% of the market utilizes two primary compressor drive methods: auxiliary diesel engines with direct drive, where the engine powers the compressor and fans directly, and auxiliary diesel engines that operate a generator to supply electrical power to the compressor and fans. Depending on unit size, these engines usually run from 1 to 5 L/h in fuel consumption. Either via a direct belt drive or an alternator unit, some TRUs also have systems that run the compressor using the primary vehicle engine. However, these systems have a small market share, especially in long-distance travel (Engineering Guidelines & Minimum System Standards for Implementation in Cold Chain Components 2025; Panda et al. 2024a,b).

13.2.1 Cryogenic Transport Refrigeration Systems

Cryogenic transport refrigeration systems operate on comparable principles, but they utilize LN_2 or LCO_2 as refrigerants. The storage capacities of these systems typically range from 420 to 700 kg, and they are characterized by a sizeable vacuum-insulated cylinder that is affixed beneath the vehicle chassis. The storage pressures of these cryogens are determined by their thermophysical properties. For example, LN_2 is stored at 3 bar, while LCO_2 is maintained at 8.6 bar. However, these substances are stored at fuelling stations at substantially higher pressures and lower temperatures – LN_2 at 18 bar and −196 °C and LCO_2 at approximately 22 bar and −57 °C. Direct, indirect, and hybrid are the three categories into which cryogenic refrigeration systems can be classified (Rai and Tassou 2017a).

13.2.2 Direct and Indirect Cryogenic Systems

Direct cryogenic refrigeration systems introduce the cryogenic fluid straight into the cargo area, allowing it to evaporate and absorb heat before being released into the atmosphere when the cargo doors are opened. The boiling points of LN_2 and LCO_2 at their respective storage pressures are −185.24 °C and −44.07 °C. In a mere 30 minutes, the cargo temperature is reduced from an ambient 30 °C to subzero levels as the liquid cryogen rapidly converts to gas upon contact with the warming air within the trailer. Although this procedure is efficient, it does pose safety concerns regarding oxygen depletion in the cargo area. Contemporary direct systems incorporate safety features, such as monitoring oxygen levels and instituting access restrictions when oxygen concentrations fall below 19.5%, to mitigate risks (Barta et al. 2021; Thermoking n.d.).

Indirect cryogenic refrigeration systems address the safety concerns of direct systems by expanding the cryogen within a heat exchanger before releasing it into the atmosphere. Although LN_2 and LCO_2 maintain the same boiling temperatures as indirect systems, the cooling effect is transferred to the cargo space via air circulated by a fan across the heat exchanger coil. The chilling capacity of these systems is approximately 0.101 kWh/kg of cryogenic fluid. Nevertheless, the rate of their pulldown is weaker than that of direct systems. The design of indirect systems is subject to variation among manufacturers due to factors such as the integration of additional cooling units, cargo size, and cooling capacity.

The food industry relies on cryogenic refrigeration to swiftly freeze, calm, and convey perishable goods, including meat, fish, dairy, fruits, and vegetables. Cryogenic freezing prevents ice crystals from forming, preserving food's structure, flavour, and nutrients. Unlike mechanical freezing, cryogenic freezing cools food evenly and extends shelf life, decreasing food waste. LN_2 and LCO_2 are ideal for producing high-quality frozen food due to their rapid cooling capabilities. Because it creates fewer greenhouse emissions than vapour compression, cryogenic processing in the food business has a significant impact on the environment. Cryogenic cooling doesn't utilize HFC refrigerants, which cause global warming (GWP).

Furthermore, enhanced food preservation contributes to decreased food waste, which subsequently reduces methane emissions resulting from decomposing food in landfills. The energy necessary for cryogen production should be taken into account, as it impacts the overall carbon footprint. The production and distribution of cryogens through renewable energy sources can improve the sustainability of cryogenic refrigeration in food processing. Adherence to food safety and refrigeration standards is essential for the extensive implementation of cryogenic refrigeration within the food industry. Regulations related to temperature control, energy efficiency, and emissions impact the design and operation of cryogenic TRUs. The advancement of renewable energy-driven cryogen production can markedly improve sustainability.

The ongoing investigation focuses on enhancing cryogenic refrigeration systems by improving insulation, minimizing energy usage, and creating hybrid cooling solutions that combine cryogenic and mechanical refrigeration to boost efficiency. Cryogenic refrigeration systems offer notable environmental and operational advantages over conventional diesel-powered TRUs, particularly in the food industry. Their ability to quickly cool and preserve perishable items, along with lower carbon emissions, makes them a practical substitute for traditional refrigeration methods. It is essential to tackle challenges such as energy demands for cryogen production, infrastructure requirements, and operational costs. As sustainability becomes increasingly important in food logistics, advancements in cryogenic refrigeration technology are expected to promote its broader application in food processing and transportation (Kumar et al. 2024a,c; Refrigeration and Freezing of Foods n.d.).

13.3 Environmental Impact Analysis

13.3.1 Greenhouse Gas Emissions of Cryogenic Systems

Cryogenic transport refrigeration systems using LCO_2 and LN_2 have been investigated as substitutes for traditional diesel-powered vapour-compression refrigeration units. The research emphasizes that (Kumar et al. 2011; Kumar and Gangawane 2024c; Rai and Tassou 2017b):

- The production phase of cryogenic fluids results in significantly higher GHG emissions compared to the production of diesel fuel and refrigerants.
- The renewable energy division and vaporization procedures raise LCO_2 output emissions over LN_2.
- Though LN_2-based systems exhibit reduced total emissions when compared to LCO_2-based systems, the variations are minimal and are within the margin of error according to the perspectives expressed in the research.
- Cryogenic systems have a higher mass intensity (20–60 kg/h) than diesel systems (2.0–4.0 L/h), meaning they consume significantly more material to achieve equivalent refrigeration effects.
- When evaluating entire WTW emissions, cryogenic refrigeration systems are projected to have 50–60% fewer emissions compared to diesel-driven systems, assuming zero operating emissions from cryogenic fluids.

13.3.2 Emissions from Diesel-Driven Systems

Vapour-compression refrigeration systems driven by diesel auxiliary engines contribute to GHG emissions through two primary mechanisms (Blowers and Lownsbury 2010; Chaubey et al. 2024; Stettler et al. 2016): Direct emissions result from refrigerant leakage: The research approximates yearly refrigerant leakage rates between 5% and 25%, with UK-specific rates averaging 8% annually. Refrigerants like R-404A and R-134a possess elevated Global Warming Potentials (GWP) of 3,922 and 1,430, respectively. Indirect emissions from diesel combustion: The fuel combustion process results in CO_2 emissions of approximately 2.676 kg CO_2/L of diesel. Additionally, diesel engines produce particulate matter (PM) and nitrogen oxides (NO_x), which have adverse environmental and health impacts. Larger refrigerated vehicles (38-tonne trucks) exhibit more than 50% lower emissions per kilogram of transported food compared to smaller vehicles (18-tonne trucks), making them more environmentally efficient for large-scale distribution.

13.3.3 Comparative Analysis of Production and Operational Emissions

The environmental impact of refrigeration systems depends on production and operation emissions. Research concludes that the emissions associated with the production of diesel and refrigerants are significantly reduced by as much as 66% compared to those of cryogenic fluids (McCulloch and Lindley 2003; Yalcin et al. 2019). The emissions associated with the production of diesel and refrigerants are significantly reduced by as much as 66% compared to those of cryogenic fluids. This reduction is primarily attributable to the substantial energy requirements involved in the production of LCO_2 and LN_2. Operational emissions of cryogenic systems are negligible, as cryogenic fluids do not generate GHGs upon release. Total emissions from diesel and cryogenic systems, encompassing both production and operation, are comparable. LN_2 refrigeration demonstrates marginally lower total emissions than LCO_2 and diesel; however, the difference is not substantial enough to definitively favour one system over the others.

13.4 Challenges in Cryogenic Processing

Cryogenic processing has emerged as a viable alternative to traditional vapour-compression refrigeration for food preservation and transportation. While cryogenic refrigeration offers benefits such as rapid cooling, enhanced food quality retention, and reduced operational emissions, its broad implementation encounters considerable technical, logistical, and economic obstacles. This section examines the three main obstacles to the implementation of cryogenic refrigeration systems in the food industry (Neittaanmäki et al. 2024; Kumar 2024; Kumar et al. 2025; Pal and Sen 2025a,c).

- High mass intensity and distribution range limitations
- Infrastructure for replenishing cryogens
- Operational and economic barriers

Each challenge is discussed in detail with supporting data and insights into possible mitigation strategies.

13.4.1 High Mass Intensity and Distribution Range Limitations

The high mass intensity of cryogenic fluids makes food distribution and transportation difficult in cryogenic food refrigeration (Maiorino et al. 2021; Kumar and Gangawane 2024a; Pal and Sen 2024b, 2025b).

13.4.1.1 Comparison with Diesel-Driven Refrigeration Systems

Traditional vapour-compression refrigeration systems use diesel-powered auxiliary engines, consuming approximately 2.0–4.0 L of diesel per hour. In contrast, cryogenic refrigeration systems require 20–60 kg of liquid cryogens per hour for the same cooling effect. This means cryogenic refrigeration has 10 times higher mass intensity than diesel-driven refrigeration units. The higher mass requirement reduces the effective transportation range, necessitating frequent refuelling of cryogenic fluids. For example, a 38-tonne refrigerated truck using diesel refrigeration can operate for several days without refuelling, whereas a cryogenic refrigeration truck may require replenishment within 10–15 hours.

13.4.1.2 Cryogenic Fluid Consumption and Its Impact on Logistics

Cryogenic fluids such as LCO_2 and LN_2 are widely used in cold chain logistics due to their excellent cooling capacity. LCO_2 (O=C=O, molar mass: 44.01 g/mol) sublimates at −78.5 °C, while LN_2 (N≡N, molar mass: 28.01 g/mol) boils at −195.79 °C. Their consumption depends on ambient conditions, food type, and insulation quality. Typical cryogen usage rates are:

- LCO_2: 0.4–1.2 kg/m^3 per delivery cycle
- LN_2: 0.6–1.5 kg/m^3, owing to higher latent heat demands

Thermal ingress, calculated via Fourier's Law, drives vaporization loss and increases refill frequency. For example, with 5 m^2 surface area and 0.05 m insulation, heat load ≈140 W, leading to ~2.5 kg LN_2/h. Efficient insulation and route optimization can significantly reduce cryogen loss, cost, and environmental impact (Cheng et al. 2008).

13.4.2 Infrastructure for Replenishing Cryogens

The lack of refuelling stations for cryogenic fluids is another critical barrier preventing the widespread adoption of cryogenic refrigeration in food industries (Leachman et al. 2017; Kumar and Gangawane 2024b; Pal and Sen 2023a,b, 2024a).

13.4.2.1 Current Status of Cryogen Replenishment Infrastructure

Unlike diesel, which benefits from a well-established global distribution network, cryogenic fluids such as LN_2 and LCO_2 require specialized refuelling stations equipped with vacuum-insulated storage tanks. Currently (Aghazadeh Dokandari et al. 2014):

- Diesel refuelling stations are widely available, ensuring the seamless operation of diesel-driven refrigerated trucks.
- Cryogenic refuelling stations are limited to industrial zones, making it difficult for long-haul refrigerated trucks to operate efficiently using LN_2 or LCO_2.
- Cryogen storage tanks require specialized handling, as LN_2 must be stored at −196 °C at 3 bar, while LCO_2 requires −57 °C at 22 bar.

13.4.2.2 Economic and Logistical Constraints of Expanding Infrastructure

The cost of building a cryogenic refuelling station is much more than that of traditional diesel stations because of the:

1) High capital investment required for vacuum-insulated cryogen storage tanks
2) The energy-intensive process of liquefying and transporting LN_2 and LCO_2
3) Need for specialized safety regulations to prevent hazards such as oxygen depletion in enclosed spaces

As a result, many food distribution companies hesitate to switch to cryogenic refrigeration, as it requires both investment in new refrigeration units and access to cryogenic fuel stations (Ghorbani et al. 2025).

13.4.2.3 Potential Solutions for Infrastructure Development

Various infrastructure modifications may be considered to enhance the feasibility of cryogenic refrigeration in food distribution, including the following:

- The incorporation of cryogenic refuelling stations at key food logistics centres is essential for guaranteeing dependable access to refrigerated transport vehicles.
- Collaboration between cryogen producers and logistics companies to establish a shared distribution network.
- Development of hybrid cryogenic-diesel refrigeration systems to extend operational range while minimizing environmental impact.

If cryogenic supply chains are strategically expanded, the adoption of cryogenic refrigeration in food industries could increase significantly (Yenare et al. 2024).

13.4.3 Operational and Economic Barriers

13.4.3.1 High Capital and Operating Costs

Cryogenic refrigeration units are more expensive to install compared to conventional diesel-based systems. The cost breakdown is as follows:

- Diesel-driven refrigeration unit: £18,000–£21,000
- Cryogenic refrigeration unit: £22,000+

Additionally, while liquid CO_2 (£0.12/kg) and liquid N_2 (£0.08/kg) are cheaper per unit than diesel (£1.2/L), their higher consumption rates lead to similar total operating costs as diesel-based refrigeration systems.

13.4.3.2 Limited Economic Incentives and Regulatory Support

The existing governmental policies do not require a shift from diesel refrigeration to cryogenic refrigeration. Diesel engines face stricter regulations on NOx and particulate emissions, which may increase future adoption of cryogenic refrigeration if emission limits become more restrictive. The absence of financial incentives, including tax credits and subsidies, hinders the widespread use of cryogenic technology and impedes sector change.

13.4.3.3 Safety and Handling Challenges

Cryogenic fluids pose safety risks such as frostbite, asphyxiation, and pressure-related hazards. Liquid nitrogen storage needs continuous oversight to avert oxygen displacement in enclosed areas. Training and certification programs for cryogen handling need to be widely implemented before large-scale adoption can occur.

13.4.3.4 Strategies for Economic Viability

Several actions may be taken to help one to overcome financial obstacles: Government grants for businesses using cryogenic refrigeration are comparable to grants for energy from renewable sources. Enhanced energy efficiency in the production of cryogens diminishes the carbon footprint associated with the liquefaction of LCO_2 and LN_2. Combining diesel backup freezers with cryogenic cooling to negotiate cost and sustainability, combinations of technologies are integrated (Rai 2019).

13.5 Benefits of Cryogenic Processing

Food quality, nutritional content, and structural damage from ice crystal formation are improved by cryogenic processing over vapour-compression refrigeration. Ultra-low temperatures achieved by LN_2 and LCO_2 freezing enable quick cooling, cellular damage avoidance, and food integrity preservation (Kumar 2013; Panda et al. 2022; Kumar et al. 2023; Kumar and Gangawane 2023b). This section explores the significant benefits of cryogenic refrigeration in food processing and transportation, focusing on: Food quality preservation; impact on nutritional value, flavour, and texture; reduction of ice crystal formation.

13.5.1 Advantages in Food Quality Preservation

In food processing, cryogenic refrigeration's main benefit is its exceptional capacity to maintain food quality above traditional refrigeration techniques.

13.5.1.1 Faster Cooling and Lower Temperature Impact

Cryogenic refrigeration systems using LN_2 and LCO_2 achieve significantly lower temperatures compared to vapour-compression refrigeration, enabling faster freezing and preserving food quality, as shown in Table 13.1. Boiling at −196 °C, LN_2 quickly draws heat from food items, hence causing immediate freezing. LCO_2 boils at −44 °C, making it slightly less effective than LN_2 but still far superior to vapour-compression refrigeration. Faster freezing translates to higher retention of sensory and nutritional properties, reducing spoilage rates and improving overall product stability.

13.5.1.2 Extended Shelf Life

Cryogenic freezing minimizes microbial growth and enzymatic activity, leading to an extended shelf life for perishable food products such as meat, seafood, and dairy. Research shows that: Cryogenic freezing extends meat shelf life by 50–80% compared to

Table 13.1 Comparison of different cooling systems.

Cooling system	Cooling rate	Temperature achieved
Vapour compression (diesel)	Gradual cooling over hours	−20 to −30 °C
Liquid CO_2 refrigeration	Instant freezing in minutes	−44 to −57 °C
Liquid N_2 refrigeration	Rapid freezing within seconds	−196 °C

conventional freezing; fruits and vegetables maintain a fresher appearance and texture when frozen at ultra-low temperatures; ready-to-eat meals retain their original taste, aroma, and structure better than conventionally frozen meals. In the food distribution sector, temperature variations can impact product quality; thus, cryogenic refrigeration provides a dependable solution for ensuring consistent temperature regulation during transport (Kumar et al. 2023).

13.5.2 Impact on Nutritional Value, Flavour, and Texture

In comparison to more traditional methods of freezing, cryogenic processing is more effective in preserving the nutritional qualities of food products.

13.5.2.1 Nutrient Retention

Conventional freezing methods degrade nutrients and lose moisture, especially in foods high in vitamins. The process of cryogenic freezing

1) Reduces oxidation of vitamins such as vitamin C and B-complex
2) Minimizes protein denaturation, preserving the structural integrity of meat and seafood
3) Prevents enzymatic browning in fruits and vegetables, maintaining their fresh colour and nutritional composition

According to research, conventional freezing procedures only preserve 60–70% of the original vitamin content, while LN_2 freezing preserves up to 95%.

13.5.2.2 Flavour and Aroma Conservation

The fast-freezing procedure inhibits oxidation and moisture evaporation, therefore preserving the natural flavour and fragrance of foods. The natural oceanic flavour of seafood is more effectively preserved when frozen with LN_2. Meat products avoid freezer burn, preserving their juiciness and umami profile. Dairy products, such as ice cream, maintain creaminess without the formation of large ice crystals.

13.5.2.3 Texture Preservation

Cryogenic processing preserves the integrity of food structure. Hence, preventing crisper veggies and fruits after thawing, juicier meat and poultry, as cell membranes remain intact and smoother dairy textures that enhance the taste and texture of cheese-based goods and frozen desserts.

13.5.3 Reduction of Ice Crystal Formation

Among the most significant benefits of cryogenic freezing is its ability to prevent the formation of large ice crystals, which is a problem with slow freezing procedures.

13.5.3.1 The Formation of Ice Crystals and Their Effect on Food

When food freezes, the water inside it turns into ice crystals. The size and location of these crystals have a significant effect on how the food feels:

1) Slow freezing (−20 to −30 °C) → Large ice crystals form, damaging cell structures and causing food to lose moisture upon thawing.
2) Rapid freezing (LN_2 at −196 °C) → Small, evenly distributed ice crystals form, maintaining the original texture and juiciness of food.

13.5.3.2 Cryogenic Freezing for Delicate Food Products

Cryogenic freezing is especially beneficial for very perishable and fragile foods, including naturally hard berries and delicate fruits, preventing stale bakery and confectionery, and ensuring that high-value foods like Wagyu beef have intact marbling.

13.5.3.3 Preventing Freezer Burn and Moisture Loss

Water vapour from frozen food surfaces causes freezer burn, producing dry, discoloured areas. Freezer burn is reduced by cryogenic freezing. The ultra-low temperature locks the moisture in food and prevents dehydration, keeping the food juicy and fresh even after long-term storage.

13.6 Sustainability Considerations

Cryogenic processing presents notable environmental benefits by removing direct emissions throughout its operation. Nonetheless, the sustainability of this approach hinges on achieving a balance between energy demands, minimizing production emissions, and enhancing the overall efficiency of the system. The high energy demands of cryogen manufacture provide a significant difficulty, thereby stressing the necessity of technical breakthroughs to enhance its environmental and financial sustainability. In this part, we will discuss important aspects of sustainability, including the following: Balancing energy needs and environmental benefits; enhancing sustainability with technology and efficiency of cryogen production methods.

13.6.1 Balancing Energy Requirements and Environmental Benefits

13.6.1.1 Energy Consumption in Cryogenic Refrigeration

Although cryogenic systems lower running emissions, their energy-intensive manufacturing techniques affect general sustainability. A comparison of energy needs emphasizes this difficulty, as shown in Table 13.2: Diesel-powered systems emit GHGs from fuel combustion and refrigerant leakage, contributing to climate change; cryogenic systems eliminate

Table 13.2 Comparison of refrigeration systems by energy source, consumption and emissions.

Refrigeration system	Primary energy source	Energy consumption	Operational emissions
Diesel vapour compression	Diesel	11.83 kWh/kg of diesel	2.676 kg CO_2/L
Liquid CO_2 refrigeration	Electricity (LCO_2 liquefaction)	0.305 kWh/kg LCO_2	Zero direct emissions
Liquid N_2 refrigeration	Electricity (LN_2 liquefaction)	0.549 kWh/kg LN_2	Zero direct emissions

on-site emissions, but high energy consumption during production offsets their environmental advantage.

13.6.1.2 Well-to-Wheel (WTW) Emissions of Cryogenic vs. Diesel Systems

The WTT emissions of cryogenic systems are double those of diesel due to the high energy demand for liquefaction. WTW emissions for cryogenic refrigeration are 50–60% lower than those of diesel-driven systems because cryogenic fluids produce zero emissions during operation. The overall environmental benefit depends on reducing the carbon footprint of cryogen production through renewable energy sources and energy-efficient liquefaction methods.

13.6.1.3 Trade-Off Between Cryogen Production and Usage

The primary problem lies in reconciling the emissions generated during manufacturing with the practical advantages of cryogenic refrigeration. The use of fossil-fuel-based energy for cryogen synthesis results in a substantial GHG footprint, hence constraining sustainability. Utilizing renewable energy resources (wind, solar, or hydro) for cryogen synthesis substantially reduces the net negative impact on the environment. Optimizing transportation, such as utilizing larger refrigerated vehicles to reduce per-unit emissions, could enhance sustainability.

13.6.2 Technological Advancements for Improved Sustainability

13.6.2.1 Innovations in Cryogenic Refrigeration Systems

Recent technological developments to improve energy efficiency and reduce the environmental impact of cryogenic processing include:

- **Hybrid cryogenic refrigeration systems**: The integration of cryogenic cooling with vapour compression has the potential to enhance energy efficiency and broaden the operational range.
- **Phase-change materials (PCMs)**: The integration of phase-change materials with cryogenic systems significantly improves thermal storage capacity, leading to a reduction in overall cryogen consumption.
- **Energy recovery systems**: Technologies like LN_2 expansion turbines harness the latent energy of cryogens to generate electricity, thereby enhancing efficiency.

13.6.2.2 Renewable Energy Integration in Cryogen Production

Green hydrogen-powered liquefaction plants have the potential to substitute fossil fuel-based energy sources in cryogen production, thereby reducing emissions. The implementation of solar and wind energy for the electrification of cryogen production facilities has the potential to decrease carbon intensity significantly. Carbon capture and storage (CCS) technologies can be utilized in the production of LCO_2 to mitigate CO_2 emissions, thereby enhancing the sustainability of the process.

13.6.2.3 Advances in Cryogenic Fuel Infrastructure

1) Decentralized cryogenic fuel production units have the potential to minimize transportation emissions linked to cryogen delivery.
2) Advanced cryogenic storage systems utilizing innovative insulation materials can reduce evaporative losses, thereby enhancing sustainability.
3) On-demand cryogen generation systems have the potential to minimize the requirement for bulk storage while enhancing operational efficiency in food processing facilities.

13.6.3 Efficiency of Cryogen Production Methods

13.6.3.1 Energy Intensity of Cryogen Production

Cryogen production is a process that demands significant energy input, necessitating the use of advanced liquefaction techniques to improve efficiency, as tabulated in Table 13.3. LN_2 production requires more energy than LCO_2 due to the air separation process; however, it results in a lower GHG footprint. LCO_2 production encompasses the processes of CO_2 capture and purification, positioning it as a viable CO_2 recycling solution for various industries.

13.6.3.2 Improving Liquefaction Efficiency

Innovations have been suggested to minimize energy consumption while improving environmental sustainability. Magnetic refrigeration employs solid-state cooling rather than gas compression, leading to a substantial reduction in energy consumption during liquefaction. Advanced cryogenic heat exchangers enhance heat transfer effectiveness, thereby decreasing power consumption in air separation plants. Superconductive liquefiers: Employ superconducting substances to minimize electrical losses in the liquefaction procedure. Cryogenic carbon capture (CCC) captures CO_2 directly from industrial emissions and produces LCO_2, resulting in a carbon-neutral process.

Table 13.3 Energy requirements and GHG emissions of cryogen production.

Cryogen	Energy required for production	GHG emissions (kg CO_2e/kg)
LCO_2	0.305 kWh/kg (Capture + liquefaction)	0.305 kg CO_2e/kg
LN_2	0.549 kWh/kg (Air separation + liquefaction)	0.254 kg CO_2e/kg

13.6.3.3 Cryogen Storage and Distribution Optimization
High-efficiency cryogenic tanks with multi-layer vacuum insulation may cut boil-off losses by 80%. Using smart logistics for cryogen shipping, such as route optimization and demand forecasting, may reduce delivery fleet emissions (Bose et al. 2024; Jeje et al. 2024).

13.7 Case Studies and Applications

13.7.1 Examples of Cryogenic Systems in the Food Industry

Cryogenic processing has been extensively used in several sectors of the food business owing to its capacity to elevate product quality, boost processing efficiency, and prolong shelf life. This section examines significant instances of cryogenic systems used in food processing.

13.7.1.1 Cryogenic Freezing for Perishable Foods
Cryogenic freezing, typically using LN_2 and CO_2, is used for the quick preservation of perishable food items, including fish, meat, and prepared meals. The very low temperatures (−196 °C for LN_2) provide rapid phase transitions, minimizing ice crystal formation and preserving product texture and nutritional integrity. Companies like Air Products, Linde, and Messer have developed cryogenic freezing methods that provide uniform freezing efficacy while reducing processing duration.

13.7.1.2 Cryogenic Grinding for Spice and Herb Processing
Cryogenic grinding is a unique method used in the preparation of spices and herbs to preserve essential oils and active chemicals that may deteriorate at room temperature. Employing LN_2 in grinding mills preserves volatile components, hence boosting the taste, fragrance, and shelf life of spices like black pepper, cardamom, and turmeric. Industries using this technique include those manufacturing pharmaceutical-grade botanical extracts and food-grade spices.

13.7.1.3 Cryogenic Sample Preparation in Food Safety Analysis
Cryogenic processing is essential for ensuring food safety and quality control, especially in the study of pesticide residues. The use of LN_2 in sample preparation guarantees uniform mixing of food matrices, averting thermal destruction of sensitive analytes. Research has shown that cryogenic comminution enhances the precision and consistency of pesticide detection in fruits, vegetables, and cereals, thereby facilitating regulatory compliance and ensuring consumer safety.

13.7.2 Real-World Performance and Benefits

The use of cryogenic processing in the food industry has shown several benefits compared to traditional approaches. This section examines the practical implications of cryogenic technology via industry case studies and scientific assessments.

13.7.2.1 Enhanced Product Quality and Shelf Life

Cryogenic processing substantially enhances the nutritional value and longevity of perishable food items by reducing microbial proliferation, enzymatic activity, and oxidation. Research indicates that cryogenic freezing of fish enhances texture preservation and minimizes drip loss relative to traditional freezing techniques. Likewise, cryogenic grinding improves the stability of bioactive components in spices, inhibiting oxidation-induced breakdown.

13.7.2.2 Energy Efficiency and Reduced Environmental Footprint

Cryogenic systems, while requiring substantial initial capital, provide long-term energy efficiencies by decreasing processing duration and minimizing refrigeration energy demands. The use of LN_2, often obtained as a by-product of industrial gas manufacturing, enhances sustainability by recycling otherwise discarded resources. Moreover, cryogenic freezing systems have been shown to reduce carbon emissions relative to mechanical freezing options by decreasing energy usage.

13.7.2.3 Food Waste Reduction and Sustainability

Cryogenic processing mitigates food waste at both industrial and consumer levels by improving preservation. The capacity to freeze and preserve food items for prolonged durations without substantial degradation mitigates premature waste. Moreover, cryogenic grinding of food by-products, including fruit peels and herbs, enables their transformation into value-added goods such as natural food colorants and extracts, hence advancing circular economy concepts within the food business (Joy and Vanapalli 2023).

13.8 Future Directions in Cryogenic Processing

Cryogenic processing in the food business has shown considerable benefits in food preservation, logistical efficiency, and environmental sustainability. Nonetheless, extensive adoption requires technology advancements, policy endorsement, and infrastructural enhancement. This section examines significant potential trajectories for cryogenic refrigeration, encompassing: Advancements and novel technologies; possibility for expanded utilization; and recommendations for industrial practice.

13.8.1 Innovations and Emerging Technologies

Recent breakthroughs in cryogenic refrigeration, artificial intelligence (AI), and sustainable energy integration are anticipated to improve the efficiency and feasibility of cryogenic food processing.

13.8.1.1 Hybrid Cryogenic Refrigeration Systems

Hybrid cryogenic-compression refrigeration innovations seek to integrate the swift cooling capabilities of cryogenic systems with the efficiency of vapour-compression refrigeration, resulting in (Popov et al. 2021):

1) Decreased operating expenses by minimizing dependence on LN_2 or LCO_2.
2) Enhanced operating range by reducing cryogen refill needs.
3) Adaptive temperature regulation, allowing adequate refrigeration under diverse load situations.

13.8.1.2 AI-Powered Predictive Maintenance and Process Optimization

Cryogenic processing will become more automated and efficient as a result of the combination of AI and machine learning (ML). Those relying on AI are able to (Benes et al. 2024):

1) Enhance freezing rates instantaneously, minimizing cryogen use.
2) Anticipate maintenance requirements in cryogenic refrigeration equipment to minimize downtime and operating expenses.
3) Enhance logistical coordination to guarantee just-in-time delivery of cryogenic fuel supply.

13.8.1.3 Renewable Energy Integration in Cryogen Production

A significant limitation of cryogenic refrigeration is the substantial energy requirement for the liquefaction of cryogens, with LN_2 needing 0.549 kWh/kg and LCO_2 requiring 0.305 kWh/kg. Future advancements will concentrate on (Ghorbani et al. 2023): Employing green hydrogen for liquefaction to reduce carbon emissions; making air separation units (ASUs) with renewable energy sources (solar, wind) to decarbonize the generation of LN_2 and LCO_2; and energetically efficient liquefaction techniques, including magnetic refrigeration and sophisticated cryogenic heat exchangers.

13.8.1.4 Superconducting Cryogenic Technologies

Systems of superconducting magnetic energy storage (SMES) could reduce cryogenic refrigeration energy losses. Advanced insulating materials for cryogenic tanks might cut cryogen boil-off losses by as much as 80% (Silva Elipe et al. 2024).

13.8.2 Potential for Broader Adoption

13.8.2.1 Expanding Cryogenic Processing Beyond High-End Applications

At present, cryogenic refrigeration is extensively used for high-value perishables (such as seafood, medicines, and gourmet foods) owing to its elevated operating costs relative to vapour-compression refrigeration. To facilitate wider acceptance, progress must minimize cryogen use per food unit with intelligent injection and energy recuperation technologies; enhance the bulk cryogen delivery infrastructure to provide accessibility for small- and medium-sized food industries; and formulate cost-sharing solutions, such as leasing cryogenic equipment, to reduce upfront capital expenditure.

13.8.2.2 Infrastructure Development for Cryogenic Refuelling

A key impediment to adoption is the insufficient availability of cryogen recharging stations. Subsequent projects need to:

- Establish decentralized LN_2/LCO_2 refuelling stations in conjunction with food distribution facilities.

- Encourage investment in cryogenic transportation networks to facilitate broader implementation in food logistics.
- Implement adaptable hybrid systems that enable food processors to alternate between vapour compression and cryogenic chilling depending on cost and availability.

13.8.2.3 Policy and Regulatory Incentives

Government and business assistance may facilitate the broader use of cryogenic refrigeration by:

1) Offering tax incentives for the use of cryogenic refrigeration, akin to subsidies for renewable energy
2) Regulating diesel emissions from vapour-compression systems to promote a transition to cleaner cryogenic alternatives
3) Formulating safety standards for extensive LN_2 and LCO_2 distribution to guarantee secure handling and logistics (Joy and Vanapalli 2023)

13.8.3 Recommendations for Industry Practices

13.8.3.1 Optimizing Cryogen Consumption for Cost Efficiency

Industries may minimize cryogen use and operating expenses by:

1) Implementing AI-driven cryogenic process control to minimize overcooling losses
2) Utilizing phase-change materials (PCMs) in cryogenic systems to improve thermal efficiency
3) Implementing closed-loop cryogen recycling systems to collect and re-liquefy CO_2 and nitrogen emissions

13.8.3.2 Enhancing Cryogenic Food Logistics

Food corporations need to:

1) Implement real-time monitoring sensors in cryogenic food storage systems.
2) Implement hybrid cold chain systems, using cryogenic refrigeration for short-term storage and vapour compression for long-distance transportation.
3) Allocate resources for training initiatives focused on the safe management of cryogens to mitigate operational hazards.

13.8.3.3 Sustainable Supply Chain Integration

To improve sustainability and decrease carbon emissions, corporations need to:

1) Procure cryogens from low-carbon manufacturing facilities that use renewable energy sources.
2) Collaborate with AI-driven logistics companies to enhance route efficiency and diminish transportation emissions.
3) Engage in cryogenic research and development partnerships to advance next-generation food preservation technologies (Joy and Vanapalli 2023).

13.9 Conclusion

Cryogenic methodology offers important benefits in food preservation, improving product quality, extending shelf life, and improving logistics. Its quick chilling speed contributes to sustainability by lowering food waste and microbiological spoiling. Its environmental advantages, however, depend upon supplies of energy, infrastructure, and legal frameworks. Future progress requires industry-wide cooperation to improve efficiency and environmental performance. Cryogenic approach preserves food texture, nutrition, and safety more effectively than traditional procedures. The substantial energy requirements for cryogen synthesis may be mitigated with the use of renewable energy sources. Closed-loop processes and sustainable liquefaction methods reduce their environmental impact. Investment in AI-driven cryogenic control systems is used for the development of infrastructure for decentralized cryogenic refuelling systems, while policies are to be endorsed via incentives and safety rules.

References

Aghazadeh Dokandari, D., Setayesh Hagh, A., and Mahmoudi, S.M.S. (2014). Thermodynamic investigation and optimization of novel ejector-expansion CO_2/NH_3 cascade refrigeration cycles (novel CO_2/NH_3 cycle). *International Journal of Refrigeration* 46: 26–36. https://doi.org/10.1016/j.ijrefrig.2014.07.012.

Alan, F., Tim, B., and Judith, E. (2023). Carbon emissions from refrigeration used in the UK food industry. *International Journal of Refrigeration* 150: 297–303. https://doi.org/10.1016/j.ijrefrig.2023.01.022.

Armstrong, K., Dong, W., Jin, M. et al. (2025). Estimating energy consumption and GHG emissions in the U.S. food supply chain for net-zero. *npj Science of Food* 9 (1): 19. https://doi.org/10.1038/s41538-024-00346-y.

Barta, R.B., Groll, E.A., and Ziviani, D. (2021). Review of stationary and transport CO_2 refrigeration and air conditioning technologies. *Applied Thermal Engineering* 185: 116422. https://doi.org/10.1016/j.applthermaleng.2020.116422.

Benes, K.J., Porterfield, J.E., and Yang, C. (2024). AI for energy: opportunities for a modern grid and clean energy economy. US Department of Energy.

Blowers, P. and Lownsbury, J.M. (2010). Carbon dioxide emission implications if hydrofluorocarbons are regulated: a refrigeration case study. *Environmental Science & Technology* 44 (5): 1526–1529. https://doi.org/10.1021/es9023354.

Bose, D., Bhattacharya, R., Kaur, T. et al. (2024). Innovative approaches for carbon capture and storage as crucial measures for emission reduction within industrial sectors. *Carbon Capture Science & Technology* 12: 100238. https://doi.org/10.1016/j.ccst.2024.100238.

Chaubey, A.R., Chaturvedi, K.R., and Kumar, A. (2024). Aerogel applications in the oil and gas industry: opportunities, challenges, and the way ahead. In: *Hybrid Aerogels: Energy and Environmental Applications*, vol. 201. De Gruyter https://doi.org/10.1515/9783111252001-009.

Cheng, L., Ribatski, G., and Thome, J.R. (2008). Analysis of supercritical CO_2 cooling in macro- and micro-channels. *International Journal of Refrigeration* 31 (8): 1301–1316. https://doi.org/10.1016/j.ijrefrig.2008.01.010.

Clauss, R.C. (2008). Cryogenic refrigeration systems. In: *Low-noise Systems in the Deep Space Network* (ed. M.S. Reid). Pasadena: Jet Propulsion Laboratory.

Dong, Y., Coleman, M., and Miller, S.A. (2021). Greenhouse gas emissions from air conditioning and refrigeration service expansion in developing countries. *Annual Review of Environment and Resources* 46 (1): 59–83. https://doi.org/10.1146/annurev-environ-012220-034103.

National Centre for Cold-chain Development (NCCD). Engineering guidelines & Minimum System Standards for Implementation in Cold Chain Components: For schemes supported by Central Government Ministry of Farmers and Welfare (2nd print). https://nccd.gov.in/guidelines. (2025).

Flammini, A., Adzmir, H., Pattison, R. et al. (2024). Greenhouse gas emissions from cold chains in agrifood systems. *Sustainability* 16 (21): 9184. https://doi.org/10.3390/su16219184.

Ghorbani, B., Zendehboudi, S., Saady, N.M.C. et al. (2023). Strategies to improve the performance of hydrogen storage systems by liquefaction methods: a comprehensive review. *ACS Omega* 8 (21): 18358–18399. https://doi.org/10.1021/acsomega.3c01072.

Ghorbani, B., Zendehboudi, S., and Saady, N.M.C. (2025). Advancing hybrid cryogenic natural gas systems: a comprehensive review of processes and performance optimization. *Energies* 18 (6): 1443. https://doi.org/10.3390/en18061443.

Grego, S., Cottingham, R., Koottatep, T. et al. (2025). *Engineering Field Testing of Non-Sewered Sanitation Systems: Compendium of Lessons Learned*, 1–61. IWA Publishing https://doi.org/10.2166/9781789064971.

Jeje, S.O., Marazani, T., Obiko, J.O., and Shongwe, M.B. (2024). Advancing the hydrogen production economy: a comprehensive review of technologies, sustainability, and future prospects. *International Journal of Hydrogen Energy* 78: 642–661. https://doi.org/10.1016/j.ijhydene.2024.06.344.

Joy M. and Vanapalli S. (2023). Cryogenic nitrogen in the food and beverage industry. *Freezing of Foodstuffs*.

Kumar, S. (2013). Effect of cryogenic freezing on food: a review. *Journal of Postharvest Technology* 1 (01): 91–96.

Kumar, A. (2023). Superheated Drying. In: *Advanced Computational Approaches for Drying in Food Processing* (ed. K.M. Gangawane, M. Dwivedi, and R. Chandra Pradhan), 189–212. Springer International Publishing https://doi.org/10.1007/978-3-031-62550-3_9.

Kumar, A. (2024). Engineered nanomaterials and associated threats in the environment risk assessment strategies.

Kumar, A. and Gangawane, K.M. (2022a). Effect of precipitating agents on the magnetic and structural properties of the synthesized ferrimagnetic nanoparticles by co-precipitation method. *Powder Technology* 401: 117298. https://doi.org/10.1016/j.powtec.2022.117298.

Kumar, A. and Gangawane, K.M. (2022b). Synthesis and effect on the surface morphology & magnetic properties of ferrimagnetic nanoparticles by different wet chemical synthesis methods. *Powder Technology* 410 (June): 117867. https://doi.org/10.1016/j.powtec.2022.117867.

Kumar, A. and Gangawane, K.M. (2023a). *Advanced Computational Approaches for Water Treatment*, 1e, 1–12. New York: CRC Press https://doi.org/10.1201/9781003325147-1.

Kumar, A. and Gangawane, K.M. (2023b). Molecular dynamics for the membrane process. In: *Advanced Computational Approaches for Water Treatment*, 1e, 13–38. New York: CRC Press https://doi.org/10.1201/9781003325147-2.

Kumar, A. and Gangawane, K.M. (2024a). Green nanomaterials in energy storage: advancements and challenges [chapter]. In: *Materials for Boosting Energy Storage. Volume 1: Advances in Sustainable Energy Technologies, Part 12 - Green Nanomaterials in Energy Storage: Advancements and Challenges*, vol. 1, 275–291. American Chemical Society https://doi.org/10.1021/bk-2024-1477.ch012.

Kumar, A. and Gangawane, K.M. (2024b). Nanoparticle-modified multifunctional nano carbons — advances in energy storage [chapter]. In: *Materials for Boosting Energy Storage. Volume 3: Advances in Sustainable Energy Technologies, Part 7 - Nanoparticle-Modified Multifunctional Nano Carbons—Advances in Energy Storage*, vol. *3*, 1–30. https://doi.org/10.1021/bk-2024-1488.ch007.

Kumar, A. and Gangawane, K.M. (2024c). Recent advances in organic/composite phase change materials (PCM) for energy storage. In: *Materials for Boosting Energy Storage. Volume 3: Advances in Sustainable Energy Technologies*, vol. 1488, 309–334. American Chemical Society https://doi.org/10.1021/bk-2024-1488.ch013.

Kumar, S., Kwon, H.-T., Choi, K.-H. et al. (2011). LNG: an eco-friendly cryogenic fuel for sustainable development. *Applied Energy* 88 (12): 4264–4273. https://doi.org/10.1016/j.apenergy.2011.06.035.

Kumar, A., Gangawane, K.M., and Ramanjaneyulu, B. (2022). Ferrofluids for waste-water treatment. In: *International Conference on Chemical, Bio and Environmental Engineering*, 723–744. Springer https://doi.org/10.1007/978-3-030-96554-9_48.

Kumar, A., Jena, H.M., and Gangawane, K.M. (2023). Effect of ferrimagnetic nanoparticles as an oxygen absorber on the shelf life of tomatoes. *9th International Food Convention (IFCoN).*

Kumar, A., Panda, D., and Gangawane, K.M. (2024a). Chapter 3 microfabrication: techniques and technology. In: *Microfabrication and Nanofabrication*, vol. 11, 47–76. De Gruyter https://doi.org/10.1515/9783111335537-003.

Kumar, A., Panda, D., Gangawane, K.M., and Jena, H.M. (2024b). Microwave-assisted drying. In: *Advanced Computational Approaches for Drying in Food Processing*, 1e, 45–73. Cham: Springer International Publishing https://doi.org/10.1007/978-3-031-62550-3_3.

Kumar, A., Panda, D., and Gangawane, K.M. (2024c). Computational modeling on the design of the morphology of aerogels. In: *Hybrid Aerogels: Energy and Environmental Applications*, vol. 1, 269. https://doi.org/10.1515/9783111252001-013.

Kumar, A., Panda, D., Jena, H.M., and Gangawane, K.M. (2024d). Electrohydrodynamic drying. In: *Advanced Computational Approaches for Drying in Food Processing* (ed. K.M. Gangawane, M. Dwivedi, and R.C. Pradhan), 167–188. Springer International Publishing https://doi.org/10.1007/978-3-031-62550-3_8.

Kumar, A., Jena, H.M., and Gangawane, K.M. (2025). Iron oxide nanoparticles as efficient adsorbents for malachite green dye removal: isotherms, kinetics, and thermodynamics analysis. *Environmental Science and Pollution Research* 32: 9638–9654. https://doi.org/10.1007/s11356-025-36291-w.

Leachman, J.W., Jacobsen, R.T., Lemmon, E.W., and Penoncello, S.G. (2017). *Thermodynamic Properties of Cryogenic Fluids*. Springer International Publishing https://doi.org/10.1007/978-3-319-57835-4.

Lin, H.-J., Chen, P.-C., Lin, H.-P., and Hsieh, I.-Y.L. (2025). Quantifying carbon emissions in cold chain transport: a real-world data-driven approach. *Transportation Research Part D: Transport and Environment* 142: 104679. https://doi.org/10.1016/j.trd.2025.104679.

Maiorino, A., Petruzziello, F., and Aprea, C. (2021). Refrigerated transport: state of the art, technical issues, innovations and challenges for sustainability. *Energies* 14 (21): 7237. https://doi.org/10.3390/en14217237.

Marks, A.B. (2024). Transitioning to regenerative agriculture one French fry at a time. *Nevada Law Journal* 24 (3): 8.

McCulloch, A. and Lindley, A.A. (2003). From mine to refrigeration: a life cycle inventory analysis of the production of HFC-134a. *International Journal of Refrigeration* 26 (8): 865–872. https://doi.org/10.1016/S0140-7007(03)00095-1.

Meneghetti, A. and Ceschia, S. (2020). Energy-efficient frozen food transports: the refrigerated routing problem. *International Journal of Production Research* 58 (14): 4164–4181. https://doi.org/10.1080/00207543.2019.1640407.

Mustafa, M.F.M.S., Navaranjan, N., and Demirovic, A. (2024). Food cold chain logistics and management: a review of current development and emerging trends. *Journal of Agriculture and Food Research* 18: 101343. https://doi.org/10.1016/j.jafr.2024.101343.

Neittaanmäki, P., Akimov, K., Akimova, V., and Tuovinen, R. (2024). Challenges and current solutions of refrigerated transportation. In: *Advanced Computational Methods and Design for Greener Aviation*, 241–256. Cham: Springer https://doi.org/10.1007/978-3-031-61109-4_16.

Pal, D. and Sen, S. (2023a). Emerging petroleum pollutants and their adverse effects on the environment. In: *Impact of Petroleum Waste on Environmental Pollution and its Sustainable Management Through Circular Economy*, 103–137. https://doi.org/10.1007/978-3-031-48220-5_5.

Pal, D. and Sen, S. (2023b). In-depth coverage of petroleum waste sources, characteristics, environmental impact, and sustainable remediation process. In: *Impact of Petroleum Waste on Environmental Pollution and its Sustainable Management Through Circular Economy*, 1–38. https://doi.org/10.1007/978-3-031-48220-5_1.

Pal, D. and Sen, S. (2024a). Optimal synthesis of dolochar derived faujasite zeolite X for highly effective Cd(II) removal. *Environmental Research* 240: 117494. https://doi.org/10.1016/j.envres.2023.117494.

Pal, D. and Sen, S. (2024b). Valorization of sponge iron industrial waste into iron-modified zeolite X for ciprofloxacin removal: a multi-parameter optimization study. *Environmental Science and Pollution Research* 31 (40): 53381–53398. https://doi.org/10.1007/s11356-024-34730-8.

Pal, D. and Sen, S. (2025a). Circularity and sustainable economy model towards supply chain management and zero waste. In: *Environmental Science and Engineering*, 353–383. https://doi.org/10.1007/978-3-031-81615-4_15.

Pal, D. and Sen, S. (2025b). Sustainable technologies in the development of novel processes and products from environmental hydrocarbon pollutants. In: *Environmental Science and Engineering*, 201–240. https://doi.org/10.1007/978-3-031-81615-4_10.

Pal, D. and Sen, S. (2025c). Toxicity and health impacts of emerging pollutants. In: *Interdisciplinary Biotechnological Advances*, 55–79. https://doi.org/10.1007/978-981-97-9922-0_4.

Panda, D., Kumar, A., Gangawane, K.M., and Mohamad, A.A.A. (2022). Overview of different computational approaches for heat and mass transfer in food processing. In: *Advanced*

Computational Techniques for Heat and Mass Transfer in Food Processing: Vol. M, 1–20. Boca Raton: CRC Press. https://doi.org/10.1201/9781003159520-1.

Panda, D., Kumar, A., Gangawane, K.M., and Dwivedi, M. (2024a). Overview of computational techniques for drying. In: *Advanced Computational Approaches for Drying in Food Processing* (ed. K.M. Gangawane, M. Dwivedi, and R.C. Pradhan), 25–43. Springer International Publishing https://doi.org/10.1007/978-3-031-62550-3_2.

Panda, D., Kumar, A., Sahu, A.K., and Gangawane, K.M. (2024b). Computational Techniques on Freeze Drying. In: *Advanced Computational Approaches for Drying in Food Processing* (ed. K.M. Gangawane, M. Dwivedi, and R. Chandra Pradhan), 213–230. Springer International Publishing https://doi.org/10.1007/978-3-031-62550-3_10.

Popov, D., Akterian, S., Fikiin, K., and Stankov, B. (2021). Multipurpose system for cryogenic energy storage and tri-generation in a food factory: a case study of producing frozen French fries. *Applied Sciences* 11 (17): 7882. https://doi.org/10.3390/app11177882.

Protocol, M. and Layer, O. (2022). Refrigeration, air conditioning and heat pumps technical options committee. *UN Environment Programme*.

Rai, A. (2019). Energy demand and environmental impacts of food transport refrigeration and energy reduction methods during temperature-controlled distribution. Brunel University London.

Rai, A. and Tassou, S.A. (2017a). Energy demand and environmental impacts of alternative food transport refrigeration systems. *Energy Procedia* 123: 113–120. https://doi.org/10.1016/j.egypro.2017.07.267.

Rai, A. and Tassou, S.A. (2017b). Environmental impacts of vapour compression and cryogenic transport refrigeration technologies for temperature controlled food distribution. *Energy Conversion and Management* 150: 914–923. https://doi.org/10.1016/j.enconman.2017.05.024.

Saeed, M.Z., Contiero, L., Blust, S. et al. (2023). Ultra-low-temperature refrigeration systems: a review and performance comparison of refrigerants and configurations. *Energies* 16 (21): 7274. https://doi.org/10.3390/en16217274.

Silva Elipe, M.V., Ndukwe, I.E., and Murray, J.I. (2024). Cryogen-free 400-MHz nuclear magnetic resonance spectrometer as a versatile tool for pharmaceutical process analytical technology. *Magnetic Resonance in Chemistry* 62 (7): 512–534. https://doi.org/10.1002/mrc.5434.

Stettler, M.E.J., Midgley, W.J.B., Swanson, J.J. et al. (2016). Greenhouse gas and noxious emissions from dual fuel diesel and natural gas heavy goods vehicles. *Environmental Science & Technology* 50 (4): 2018–2026. https://doi.org/10.1021/acs.est.5b04240.

Tassou, S.A., De-Lille, G., and Ge, Y.T. (2009). Food transport refrigeration – approaches to reduce energy consumption and environmental impacts of road transport. *Applied Thermal Engineering* 29 (8–9): 1467–1477. https://doi.org/10.1016/j.applthermaleng.2008.06.027.

Thermoking (2024). CryoTech refrigeration precise temperature control. [Brochure]. Thermo King. https://www.thermoking.com/brochures/cryotech.pdf

UNEP Technical Options Committee (2019). Refrigeration, air conditioning and heat pumps technical options committee. *Assessment*.

Wu, J., Liu, G., Marson, A. et al. (2022). Mitigating environmental burden of the refrigerated transportation sector: carbon footprint comparisons of commonly used refrigeration systems and alternative cold storage systems. *Journal of Cleaner Production* 372: 133514. https://doi.org/10.1016/j.jclepro.2022.133514.

Yalcin, E., Sogut, M.Z., Altuntas, O., and Karakoc, T.H. (2019). Comparative analysis of various refrigerants used in transport refrigeration based on thermodynamics and environmental performances and cold chain management. *International Journal of Global Warming* 19 (4): 407. https://doi.org/10.1504/IJGW.2019.104262.

Yenare, R.R., Sonawane, C.R., Sur, A. et al. (2024). A comprehensive review of portable cold storage: technologies, applications, and future trends. *Alexandria Engineering Journal* 94: 23–33. https://doi.org/10.1016/j.aej.2024.03.014.

14

Advancements and Future Avenues in Cryo-Processing Technologies in Food

Keerthivasan Ragupathi[1], *Sankar Manokaran*[2], *and Krishnakumar Panneerselvam*[2]

[1] *Department of Floriculture and Landscape Architecture, Horticultural College and Research Institute, Tamil Nadu Agricultural University, Coimbatore, Tamil Nadu, India*
[2] *Department of Food Process Engineering, National Institute of Food Technology, Entrepreneurship & Management, Thanjavur, Tamil Nadu, India*

14.1 Introduction

Cryogenics is a field of engineering that focuses on the generation and utilization of low temperatures, ranging from absolute zero (−273.15 °C) to −150 °C (Goswami 2017). Cryo-processing and preservation are the most advanced technologies used in the food sector to extend shelf life, preserve nutrients, retain flavour and aroma, and provide a customer-friendly product (Jha et al. 2024). The Greek words 'Kryos' (frost) and 'Genic' (create) are the basis of the word 'cryogenics'. The field of cryogenic research most likely began with gas liquefaction (Balasubramanian et al. 2012).

The product's surface may be subjected to a liquid N_2 spray, a solid ('snow') and gaseous CO_2 mixture, or a direct immersion in the liquid cryogen during the cryogenic processing procedure. The cryogenic material is typically delivered to the food manufacturing facility as a high-pressure liquid, making handling the refrigerant far easier than with mechanical refrigeration systems. After that, it is stored in a tank that requires chilling or insulation. A pipe assembly is used for spraying with a nozzle and valve. For both cryogens, the equipment is the same, but the design parameters are different (Estrada-Flores 2012). Cryogenic applications have revolutionized the food sector by enabling low-temperature uses in food processing, rapid freezing, and preservation. Cryogenic grinding is a practical technique used in the processing of spices.

14.1.1 Significance of Cryogenic Technology

When compared to traditional methods, cryogenic freezing significantly reduces the time required to freeze food. This quick process ensures that the food's texture, flavour, colour, and nutritional content are better retained after thawing by preventing the formation of

Cryogenic Processing: Advances in the Food Industry, First Edition. Edited by Sumit Sudhir Pathak, Rama Chandra Pradhan, and Sudarshan Ramanathan.

large frozen crystals, which can damage the cellular structure (Brown 1967; Pravarthika and Karthickumar 2025). Delivering foods that taste and appear more like their fresh counterparts is made feasible by the process, which preserves the food's organoleptic (sensory) attributes and freshness.

Cryogenic freezing significantly extends the shelf life of sensitive goods and enhances food safety by rapidly reducing temperatures and creating an environment inhospitable to microbial growth. By simply adjusting the freezing conditions, the technique can be applied to a wide range of food products, including prepared meals, meats, shellfish, and baked goods. Larger yields, reduced product loss, lower maintenance, and a smaller system size result in operating savings, even though the original investment in cryogenic technology may be higher. Compared to mechanical refrigeration devices that employ hydrofluorocarbons, the technique is more sustainable due to its efficiency and use of naturally occurring gases (Mudgulkar 2024; Volokitina et al. 2024).

14.1.2 Overview of Recent Trends and Global Outlook

Cryogenic freezing is being revolutionized by the incorporation of innovative technologies, including dynamic airflow management, predictive analytics, and real-time monitoring (Radu et al. 2025). By identifying inefficiencies and anticipating maintenance requirements, these systems minimize downtime, improve product consistency, and optimize energy use. Automation, such as sophisticated Clean-in-Place (CIP) systems, is accelerating cleaning cycles, reducing water usage, and improving hygiene (Brown 1967). Hybrid systems that combine ammonia for high-capacity cooling and CO_2 for low-temperature freezing are gaining popularity, supporting both sustainability and performance objectives. The method prolongs shelf life without the use of chemical preservatives, supporting the clean-label movement. North America and Europe are at the forefront of premium frozen food products, automation adoption, and environmental measures (Banerjee 2021).

Urbanization, growing middle-class consumption, and the demand for prepared foods, seafood, and baked goods are driving the rapid market expansion in the Asia-Pacific and Latin America regions. In India, due to consumer and corporate demand for cutting-edge freezing solutions, the cryogenic freezers market is anticipated to expand at a compound annual growth rate (CAGR) of more than 4%, reaching around $30 million by 2033 (Kypraiou and Varzakas 2025).

14.2 Cryogens

The liquids used in the cryogenic process are referred to as 'cryogens'. Cryogens are liquids of nitrogen (N_2), oxygen (O_2), helium (He), and carbon dioxide (CO_2). Depending on their characteristics, needs, and system viability, these fluids are employed in various fields. Along with the liquefaction and separation of air gases, liquid nitrogen (LN_2) is one of the largest and fastest-growing industries in the field of cryogenic applications. The primary motivation behind cryogenics is the densification (condensation) and separation (distillation) of gases. Cryogenics utilizes a substance's latent heat of vaporization to achieve low temperatures by submerging it in a cryogenic bath (Balasubramanian et al. 2012).

14.2.1 Liquid Nitrogen

Nitrogen in its liquid state, following exposure to low temperatures (below its boiling point, which is −196 °C or −321 °F), is known as cryogenic liquid nitrogen (LN_2). Since 78% of air is nitrogen, it is a colourless liquid that is created by fractional distillation of liquid air. Due to its unique properties, including its ability to maintain freezing temperatures, cryogenic liquid nitrogen offers a wide range of applications once it is generated. Specialized Dewar containers, which are insulated vessels designed to reduce heat transfer and sustain extremely low temperatures, are required to store LN_2 (Kim et al. 2011).

Chemical inertness, a high volumetric expansion ratio linked to the liquid-to-vapour phase change, and a high enthalpy of vaporization are the primary characteristics of LN_2 that make it a desirable choice in the food and beverage industry (Bonga 2024). Because nitrogen is non-corrosive and chemically inert, it can be used in food without negatively impacting its organoleptic qualities. Notably, LN_2's superior thermophysical qualities allow it to be used as a refrigerant. Due to its high-volume expansion ratio, LN_2 can also be utilized for specific inert packaging applications, such as LN_2 dosing. When LN_2 is exposed to higher temperatures, it becomes volatile. The energy density of LN_2 is high, allowing it to absorb a large amount of heat during evaporation (Kim et al. 2011).

When a liquid cryogen is introduced into a target medium, it changes phases and becomes saturated vapour. The vapour phase then warms up to room temperature. The sensible heat difference, linked to the temperature rise of the vapours and the latent heat of vaporization for the liquid-to-saturated vapour phase transition, determines a liquid cryogen's capacity to extract heat from the target medium (Bonga 2024).

14.2.2 Carbon Dioxide (CO_2)

Compared to LN_2, liquid CO_2 sublimates at −78.5 °C (−109.3 °F) and has a milder cryogenic impact. Although it is not as quick as LN_2, CO_2 freezing is still significantly faster than traditional mechanical techniques. Snow horns or tunnels, which create and apply CO_2 snow to food products, are frequently used in CO_2 systems. In general, CO_2 is considered less expensive than LN_2 and poses fewer safety risks. It works effectively with a variety of food items, especially those that are less susceptible to changes in freezing rates. Additionally, because it can be captured from industrial operations, CO_2 is frequently regarded as a more sustainable choice; nonetheless, life-cycle assessments are essential for a comprehensive sustainability practice (Bagwan et al. 2023).

14.3 Advanced Cryogenic Processing in the Food Sector

14.3.1 Freezing and Cooling

Preservation is one of the important phenomena when it comes to storage and shelf life. The shelf life of many animal and plant-based food products is extended by low-temperature (cooling and freezing) techniques. Because low temperatures do not cause thermal damage to products, those treated using low-temperature techniques are of higher quality than those produced using traditional thermal techniques. By decreasing the product's water activity, it delays cellular metabolic responses, lowers microbiological activity, and decreases enzymatic reactions, all of which contribute to a longer shelf life (Goswami 2017; Alhamdan et al. 2018). Furthermore, since the development of low-temperature processing techniques, the need for chemical preservatives in foods has significantly decreased. Air-blast, plate, contact, fluidized, immersion, and cryogenic cooling are some of the cooling techniques (Bashir et al. 2025).

Frozen foods and chilled products are getting popular nowadays. Due to the changing lifestyle, diet plans, and easy-to-use products, they are attracting consumers rapidly (Çalişkan Koç et al. 2025). In the food sector, the frozen food category, which encompasses quick-frozen meat, fruits, vegetables, flour, and rice products, is experiencing rapid expansion. However, there are serious safety hazards associated with their manufacture and storage. The freezing rate significantly affects the quality of food, with the majority of published data indicating that rapid cooling is the most effective method for improving quality preservation in cellular food systems (Truonghuynh et al. 2020).

Traditionally, freezing injuries are reduced by promoting a higher freezing rate or by using pretreatment methods prior to the initiation of freezing. Modern freezing techniques, which include the use of physical fields and high-pressure processing, have proven effective in producing high-quality frozen foods. Cryogenic fluids enable ultra-rapid freezing, which results in high-speed freezing rates. By producing a large number of tiny ice crystals in the intracellular and intercellular spaces, this process reduces the negative consequences of dehydration and shrinkage-induced damage. As a result, the overall quality of frozen foods is successfully maintained (Jha et al. 2024).

14.3.1.1 Cryogenic Freezing

Cryogenic freezing beats conventional techniques in maintaining the quality of food items both during and after freezing (Table 14.1). Numerous factors, including the preservation of the original microstructure, reduced exudate loss, enhanced texture retention, decreased dehydration losses, reduced lipid oxidation, and minimal colour change,

Table 14.1 Comparison of cryogenic vs. conventional freezing techniques.

Parameter	Cryogenic freezing	Conventional freezing
Cooling rate	Very high	Moderate to low
Ice crystal formation	Small, uniform, intracellular	Large, extracellular
Product quality (texture, flavour)	Superior	Often compromised
Nutrient retention	High	Moderate
Processing time	Short (seconds to minutes)	Long (hours)
Equipment cost	High	Moderate
Operating cost	High (due to cryogen consumption)	Lower
Common applications	Premium meat, berries, seafood	General frozen goods

demonstrate that this is one of the highest-quality preservation methods (Truonghuynh et al. 2020). The outer layer of ice crust grows on the product's surface when it is exposed to cryogens. In addition to increasing the product's mechanical strength, this crust serves as a moisture-retaining barrier. This crust's development could help to preserve and encapsulate the food's flavour, providing an extra benefit in terms of flavour retention (Jha et al. 2024). For freezing products, different types of freezers are available, based on the properties of the food being subjected to freezing.

14.3.1.2 Cryogenic Immersion Freezers

It utilizes a liquid refrigerant as a heat transfer medium. Because liquids have a high thermal coefficient, food comes into direct or indirect contact with the refrigerant during a quick cooling or freezing process. It has frequently been demonstrated in these tests that the substantial freezing rate of immersion freeze treatment maintains the original food quality. Protein structural stability is maintained, and cell damage is reduced in immersion freeze-treated samples due to the small size of the ice crystals, which significantly improves product quality. Food preservation is favoured by the increased quantity of tiny ice crystals produced by immersion freezing. Additionally, the preservation of protein properties was enhanced by the immersion freeze technique (Ikram et al. 2025).

14.3.2 Fruits and Vegetables

Fruits offer higher nutritional benefits, but at the same time, they are highly perishable, resulting in shorter shelf life. Additionally, fruits are seasonal; to maintain year-round availability, proper preservation techniques are necessary. The final quality of frozen meals depends on the size and structure of the ice crystals, with the rate of heat removal being a primary parameter that influences the rate of crystal development. Rapid freezing encourages intense nucleation and the development of intracellular small ice crystals, whereas slow freezing facilitates the formation of large extracellular ice crystals that harm vegetative tissues. The rapid freezing of berries and other food items using liquid nitrogen spray has gained interest recently (Dalmau et al. 2024).

During the frozen storage period, measurements were made of the colour values, textural characteristics (hardness, elasticity, chewiness, and resilience), and nutritional characteristics (enzymes and sugars) of the fresh and frozen dates. The freezing process and the frozen storage time had an impact on the colour value of the frozen dates. The quality of frozen fruits produced by cryogenic freezing differs significantly from that of those produced using conventional slow freezing. The findings revealed a significant difference in the freezing times between the two approaches. In the cryogenic freezing approach, the freezing period was 10 minutes, whereas in the standard slow-freezing system, it was 30 hours (Alhamdan et al. 2018). By presenting a novel approach to reducing fruit cracking during freezing without the need for packaging, this technology offers fresh perspectives on the field of food freezing technology. The results also suggest potential benefits for the food processing sector, including reduced nitrogen consumption, shorter freezing times, and lower operating expenses (Dalmau et al. 2024).

14.3.3 Meat and Seafood

Due to microbial deterioration during storage, high-nutrient meat products degrade rapidly. Low temperatures are necessary to preserve the meat's quality. Freezing is a common technique for preserving meat and meat products (Ikram et al. 2025). Meat that is frozen and then thawed can undergo significant changes, affecting the nutritional profile, texture, colour, flavour, and microbiological stability of the end product (Pesce et al. 2025).

Seafood quality can be well preserved during frozen storage by pretreatment with liquid nitrogen freezing (LNF). A study investigated the effect of LNF temperatures on the meat quality of gazami crabs (*Portunus trituberculatus*) to improve their flesh quality during frozen storage. The findings demonstrated that the LNF temperature of −100 °C, which is lower than the ideal LNF temperature for shell-free aquatic items like fish, more successfully prevented the oxidative deterioration of crab meat during storage, increasing the shelf life of frozen gazami crab to four months. By offering recommendations for determining the optimal LNF temperature for aquatic products, this study enhances the application parameters of LNF in the freezing sector of crustacean aquatic products (Ren et al. 2025). When compared to conventional procedures, Atlantic salmon treated with LN_2 supercooling showed better gaping scores, a lower incidence of blood spots, and increased firmness after smoking (Chan et al. 2020). LNF decreased thawing loss and water migration in crucian carp and catfish, assisting in maintaining the fillet's succulence by retaining its water-holding capacity (Jin et al. 2025).

14.3.3.1 Changes to the Physical Characteristics of Frozen Meat Products

- **Microstructure:** Monitoring the development of ice crystals throughout the freezing process is essential. The time allotted for crystal formation determines how ice crystals develop.
- **Colour:** It is the primary consideration in consumer preference. During freezing and after thawing, the colour of the meat will change.
- **Tenderness:** Shear force, a sensory characteristic used to evaluate the quality of raw meat, is frequently used to objectively measure tenderness. Meat becomes less tender as the shear force rises. The enzymatic breakdown of muscle fibres and the tenderization of beef tissue were caused by protein hydrolysis and the aging process.
- **Drip Loss and Water Holding Capacity:** Drip loss, snow recrystallization, sublimation, and the fixing and reorganization of dampness can all result in moisture damage. Food's weight, surface, juiciness, and appearance are all impacted by moisture migration.

14.4 Cryogenic Grinding

India is the world's top producer and exporter of spices. Ground powders of various spices, either alone or in the form of spice mixes, have demonstrated enormous potential for commercial exploitation as value-added products (Saxena et al. 2018a). Although there are other ways to add value to spices, ground powders are the most widely used and significant. In addition to adding taste and flavour, powdered spices of different types provide a preservative benefit to prepared meals.

Grinding is a unit operation used in the food industry to reduce the size of materials and produce powders. For the grinding process, different types of grinding mills are available. Each mill has a specific application. Since ancient times, spices have been valued for their distinct aroma and unique flavour. Spice grinding has long been accomplished with various grinding tools tailored to the specific situation. The demand for high-quality processed spices has increased in recent years. Thermal damage is one of the primary drawbacks of the traditional grinding method. Grinding under regulated temperature conditions is one effective way to minimize thermal damage. Therefore, it is particularly crucial to grind at lower temperatures. The basic idea of grinding remains the same in both ordinary and cryogenic settings.

The intrinsic quality of the spices is determined by their moisture, volatile oil, and fixed oil content. A premium spice powder with a distinctive flavour and high essential oil content is always preferred by consumers in the culinary sector. Consequently, it is preferable to grind in a way that preserves the most essential volatile oils while minimizing quality loss. A well-known method for producing high-quality spice grinding is cryogenic grinding, which involves low temperatures (below −153 °C). By absorbing the heat produced during the grinding process, the LN_2 in this method provides the refrigeration required to pre-cool the spices and maintain the appropriate low temperature (Saxena et al. 2018b).

For materials that contain large amounts of volatile oils and fats that are readily oxidized, changes in the material's quality during grinding are incredibly complicated. There are two reasons to use cryogen, such as liquid nitrogen (LN_2). It lowers the grinding temperature and creates an atmosphere that prevents the material from undergoing oxidative changes. In the food industry, cryogenics plays a crucial role. Among all the cryogens available, liquid nitrogen is the most recommended for achieving the required low temperature in food grinding. Superior grade ground powder with improved hygiene was generated by grinding spices at a very low temperature (Saxena et al. 2018a).

14.4.1 Effects of Cryogenic Grinding on Spices

The impact of cryogenic grinding on cumin was assessed at different temperatures (230 to −108 °C and ambient), feed rates (5–7 kg/h), and sieve sizes (0.8–1.5 mm) on mean particle size, grinding duration, and volatile oil content. Cryogenic grinding preserved 62.56% more volatile oil compared to grinding at ambient temperature. Based on the retention of volatile oil, the highest-quality cumin powder was achieved at a grinding temperature of 230 °C, with a feed rate of 7 kg/h and a sieve size of 0.8 mm. Storage tests on cryogenically ground cumin powder indicated that refrigeration provided benefits over ambient storage regarding freshness and volatile oil retention over extended periods (Saxena et al. 2018b).

Cryogenic grinding can increase the velocity and yield of essential oils (Yan et al. 2025). In comparison to a simple grinding pretreatment technique, cryogenic grinding successfully increased the variety of compounds in nutmeg essential oils (NEO), thereby improving the quality of the essential oil, as determined by gas chromatography-mass spectrometry (GC-MS) analysis conducted to assess the composition of the extracted essential oils. Cryogenic grinding in conjunction with ultrasonic-assisted pretreatment and hydrodistillation (CG-UAP-HD) is a successful extraction technique that improves the quantity and

Table 14.2 Summary of cryogenic grinding benefits on spices.

Spice	Key benefits of cryogenic grinding	References
Cumin	+62.56% volatile oil retention, better shelf life	Saxena et al. (2018b)
Nutmeg	Enhanced essential oil diversity (via GC-MS)	Yan et al. (2025)
Coriander	Higher antioxidant content and oleoresin yield	Saxena et al. (2015)
Pepper	Superior aroma retention vs. hammer milling	Liu et al. (2013)

quality of essential oils, supporting research in the field of nutmeg essential oil extraction with theoretical data.

The colour, taste, and sensory qualities of the spices were barely impacted by cryogenic grinding (Table 14.2) (Liu et al. 2013). Although the concentrations of the primary scent ingredients were significantly decreased after holding the samples at 4 °C for six months, cryogenic grinding was still superior to hammer milling in terms of maintaining the major potent aroma compounds. Cryogenic grinding outperformed hammer milling in maintaining the pepper's taste and sensory qualities without appreciably lowering its quality.

Cryogenic grinding has outperformed ambient grinding in post-harvest processing, producing higher-quality coriander powder (Hiremath et al. 2025). An analysis has been conducted to investigate the impact of cryogenic grinding on the volatile oil, oleoresin content, total phenolics, flavonoid content, and antioxidant properties of seed extracts from nine genotypes of coriander (*Coriandrum sativum* L.). The amount of volatile oil and oleoresin in cryogenically ground samples was considerably higher. All genotypes' cryoground samples consistently produced large yields of methanol crude seed extract, and all genotypes also had high levels of total phenolic content and total flavonoid content. The total antioxidant and 2,2-diphenyl-1-picrylhydrazyl (DPPH) scavenging percentages are also higher in cryogrind samples of all genotypes (Sharma et al. 2016).

The impact of cryogenic grinding on the recovery of fatty oil and volatile oil, as well as their contents in two cumin genotypes (*Cuminum cyminum* L.), has been examined. In addition to preserving the volatiles, cryogenic grinding increased recovery by 33.9% in GC 4% and 43.5% in RZ 209. Additionally, genotype RZ 209 showed a notable increase (29.9%) in oil percentage above typical grinding. The essential oil of both genotypes contained 19 main components. The primary component in both genotypes was cumin aldehyde, whose quantity rose from 48.2% to 56.1% in GC 4 upon cryogrinding (Sharma et al. 2016).

14.5 Advantages of Cryogenic Technology

There are potential commercial applications for cryogenic grinding of spices and herbs, as it preserves volatiles without compromising flavour quality or uniqueness. Another significant application of this technique for grinding herbs in Ayurveda and other traditional medical systems is made possible by the improvement of its pharmacological properties. Cryogenic grinding can produce particles up to 50 μm in size, thereby increasing their efficacy in therapeutic or medical applications. It is possible to conclude that, regardless of genotype, cryogenic grinding technology preserves the flavour and antioxidant qualities of spices. It can also increase the yield and recovery (Saxena et al. 2015).

It can increase the shelf life of the products by reducing the use of chemical preservatives. When compared to samples frozen using an air-blast freezer, the overall quality of cryogenic immersion-frozen products was superior (Qian et al. 2018). Cryogenics helps create a more sustainable food supply chain by lowering food waste through improved preservation and shelf life. Significantly lowering the risk of foodborne illness, cryogenics quickly stops microbial activity (Pravarthika and Karthickumar 2025).

14.6 Challenges of Cryogenic Processing

Despite being a proven technology, its high initial cost has made it unpopular in developing nations (Pravarthika and Karthickumar 2025). Cryogenic freezers and other equipment needed for cryogenic processing are not always accessible. Cryogenic system operation requires highly qualified staff and strict adherence to safety regulations. The creation of cryogenic gases may have an impact on the environment, even though cryogenics reduces waste.

14.7 Integration of Artificial Intelligence in Cryogenic Processing

The development of the frost layer on cryogenic surfaces is a significant subject, and studies on heat and mass transport may benefit from a precise forecast of this growth. It is beneficial and worthwhile to employ artificial intelligence (AI) approaches for estimation since they are affordable, dependable, and easy to apply. The frost layer thickness, δ, on vertical cryogenic surfaces was estimated using four models: multiple linear regression (MLR), least squares support vector machine (LSSVM), artificial neural network (ANN), and adaptive neuro fuzzy inference system (ANFIS) (Zendehboudi et al. 2017).

The freezing procedure has a significant impact on the quality of both frozen and thawed meat. Meat that is frozen at will may lose some of its nutritional value and suffer from a significant decline in sensory qualities (Qiao et al. 2024). AI is anticipated to support the advancement of freezing technology and oversee the freezing procedure to preserve the nutritional value and quality of meat products as much as possible. The method was developed to track the oxidative damage to pork's lipids and proteins during the freeze–thaw cycle. This approach features end-to-end multitask analysis, combining convolutional neural network (CNN) and hyperspectral imaging, which provides the advantage of quick freezing. The automation also facilitates a simple optimization process.

14.8 Hybrid Cryogenic Systems

To improve the quality of meat, radiofrequency-assisted nitrogen cryogenic freezing (RF-CCF) has been suggested. Both pulsed cryogenic freezing (PCF, 3 seconds pulse with a 10 seconds interval for 10 minutes) and continuous cryogenic freezing (CCF, for 2.5 minutes) freezing methods were employed. RF-CCF (10 seconds 2 kV RF pulse with 20 seconds interval and 2.5 minutes of continuous N_2 flow) and RF-PCF (30 seconds 2 kV RF pulse with 60 seconds intervals, with 3 seconds nitrogen pulse with 10 seconds interval for 10 minutes) were subsequently created by combining CCF and PCF with radiofrequency (RF) pulses. These procedures have been compared with blast freezing (BF) and slow freezing (SF) (Manzocco et al. 2022).

It is challenging to effectively break down the plant cell wall using simple grinding as a pretreatment, which slows the release of essential oils. Additionally, the machine's localized high temperature will cause the loss of heat-sensitive chemicals. The nutmeg was pretreated using cryogenic grinding and ultrasonication, and the addition of sodium chloride (NaCl) was intended to minimize the loss of essential oil during the extraction and separation procedures. The essential oil from nutmeg was then extracted using hydrodistillation. Following a methodical optimization of the extraction conditions, the yields for cryogenic grinding and simple grinding of nutmeg were 7.0409% and 5.7084%, respectively (Yan et al. 2025).

Membrane technology and membrane-based hybrid technology are strongly driven by the high CO_2 concentration and high pressure of the feed used in pre-combustion carbon capture (Ni et al. 2025). The membrane–cryogenic hybrid method offers a novel approach, complementing conventional separation techniques, as it can economically achieve high CO_2 purity and recovery.

14.9 Quantum Cryogenics in Food: A Theoretical Approach

At the most advanced field of study, known as quantum cryogenics, ultra-low temperatures made possible by cryogenic methods are integrated with the concepts of quantum mechanics. It focuses on utilizing the unique behaviours of quantum systems at extremely low temperatures to investigate fundamental scientific processes and develop new technologies (Braga et al. 2021).

14.9.1 Benefits of Quantum Cryogenics

The delicate quantum states required for quantum computing, which hold the potential to transform computation, cryptography, and optimization issues, must be created and maintained at extremely low temperatures. Cryogenic temperatures enable quantum sensors and detectors to achieve lower noise levels and higher sensitivity. Applications such as magnetic resonance imaging (MRI), gravitational wave detection, and precise measurements depend on this. The precise control of quantum systems for simulating intricate physical processes is made possible by quantum cryogenics, which has applications in areas such as drug development and materials science.

Cryogenics is essential to the development of quantum computing because it maintains the stability of qubits, which is necessary for preserving their quantum states at extremely low temperatures. A key component of many quantum systems, superconducting qubits require sophisticated cooling techniques, as they function optimally at temperatures of a few millikelvins. Cryogenics helps to optimize other aspects of quantum systems in addition to cooling. Readout instruments and control circuits can function more efficiently in cryogenic environments, guaranteeing that all system components work together harmoniously in low-temperature settings (Braga et al. 2021).

Advanced refrigeration methods, such as dilution refrigerators, which use a sequence of cooling stages employing various helium isotopes, are necessary to maintain ultra-low temperatures. Improvements in refrigeration and materials science will have a significant impact on the advancement of cryogenics in quantum technologies. By providing more effective cooling solutions and lowering operating costs, innovations like magnetic refrigeration and new superconducting materials have the potential to transform the industry completely.

14.10 Precision Cryogenic Processing

In the food industry, precision cryogenic processing primarily refers to the rapid freezing or chilling of food products using extremely low temperatures, which are achieved with the use of liquid nitrogen or carbon dioxide. Compared to traditional mechanical freezing, this technique is advanced and superior due to its speed, effectiveness, and ability to maintain food quality. To ensure uniform and consistent freezing across batches, modern cryogenic systems allow for precise temperature and exposure duration adjustments. When food is exposed to extremely low temperatures, as low as −196 °C, with LN_2, the water in the food nearly immediately freezes. Very tiny ice crystals are formed as a result of the rapid freezing, protecting cells and preserving the food's texture, flavour, and nutritional content (Pravarthika and Karthickumar 2025).

14.11 Advances in Cryogenic Equipment and Material Science

14.11.1 Machining Technology

Cryogenics has become a cutting-edge technique in machining technology, enabling the lubrication and cooling of cutting tools during the machining process. Cryogenic machining can enhance cutting performance, decrease tool wear, lengthen tool life, and improve surface smoothness, accuracy, and precision by cooling the cutting tool using LN_2 or other cryogenic gases. Furthermore, cryogenic machining can reduce the likelihood of thermal distortion and damage, particularly when machining challenging materials. This increases the tool's wear resistance and hardness, increasing its longevity and effectiveness. It has been discovered that cryogenically treated tools can last up to four times longer than untreated ones, which lowers the frequency of tool replacement and boosts productivity (Christensen et al. 2025).

To achieve a lower total cutting time than traditional cutting procedures, the Cryogenic Cooling and Cutting System (CCCS) utilizes cryogenic cooling to decrease its impact energy absorption capabilities substantially. The effectiveness of CCCS is evaluated and compared with that of the abrasive water jet (AWJ) technology using numerical analysis. The results show that CCCS can cut at speeds up to 46.8 times faster and reduce cutting times by 87.1–97.9% across a range of monopile diameters and wall thicknesses. The purpose of this study is to demonstrate the distinct advantages of the CCCS technology over traditional approaches, such as AWJ, by significantly reducing cutting times and adverse environmental impacts, and enhancing the sustainability and economic viability of offshore wind farm decommissioning (Anand Shukla et al. 2023).

14.11.2 Storage and Transportation

There is a growing need for materials that can withstand extremely low temperatures and cryogenic conditions in various industries, such as energy storage, food processing and preservation, gas pipelines, and hydrogen energy storage tanks (Mekonnin and Waclawiak 2025). In cryogenic application domains, materials that can safely operate at extremely low temperatures are essential to providing the intended function. Additional requirements are placed on the design of storage tanks when transporting and storing hydrogen as a liquefied gas. Liquid hydrogen (LH_2) is particularly favoured due to its high storage density and effective delivery. To ensure that these containers are compatible with LH_2 at such low temperatures, resistant to hydrogen embrittlement, and possess the best mechanical and thermophysical properties, the materials used must be thoroughly examined. Although stainless steel is frequently used to make LH_2 storage tanks, hydrogen embrittlement is an issue (Qiu et al. 2023). The ability of different grades of stainless steel to withstand this problem has been evaluated. Selecting materials for these exceptionally low temperatures remains a significant challenge. Materials must retain their intended qualities at low temperatures without degrading.

The mechanical and physical characteristics of materials, including fatigue, yield strength, and tensile strength, are improved by exposure to cryogenic temperatures. In addition to improving durability and stabilizing strength properties, cryogenic temperatures enhance resistance to corrosion, erosion, wear, and abrasion. Stronger and more resilient materials are produced as a result of this process, which also stabilizes and improves the crystal lattice structure and guarantees more even distribution of carbon particles throughout the material (Mekonnin, and Waclawiak 2025).

The cryogenic storage and transportation container for LH_2 is a crucial component of equipment in the space and energy sectors, as it facilitates the large-scale storage and transportation of LH_2 (Qiu et al. 2021). The scientific application of icy materials enables the design and production of vessels for storing and transporting LH_2. A systematic and comprehensive understanding of the fundamental features of cryogenic materials for this type of container is a crucial assurance for the safe, dependable, and long-term operation of cryogenic propellant storage and transportation vessels. Austenite stainless steel, the most popular material for this container, performs well at low temperatures, making it the first material of choice for low-temperature operating circumstances. The face-centred cubic structure of austenitic stainless steel offers excellent stability at low temperatures. Its toughness and flexibility will not drastically diminish when the temperature drops.

14.11.3 Material Science

The study of metals' characteristics at low temperatures has become a significant area of interest in materials research in recent years. The scientific study of materials at extremely low temperatures, typically below 150 °C, where temperature-dependent processes and material composition lead to notable changes in their mechanical and physical properties, is known as cryogenics (Mekonnin and Waclawiak 2025). The microstructure of a material can alter at cryogenic temperatures, improving qualities such as durability, hardness, and tensile strength. Similarly, some polymers are suitable for cryogenic applications because they retain their toughness and flexibility at low temperatures. Furthermore, materials can be processed at extremely low temperatures via cryogenic treatment, such as deep cryogenic treatment, to improve the performance characteristics of conventional alloys.

In cryogenic settings, conventional metals and alloys – particularly stainless steels – display exceptional tensile properties, including resistance to oxidation, corrosion, and wear (Fujun Cao et al. 2023). They are ideal for cryogenic applications due to their specific qualities. The most crucial metric for evaluating cryogenic materials is cryogenic toughness. The following test methods are closely associated with containers used for the storage and transportation of LH_2: drop weight, full thickness, double tensile, and wide plate tests; fracture mechanics tests; and low-temperature impact toughness tests (V-notch, German Society for Testing and Materials [DVM] sample).

14.12 Start-ups, Collaborations, and Commercial Innovations

Worldwide, different start-ups are working in the area of cryogenic technology. Some of the top cryogenic engineering start-ups are given below in Table 14.3 while commercial innovations and their impacts are enlisted in Table 14.4.

Table 14.3 Top cryogenic engineering start-ups.

Company/collaboration	Innovation/focus area
P3 Technologies	High-speed cryogenic pumps/turbopumps
Intelline	Scalable, maintenance-free cryogenic coolers
Cryogenx	Portable rapid body cooling device (CGX1)
Kryotech GmbH	Ultra-low temp freezers, cryosurgery without LN_2
Air Liquide & CEA	Joint R&D in cryogenic systems, pulsed gas tubes
MSU Cryogenic Initiative	Education & research in cryogenic engineering

Table 14.4 Commercial innovation area and its impact.

Innovation area	Example/impact
Solar-powered cryogenics	Off-grid, sustainable cooling to −196 °C
Solid-state cooling	No moving parts, photonic metamaterials
AI-driven optimization	27% energy savings, predictive maintenance
Miniaturized cryocoolers	Smartphone-sized, −196 °C for portable applications
Cryogenic isotope separation	40% efficiency gains, modular reactors, medical isotopes
LNG/hydrogen equipment	Specialized joints, arms for energy transition
Hybrid membrane-cryogenic plants	Lower energy, smaller footprint

To spur scientific advancement in cryogenics, these long-time collaborators have established a collaborative laboratory. Their main goals are to improve simulation models for massive cooling systems in fusion reactors, optimize the efficiency of pulsed gas tubes for telescopic sensors, and better integrate cryogenic systems into large-scale projects. This partnership promotes developments in nuclear fusion, particle accelerator architecture, and space exploration. The Michigan State University Cryogenic Initiative, a partnership between the Facility for Rare Isotope Beams (FRIB) and MSU's College of Engineering, prepares students for careers in cutting-edge industries, such as gas liquefaction, aerospace, and quantum computing, by combining classroom instruction with practical cryogenic engineering research. The Cryogenic Engineering Conference and International Cryogenic Materials Conference (CEC/ICMC) serves as a global platform for sharing the latest research and commercial developments, fostering industry–academic partnerships, and showcasing state-of-the-art applications in hydrogen, superconductivity, medical, and space technologies.

14.13 Future Goals of Cryogenics

Advanced cryogenic technology is becoming increasingly necessary due to the growing demand for liquefied natural gas (LNG), which is driving the development of new infrastructure. Cryogenic technology is viewed as an effective method for capturing and storing carbon emissions as companies strive for decarbonization. Automation and AI, combined with cryogenic systems, will enhance operational effectiveness, reduce costs, and increase dependability. The demand for cryogenic systems is expected to increase in various industrial processes as sectors such as electronics, chemicals, and pharmaceuticals continue to expand. Creating our goal of a carbon-neutral society requires the use of CO_2 collection devices. The most developed post-combustion CO_2 capture methods to date, amine-based technologies, have primarily been used in large-scale CO_2 collection applications. Nevertheless, it has also been demonstrated that the cryogenic technique is a promising CO_2 collection technology appropriate for large-scale uses. From a system viewpoint, it has been found that using cryogenic technologies to collect 90% of CO_2 emissions by 2045 can lower the need for wind power by approximately 47%, resulting in a nearly 45% reduction in annual system costs (Asgharian et al. 2024).

14.14 Technological Gap

To make the technology more efficient, some implementation and research are needed. In cryogenic processing technology, the implementation of advanced sensors and IoT-based systems is necessary to enhance the technology's precision and efficiency (Qiao et al. 2024). For the production of cryogen, solar-based, eco-friendly production systems are important. Regulatory systems are important for managing the system and coordinating with global standards. For the efficient use of cryogenic technology, research into green cryogen, nano-based cryogenic applications, hybrid applications of cryogenic technology, and automated and robotic-controlled processing lines is considered.

14.15 Vision 2030: Strategic Roadmap for Cryogenic Food Processing

Revenue from the cryogenic technology market was assessed at USD 12.5 billion in 2024 and is projected to expand at a CAGR of 6.5% from 2026 to 2033, reaching USD 20.3 billion (Ghorbani et al. 2025). By 2030, cryogenic food processing is expected to undergo a revolutionary expansion due to advancements in technology, the growing need for sustainability, and shifting customer preferences (Chand 2022).

A strategy roadmap with important priorities and concrete actions for stakeholders is provided below:

1) Utilize AI algorithms to optimize freezing settings (temperature, LN_2 flow) in real time, enhancing product uniformity and reducing energy consumption by 20–30%. For high-end snacks, medications, and probiotics, combine cryogenic pelletizing and freeze-drying, using sublimation to preserve flavour and nutritional characteristics.
2) To cut energy use over traditional procedures, combine LN_2 freezing with microwave drying or pulsed electric fields.
3) In large plants, use closed-loop recovery equipment (like Demaco Phase Separators) to cut LN_2 waste by 90%.
4) In order to meet worldwide phasedown requirements, replace 30% of HFC-dependent equipment with transcritical CO_2 or ammonia refrigerators by 2028.
5) Aim for 15% yearly growth in the high-protein alternative meat and freeze-dried probiotics markets by optimizing cryogenic grinding.
6) Collaborate with International Organization for Standardization (ISO)/WHO to standardize cryogenic handling procedures, which will result in a 25% reduction in compliance expenses for multinational operators.
7) Connect blockchain traceability to Internet of Things (IoT)-enabled cryogenic tanks for real-time supply chain quality monitoring.

14.16 Conclusion

Cryogenic food freezing is a technique that preserves food at its best, increasing its availability and reducing waste. This technique, grounded in sound science, offers tangible benefits for both food manufacturers and consumers. In food processing, cryogenic technology is a revolutionary development. Its capacity to quickly and consistently freeze or cool food maintains its nutritional content, safety, and quality, satisfying the exacting requirements of both customers and the modern food business. Cryogenic systems will become increasingly important as the industry innovates to ensure effective, sustainable, and superior food production globally.

References

Alhamdan, A., Hassan, B., Alkahtani, H. et al. (2018). Cryogenic freezing of fresh date fruits for quality preservation during frozen storage. *Journal of the Saudi Society of Agricultural Sciences* 17: 9–16.

Anand Shukla, O., Rawat, R., and Vidya, S. (2023). Transforming machining technology with cryogenics: advancements and innovations. In: *International Conference on Scientific and Technological Advances in Materials for Energy Storage and Conversions*, 273–283. Singapore: Springer Nature Singapore.

Asgharian, H., Marques, D.L., Iov, F. et al. (2024). The role of cryogenic carbon capture in future carbon-neutral societies. *International Journal of Greenhouse Gas Control* 135: 104161.

Bagwan, N.S., Gevaers, R., and Dewulf, W. (2023). Cooling technologies in cooled supply chains. about the suitability and sustainability of dry ice as a cooling medium. An exhaustive review. In: *Interdisciplinary Conference on Production, Logistics and Traffic*, 274–290. Springer.

Balasubramanian, S., Gupta, M.K., and Singh, K. (2012). Cryogenics and its application with reference to spice grinding: a review. *Critical Reviews in Food Science and Nutrition* 52: 781–794.

Banerjee, S. (2021). A comprehensive review on the economic status of the global convenience food industry. *International Journal of Business, Management and Economics* 2: 43–52.

Bashir, B., Rafiq, S., Fayaz, I., and Makroo, H. (2025). Food and dairy freezing and cooling techniques. In: *Engineering Solutions for Sustainable Food and Dairy Production: Innovations and Techniques in Food Processing and Dairy Engineering*. Springer.

Bonga, K.B. (2024). Tuning of self-growing natural composite materials using non-pathogenic fungal strains-towards 3D constructs.

Braga, D., Li, S., and Fahim, F. (2021). Cryogenic electronics development for high-energy physics: an overview of design considerations, benefits, and unique challenges. *IEEE Solid-State Circuits Magazine* 13: 36–45.

Brown, D. (1967). The application of cryogenic fluids to the freezing of foods. In: *Advances in Cryogenic engineering: Proceedings of the 1966 Cryogenic Engineering Conference University of Colorado Engineering Research Center and Cryogenics Division NBS Institute for Materials Research Boulder, Colorado June 13–15, 1966*, 11–22. Springer.

Çalişkan Koç, G., Özkan Karabacak, A., Süfer, Ö. et al. (2025). Thawing frozen foods: a comparative review of traditional and innovative methods. *Comprehensive Reviews in Food Science and Food Safety* 24: e70136.

Chan, S.S., Roth, B., Skare, M. et al. (2020). Effect of chilling technologies on water holding properties and other quality parameters throughout the whole value chain: from whole fish to cold-smoked fillets of Atlantic salmon (Salmo salar). *Aquaculture* 526: 735381.

Chand, R. (2022). Indian agriculture towards 2030—the need for a transformative vision. *Indian Agriculture Towards* 2030: 1–7.

Christensen, K.B., Maheri, A., Siddiq, M.A., and Jalili, S. (2025). A study on the potential of cryogenic cooling and cutting technique in reducing the decommissioning cost of offshore monopiles. *Cryogenics* 145: 103991.

Dalmau, E., Araya-Farias, M., and Ratti, C. (2024). Cryogenic pretreatment enhances drying rates in whole berries. *Food* 13: 1524.

Estrada-Flores, S. (2012). Chilling and freezing by cryogenic gases and liquids (static and continuous equipment). In: *Operations in Food Refrigeration* (ed. R.H. Mascheroni), 233–251. CRC Press, capítulo, 12.

Cao, F., Sun, T., Jinpeng, H. et al. (2023). Enhanced mechanical and anticorrosion properties in cryogenic friction stir processed duplex stainless steel. *Materials & Design* 225: 111492.

Ghorbani, B., Zendehboudi, S., and Saady, N.M.C. (2025). Advancing hybrid cryogenic natural gas systems: a comprehensive review of processes and performance optimization. *Energies* 18: 1443.

Goswami, T. (2017). Recent trends of application of cryogenics in food processing and preservation. *Journal of Food, Nutrition and Population Health* 1: 27.

Hiremath, R., Hosamath, J., Srinivasulu, B., and Hegde, R. (2025). Trends, innovations, and sustainable practices in coriander production: a global perspective. *Journal of Scientific Research and Reports* 31: 891–907.

Ikram, A., Arshad, M.T., Maqsood, S. et al. (2025). Modern approaches for ensuring the safety and integrity of frozen foods: an overview. *Food and Agricultural Immunology* 36: 2491594.

Jha, P.K., Chapleau, N., Meyers, P.-E. et al. (2024). Can cryogenic freezing preserve the quality of fruit matrices during long-term storage compared to the mechanical method? *Applied Food Research* 4: 100374.

Jin, H., Zhang, Y., Shi, M. et al. (2025). Impact of freezing technologies on the quality of catfish (*Ictalurus punctatus*) fillets: insights from protein properties and ice crystal analysis. *International Journal of Refrigeration* 173: 44–54.

Kim, T., Kim, Y., and Kim, S.-K. (2011). Numerical study of cryogenic liquid nitrogen jets at supercritical pressures. *The Journal of Supercritical Fluids* 56: 152–163.

Kypraiou, C. and Varzakas, T. (2025). Evolution and evaluation of ultra-low temperature freezers: a comprehensive literature review. *Food* 14.

Liu, H., Zeng, F., Wang, Q. et al. (2013). The effect of cryogenic grinding and hammer milling on the flavour quality of ground pepper (*Piper nigrum* L.). *Food Chemistry* 141: 3402–3408.

Manzocco, L., Alongi, M., Cortella, G., and Anese, M. (2022). Optimizing radiofrequency assisted cryogenic freezing to improve meat microstructure and quality. *Journal of Food Engineering* 335: 111184.

Mekonnin, A.S. and Waclawiak, K. (2025). Investigation of material properties under cryogenic conditions: a review. *Scientific Papers of Silesian University of Technology. Organization & Management Series* 216.

Mudgulkar, S. (2024). Food processing and preservation. In: *Frontiers in Food Biotechnology*. Springer.

Ni, Z., Li, S., Zhang, X. et al. (2025). Analysis and comparison of the membrane-cryogenic hybrid process and multistage membrane process for pre-combustion carbon capture based on the superstructure method. *Separation and Purification Technology* 353: 128636.

Pesce, F., Parafati, L., Fallico, B., and Palmeri, R. (2025). Use of liquid nitrogen in food products: a review. *Food Frontiers* 6: 1617–1644.

Pravarthika, R. and Karthickumar, P. (2025). Exploring cutting-edge technologies for sustainable seafood freezing. *Signs (Chicago).*

Qian, P., Zhang, Y., Shen, Q. et al. (2018). Effect of cryogenic immersion freezing on quality changes of vacuum-packed bighead carp (*Aristichthys nobilis*) during frozen storage. *Journal of Food Processing and Preservation* 42: e13640.

Qiao, J., Zhang, M., Wang, D. et al. (2024). AI-based R&D for frozen and thawed meat: research progress and prospects. *Comprehensive Reviews in Food Science and Food Safety* 23: e70016.

Qiu, Y., Yang, H., Tong, L. et al. (2021). Research progress of cryogenic materials for storage and transportation of liquid hydrogen. *Meta* 11: 1101.

Qiu, J., Matsukawa, K., and Edashige, K. (2023). Equilibrium vitrification of oocytes using low concentrations of cryoprotectants. *Cryobiology 113*: 104586.

Radu, I.-C., Vadureanu, A.-M., Cozorici, D.-E. et al. (2025). Advancing sustainability in modern polymer processing: strategies for waste resource recovery and circular economy integration. *Polymers* 17: 522.

Ren, Y., Liang, R., and Mao, X. (2025). Effect of liquid nitrogen freezing pretreatment on the meat quality of gazami crab (*Portunus trituberculatus*) during frozen storage. *Food Chemistry* 468: 142367.

Saxena, S., Barnwal, P., Balasubramanian, S. et al. (2018a). Cryogenic grinding for better aroma retention and improved quality of Indian spices and herbs: a review. *Journal of Food Process Engineering* 41: e12826.

Saxena, S., Sharma, Y., Rathore, S. et al. (2015). Effect of cryogenic grinding on volatile oil, oleoresin content, and anti-oxidant properties of coriander (*Coriandrum sativum* L.) genotypes. *Journal of Food Science and Technology* 52: 568–573.

Saxena, V., Patel, B.B., Sutar, R., and Joshi, D. (2018b). Improving the quality of cumin powder through cryogenic grinding technology. *Journal of Food Processing and Preservation* 42: e13371.

Sharma, L., Agarwal, D., Rathore, S. et al. (2016). Effect of cryogenic grinding on volatile and fatty oil constituents of cumin (*Cuminum cyminum* L.) genotypes. *Journal of Food Science and Technology* 53: 2827–2834.

Truonghuynh, H.T., Li, B., Zhu, H. et al. (2020). Freezing methods affect the characteristics of the large yellow croaker (*Pseudosciaena crocea*), particularly when using cryogenic freezing for long-term storage. *Food Science and Technology* 40: 429–435.

Volokitina, I., Denissova, A., Volokitin, A. et al. (2024). Application of cryogenic technologies in deformation processing of metals. *Progress in Physics of Metals* 25: 161–194.

Yan, J., Zhao, S., Liang, Z. et al. (2025). Application of cryogenic grinding combined with ultrasonic pretreatment to enhance the yield and quality of essential oils from nutmeg (*Myristica fragrans* Houtt.). *Sustainable Chemistry and Pharmacy* 44: 101979.

Zendehboudi, A., Wang, B., and Li, X. (2017). Application of innovative models for prediction of the frost layer thickness on vertical cryogenic surfaces under natural convection. *Applied Thermal Engineering* 115: 1128–1136.

Index

a

aerogel-based insulation systems 13
air-blast freezing 97–98
amorphous *vs.* crystalline ice 37–38
anharmonic contributions 31
asphyxiation 12
auxiliary systems 55

b

bakery 56
batch freezers 54
bioactive compounds, molecular basis of 42–43
  antioxidants and polyphenols 43
  encapsulation and retention 43
  essential oils 43
biomolecules 34–37
  carbohydrate glass formation and mobility 36–37
  lipid crystallization 36
  oxidative stability 36
  protein denaturation and refolding 35
Biot number (B_i) 27

c

cabinet freezers 83
carbohydrates 34, 36–37
carbon dioxide (CO_2)
  physical and chemical properties 38–39
  reaction with food constituents 39–40
  safety and handling considerations 40
cellular integrity 57
Celsius scale 6, 7
chemical reaction rates 32
chromatographic techniques 62
commercial innovations 279–280
confectionery 55
confocal laser scanning microscopy (CLSM) 60–61
cryo-assisted drying 54
cryo-electron microscopy (Cryo-EM) 61
cryo-encapsulation 54
cryogenics 227
  advanced characterization techniques
    chromatographic techniques 62
    microscopy techniques 60–61
    spectroscopic techniques 61–62
    thermal analysis 62
  advantages of 55, 274–275
  artificial intelligence 275
  challenges 275
  concept 2–10
  crystalline structures and phase transitions
    glass transition and vitrification 58–59
    ice crystallization 58
    phase separation and recrystallization 59–60
  defined 1, 168–169
  development 228
  equipment and techniques 54–55

Cryogenic Processing: Advances in the Food Industry, First Edition. Edited by Sumit Sudhir Pathak, Rama Chandra Pradhan, and Sudarshan Ramanathan.

cryogenics (*cont'd*)
evolution 52
food preservation 65–66
food processing 15
food quality
bioactive compounds 64
bioavailability 64
essential nutrients 64
flavour improvements 63–64
safety 65
shelf stability 65
texture improvements 63
food systems 52–53
frozen desserts and baked goods 66
functional foods and nutraceuticals 67
future directions and challenges 67–68
goals of 280–281
health and medicine 14–15
heat transfer mechanism 75–76
historical developments 2–3
history 52
impacts 13, 14
industrial role 76–78
limitations and challenges 18–19
materials 51
meat and poultry processing 66
microstructural changes 57–58
molecular changes 56–57
physics and material science 14
preservation 181
principles 53–54, 74–75
safety 12–13, 16–18, 156–157
sensory analysis 62–63
shelf-life extension 65–66
significance 2
in space science 13
structural alterations 56
thermodynamic applications 76
cryogenic encapsulation systems 55
cryogenic fluids
challenges 12
hazards 16–17
phase transitions 8–10
properties 11
transporting 17
cryogenic grinders 54
cryogenic refrigerant 96–97
cryogens 2, 181, 268
carbon dioxide 269
liquid nitrogen 269
cryogen-specific energy consumption 30
cryo-grinding 54
cryo-mechanochemistry 54
cryopreservation 87
plant-based food preservation 134–135
cryo-processing technologies
freezing and cooling 269–271
fruits and vegetables 271
recent trends 268
significance 267–268
cryoprotectants 35
cryoprotection 33–34
crystalline structures and phase transitions
ice crystallization 58

d

dairy production and preservation 107–108
applications 110–111
challenges 111–112, 125–126
cheese processing 112–113
cryogenic freezing technology 108–109
cultures and enzymes
benefits 122
storage 120, 121
sustainability 122
transportation 120, 121
energy efficiency 111
environmental benefits 111
milk processing and preservation 109–110
starter culture preservation 113
technological evolution 125
value-added dairy products
colostrum 124
environmental and economic considerations 124–125
nitrogen-infused milk beverages 122–123
nutrient encapsulation 123–124
desserts 55

Dewar flasks 10, 16
direct-contact cryogenic freezing 99–101
DNA barcoding 194–195

e

effective cooling techniques 5
electron microscopy 14

f

Fahrenheit scale 6, 7
fermented milk processing 118
  challenges and future prospects 120
  energy efficiency, sustainability, and operational benefits 119–120
  microbial viability and probiotic efficacy 119
  product quality preservation 119
fluidized bed freezers 84
food 15–16
food industry 25–26
  amorphous *vs.* crystalline ice 37–38
  application 45–46, 55–56
  biomolecules 34–37
  carbohydrates 36–37
  cell effects 38
  composite 42
  cryogenic processing
    AI-powered predictive maintenance 258
    high-end applications 258
    hybrid cryogenic refrigeration systems 257–258
    infrastructure development 258–259
    policy and regulatory incentives 259
    renewable energy integration 258
    superconducting magnetic energy storage 258
  cryoprotectants 35
  environmental impact
    challenges 248–251
    comparative analysis 244–245
    cryogenic systems 256
    diesel-driven systems 248
    greenhouse gas emissions 247
    production and operational emissions 248
    real-world performance and benefits 256–257
    refrigeration 244
    sustainability 253–255
    vapour-compression refrigeration systems 243
  flavour and volatile compound retention 42
  food structure and quality 40–41
  functional foods 45
  industrial relevance 45–46
  lipid crystallization 36
  nanostructured cryogenic interfaces 44
  nucleation and growth of ice crystals 37
  nutritional stability and enzyme inactivation 42
  protein denaturation and refolding 35
  quality preservation
    crystal formation 253
    extended shelf life 251–252
    faster cooling and lower temperature impact 251
    nutritional value, flavour, and texture 252
  smart packaging and sensors 44
  technologies and mechanisms 245–246
    direct and indirect cryogenic systems 246–247
    transport refrigeration systems 246
  texture modification and structural integrity 41
  thermodynamic principles 27–30
  tissue integrity 38
  water behaviour 37–38
food microbiology
  challenges, limitations, and future direction 237–240
  cryogenics 229
  freezing 233–234
    advantages 234
    viability 234
  mechanisms of inactivation 232–233
  microbial populations 229–231
  safety 234–237
  survival and proliferation 231–232

food preservation 55
food processing 281
  challenges and future trends 86–87
  industrial design and engineering 78–79
    encapsulation 86
    freezers 83–84
    grinding 85
    liquid carbon dioxide 82–83
    liquid nitrogen 82–83
    material selection 79–80
    safety measures and standards 81–82
    storage tanks 84–85
    thermal insulation technique 80–81
food quality
  bioactive compounds 64
  bioavailability 64
  essential nutrients 64
  flavour improvements 63–64
  safety 65
  shelf stability 65
  texture improvements 63
food systems 52–53
Fourier number (F_0) 28
Fourier's law of heat conduction 27
Fourier Transform Infrared Spectroscopy (FTIR) 61
freeze-drying process 33
freezers 54
freezer tunnel 83–84
freezing point depression 28
freezing technology 15
frozen foods 55
fruits and vegetables (F&Vs) 271
  air-blast freezing 97–98
  cryogenic refrigerant 96–97
  direct-contact cryogenic freezing 99–101
  freezing 95–97
    advantages 103
    case study 101–103
    economic feasibility, challenges, and limitations 103–104
  immersion freezing 98
  indirect contact cryogenic freezing method 101
  plate freezers 98
fruit tissues 205
functional and nutraceutical foods 55
futuristic technology 138

g

gas chromatography (GC) 62
gases 7–8
glass transition temperature (Tg) 32
grinding 85, 272–274
  plant-based food preservation 135–136

h

heat transfer 27–28
heat transmission 7
helium 6, 7
herbal medicine processing 183–184
  cryogenic applications 193–194
  phytomedicines 184–186
    good manufacturing practices 188–189
    process control 189–190
    production controls 187–188
    purity management 191–193
    work-in-progress testing 190–191
high-entropy alloys (HEAs) 5
high-performance liquid chromatography (HPLC) 62
hybrid cryogenic refrigeration systems 257–258
hybrid cryogenic systems 276
hydrogen bonding 33–34
hydrophobic interactions 34

i

ice cream and frozen desserts 116–117
  challenges and innovations 118
  energy efficiency and operational flexibility 118
  product quality and flavours 117
  texture and structure preservation 117
ice crystallization 37–38
immersion freezers 84
immersion freezing 98
indirect contact cryogenic freezing method 101

industry practices
cost efficiency 259
food logistics 259
sustainable supply chain integration 259
International Institute of Refrigeration 181

k
Kelvin scale 5, 7
kinetic consequences 32–33

l
lipid crystallization 36
lipids 34
liquefaction efficiency 255
liquid helium 9
liquid nitrogen (LN_2) 6, 7, 17
physical and chemical properties 38–39
reaction with food constituents 39–40
safety and handling considerations 40
liquid oxygen (LOX) 73
liquids 7–8
low-temperature
environment 1
thermodynamics 8

m
machining technology 277–278
marine/aquatic products 144, 149–150
cabinet freezers 146
combined freezers 146
drip loss 151
freezing loss 150
immersion freezers 146
lipid oxidation 152–153
morphological characteristics 153–154
pellet freezers 146
pH and colour 153
protein denaturation 152
rotary freezers 145–146
sensory attributes 154
tunnel freezers 144–145
water content 151–152
weight loss during storage 150–151
materials
behaviour of 3, 5
selection 12
material science 279
meat 56, 144, 272
biochemical properties 148–149
cabinet freezers 146
colour 148
combined freezers 146
immersion freezers 146
lipid and protein oxidation 149
loss of moisture 147–148
microbial properties 149
microstructure 147
pellet freezers 146
physical properties 147–148
protein denaturation 148–149
quality changes 146–147
rotary freezers 145–146
tenderness 148
tunnel freezers 144–145
microbial viability 234
microbiological analysis 190–191
micronization 57–58
microscopy techniques 60–61
milk powder processing 114
challenges and innovations 116
energy efficiency and environmental sustainability 115–116
preservation of nutritional and bioactive compounds 115
mobility 36–37
molecular diffusion 31–32
molecular dynamics 30

n
non-Arrhenius behaviour 30–31
non-covalent interactions 33
nuclear magnetic resonance (NMR) spectroscopy 61
nucleation dynamics 30

o
Occupational Safety and Health Administration (OSHA) 13
oxidative stability 36

p

personal protective equipment (PPE) 12
phase diagrams of water 28
phytomedicines
  good manufacturing practices 188–189
  process control 189–190
  production controls 187–188
  purity management 191–192
  work-in-progress testing 190–191
plant-based food preservation 129–130
  cryopreservation 134–135
  emerging trend 130–131
    cheese 132
    eggs 132
    fermented mycoprotein 131–132
    microscopic algae 132–133
    milk 132
    plant-based protein products 131
    seafood 132
  engineering challenges 136–137
  grinding 135–136
  principles 133–134
plant-based foods 206
plate freezers 98
poultry 56, 144
  cabinet freezers 146
  combined freezers 146
  immersion freezers 146
  pellet freezers 146
  rotary freezers 145–146
  tunnel freezers 144–145
precision cryogenic processing 277
proteins 34
purity management 191–193

q

quantum cryogenics 276–277
quantum mechanical processes 7

r

Raman spectroscopy 61
rapid freezing 54
refrigeration 10
  energy consumption in 253–254
  innovations 254
  renewable energy integration 255
rotational motion 31

s

scanning electron microscopy (SEM) 60
seafood 272
self-propagating reactions 32
sensible heat 28
sensory analysis 62–63
shrimp
  freezing 155
  frozen storage 155
  glazing 155
  harvesting 154
  holding 154–155
  packaging 155
  peeling 155
snacks 56
solids 7–8
spectroscopic techniques 61–62
spice processing
  grinding process 163
  grinding technology 168–169
    benefits 169–170
    challenges and limitations 175
    construction methods 172
    design and engineering 171–173
    grinder 172–173
    liquid nitrogen 172
    pre-coolers 172
    sensors 173
  harvest and the post-harvesting 167
  India 163
  limitations 168
  particle size 167–168
  post-harvest practices 164–166
  powders 164
  quality and preservation
    flavour and aroma retention 173–174
    traditional method 174
  raw material 167
  storage and packaging 168
spiral freezers 84
superconducting magnetic energy storage (SMES) 258

superconductivity 6
supercritical fluids 7–8
superfluidity 6
superinsulation technology 11

t
technological gap 281
temperature scales 5
thermal analysis 62
thermal insulation technique 80–81
thermodynamic arrest 32
tissue integrity 38
total energy 29
transformation-induced plasticity (TRIP) 5
translational mobility reduction 31–32
transmission electron microscopy (TEM) 60
transportation 278–279
tunnel freezers 54
twinning-induced plasticity (TWIP) 5

u
ultraviolet–visible (UV–Vis) spectroscopy 61–62

v
vacuum insulation 11
Van der Waals interactions 34
vegan-based products
  applications 220–221
  challenges and cost implications 222–223
  consumer attitudes 221–222
  cryogenic agents 209
  cryogenic liquids 206
  cryogenic techniques 206–207
    advantages and disadvantages 210
    cooling 214–215
    cryo-concentration 215–216
    cryo-milling 212–213
    freezing 210–211
    principle and mechanism 208–210
    pulse freezing 211–212
    ultra cryo-milling 214
  food preservation 208
  future perspectives 223
  quality characteristics 216
    bacteria 218–220
    colour 218
    concentration effects 217
    drip loss 218
    freeze cracking 217
    freezer burn 217
    moisture loss 217
    nutrient-rich and tender raw materials 217
    packaging materials 217
    rancid flavour development 217
    suppressed microbial growth 218–220
    texture 218
  sustainability and scalability 223
veganism 129
vibrational modes 31
volatile organic compounds (VOCs) 42

w
water, phase transition of 205
well-to-wheel (WTW) emissions 254
work-in-progress testing 190–191

Wiley The manufacturer's authorized representative according to the EU General Product Safety Regulation is Wiley-VCH GmbH, Boschstr. 12, 69469 Weinheim, Germany, e-mail: Product_Safety@wiley.com.

Printed and bound by CPI Group (UK) Ltd, Croydon, CR0 4YY
13/04/2026
02089254-0018